London Mathematical Society Student Texts: 2

Building Models by Games

WILFRID HODGES
School of Mathematical Sciences
Queen Mary College,
University of London

The right of the University of Cambridge to print and sell all manner of books was granted by Henry VIII in 1534. The University has printed and published continuously since 1584.

CAMBRIDGE UNIVERSITY PRESS
Cambridge
London New York New Rochelle
Melbourne Sydney

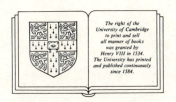

Published by the Press Syndicate of the University of Cambridge
The Pitt Building, Trumpington Street, Cambridge CB2 1RP
32 East 57th Street, New York, NY 10022, USA
10 Stamford Road, Oakleigh, Melbourne 3166, Australia

First published 1985

Printed in Great Britain at the University Press, Cambridge

British Library Cataloguing in Publication Data
Hodges, Wilfred
Building models by games. - (London Mathematical
Society student texts; 2)
1. Mathematical models
I. Title II. Series
511'.8 QA401

ISBN 0 521 26897 4 hard cover
ISBN 0 521 31716 9 paperback

CONTENTS

INTRODUCTION

Most mathematical books are about some particular structure (such as the complex numbers) or class of structures (such as Banach algebras). This book is not. Instead it is about a very general method for building structures. For a precedent one should look for a book on Cartesian Products, or perhaps on Left Adjoints of Forgetful Functors.

The method of construction is easily described. Suppose you and I want to build a group. I write down a partial specification:

$$a = b^2.$$

You add some further information:

$$b = c^2, \ bd \neq db.$$

I continue:

$$a = 1, \ df = g^2.$$

Your turn:

$$g^{-1}h^{-1}ghjh^{-1}g^{-1}hgj^{-1} \neq 1.$$

And so on. The only constraints are that we should each write down a finite amount at a time (try violating that!), and that we should never contradict anything which has been written down so far. After an infinite number of steps we shall have assembled a set S of equations and inequations. There will be a group G, unique up to isomorphism, which is generated by a, b, c, ... subject to the equations in S. Together we have built the group G. (The inequations that went into S played an indirect role in the definition of G: they prevented us putting certain equations into S later.)

Now let me ask a question. Given a property P which the group G might have, is it possible for you to make your choices so that G will have property P regardless of what I write down? If it is, we say that P is an enforceable property. Clearly the property of being non-abelian is enforceable - you can always find two new letters x, y and add $xy \neq yx$ at your next turn. Less obviously, it is enforceable that all elements of order 7 lie in the same conjugacy class of G.

Enforceable properties have one very useful feature. If P_i $(i < \omega)$ are a countable family of enforceable properties, then their conjunction $P_0 \& P_1 \& P_2 \& \ldots$ is enforceable too. It follows that if we want to show that there exists a group with some property Q, we need only show that Q is implied by some countable family of enforceable properties P_i. In practice we often find that the properties P_i are separately much easier to handle than Q, so that there is a real saving in effort. This book is full of examples.

A more whimsical feature of enforceable properties is that groups constructed in more orthodox ways (free or cartesian products, wreath products, automorphism groups etc.) tend not to have them. So we have a source of unusual and unexpected examples.

These ideas apply equally well to skew fields, boolean algebras, models of first-order arithmetic, or structures in any other reasonable class. Model theorists have studied enforceable properties in one shape or another for twenty-five years or more, sometimes under the names of omitting types or forcing. It has become steadily clearer that we have here one general method which deserves to be known in its own right. I propose to call it construction by games.

There is a difference between construction by games and most of the constructions that one meets in algebra. For most constructions one can say: the constructed object is such-and-such. But in game constructions the best we can say is: the constructed object can be guaranteed to have such-and-such properties. Sometimes this is a distinction without a difference, because the guaranteed properties determine the object up to isomorphism. But sometimes they don't, and then it is useful to think of one player as guaranteeing the properties while the other player pushes the construction this way or that. In such cases our games are a machine for producing families of pairwise non-isomorphic structures.

The book consists of a forty-minute talk which I gave at the British Mathematical Colloquium in Aberdeen in April 1983. I began with the game for building groups, as described above, and I finished with Theorem 8.2.11 at the end of the book. Of course I have added some details for the printed version. These were largely worked out in the model theory seminar at Bedford College - my thanks to the participants.

I have also added some exercises. I have no advice at all on how the reader should use them. Maybe if I had another year to think about the book I would recast them completely. But they are what they are: some of them fill small gaps in proofs in the text, some are applications of results in the text, some report major items of research and some merely indicate that certain things are known. I can't do them all myself.

My debts in writing a text of this kind are legion, but they seem to shake down into three groups.

In the first place, this whole species of mathematics owes its existence to a small circle of charmed souls. Any list would include at least B. H. Neumann, Leon Henkin, Roland Fraissé, Abraham Robinson, H. J. Keisler, Angus Macintyre. Two other mathematicians have had a vast influence on the presentation in this book, as anyone who knows their work will see at once. They are Saharon Shelah and Martin Ziegler. It was Ziegler who identified the right games to study. I learned most of what I know about the whole topic by struggling with Shelah's preprints.

Secondly I thank the many mathematicians who were asked for specific help (preprints, offprints, information, corrections) and generously gave it. The list I have is as follows, and I apologise for any omissions: R. B. J. T. Allenby, Andrew Apps, Andreas Baudisch, O. V. Belegradek, Maurice Boffa, Robert Bonnet, Kim Bruce, P. M. Cohn, Don Collins, Max Dickmann, Colin Elliott, Andrew Glass, Rami Grossberg, K. K. Hickin, Ian Hodkinson, Matt Kaufmann, Roman Kossak, Alistair Lachlan, Berthold Maier, Jerome Malitz, Jeff Paris, Vladimir Raženj, Gabriel Sabbagh, James Schmerl, Saharon Shelah, Robert Soare, Charles Steinhorn,

Simon Thomas, Sauro Tulipani, Bert Wehrfritz, Carol Wood, Mike Yates. In fairness to these many helpers I have to add that they have not all seen the final versions of the sections on which they are experts. So there are bound to be some mistakes and gaps, and of course I take full responsibility for these. The series editor Brian Davies and Rufus Neal of the Cambridge University Press were very ready with encouragement and sound advice.

And thirdly I thank my family and neighbours for putting up with a broken fence, unmended windows and constant banging noises for half a year. Tomorrow I shall buy five new fence-posts.

As I write, Bedford College is packing its bags in Regent's Park and preparing to move out of London to Egham, to join Royal Holloway College. The excellent mathematics department which I have known as home for the last dozen years or more is being disbanded, and London University has jettisoned the most beautiful of its sites. I was about to say a word about the economic policies which have brought about this vandalism (and done incomparably worse things elsewhere). But I bite my tongue. Better to use this space to say a sad goodbye to my old colleagues at Bedford, and to wish every success to the new combined college at Egham.

Wilfrid Hodges

Queen Mary College
London
Summer 1984

1 PRELIMINARIES

I know nothing about that subject except what I have heard described or seen in pictures ... my mind is virgin.
Euripides, Hippolytos 1004-1006

Mathematics is not a topic that one can easily approach with a virgin mind. It is a cumulative subject, full of prerequisites. For this book the prerequisites lie in the area of mathematical logic known as model theory. But I did not write the book just for model theorists. On the contrary, I wanted to show how a certain collection of ideas from model theory have applications throughout algebra, and maybe beyond.

So the prerequisites have to be spelt out. This is the business of section 1.2 below. I would suggest that readers take it quickly at a first reading and then come back to it again for particular facts as necessary. As for the pictures in section 1.1, I think I should let them speak for themselves. If they say nothing to you, either now or in two hundred pages' time, cast a cold eye and pass by.

1.1 *PICTURES*

When I was a student, I once asked C. C. Chang for some help with a problem. He thought for a moment and then waved his finger:

(1)

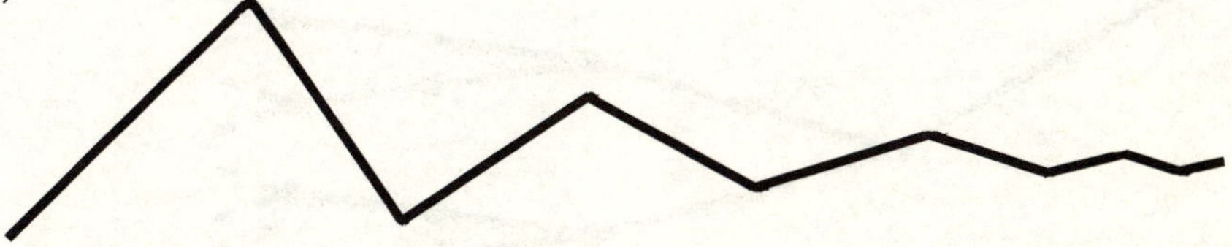

I think he only said "Yes, that should work". I tried it and it did.

Most of the ideas in this book are based on equally simple pictures. Surely 300-odd pages of mathematical text are not an efficient way of passing on pictures. So let me make a more direct attempt.

The overall picture of a construction by a game is as follows:

(2)

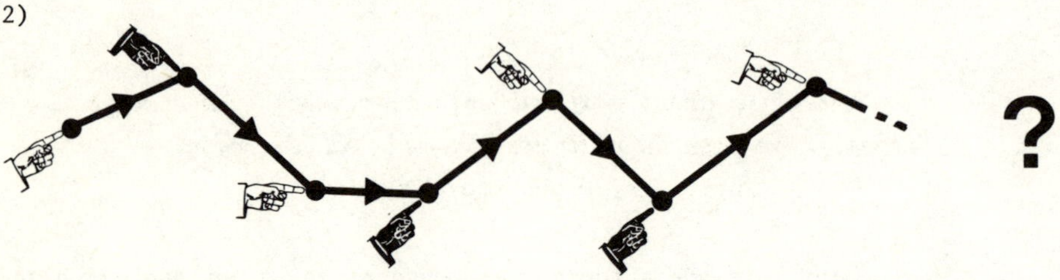

But this is not a very revealing picture, because the two players are really only a technical device for analysing possible constructions. As early as Chapter 2 we shall find one player dissolving into countably many, or vice versa. It is more helpful to picture what happens if one player follows a fixed routine and the other player tries all possibilities against him. Here there are two opposite cases. The first is the tree or perfect set case:

(3)

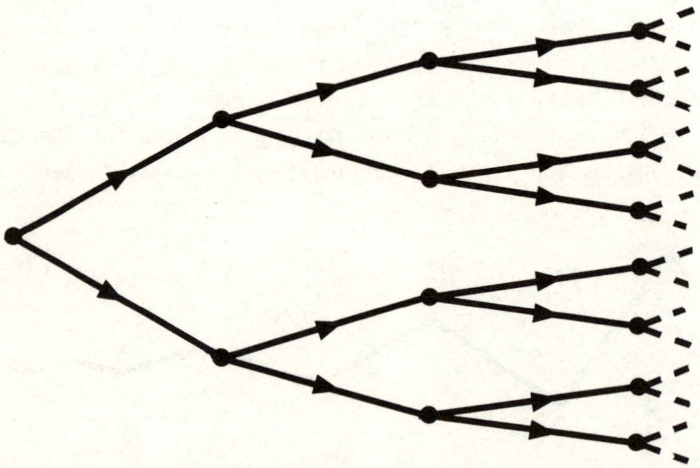

The second is the categorical case, where nothing that the fickle player does makes any difference to the outcome:

(4)

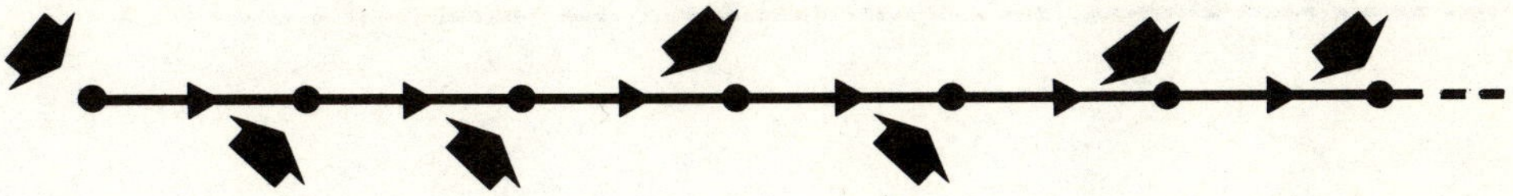

The dichotomy between (3) and (4) will occupy us a good deal, particularly in Chapter 4.

The picture (4) represents pretty well the general outcome of a categorical construction: one just keeps shovelling in bits and pieces until everything is included. But there is a better picture that expresses the fine detail:

(5)

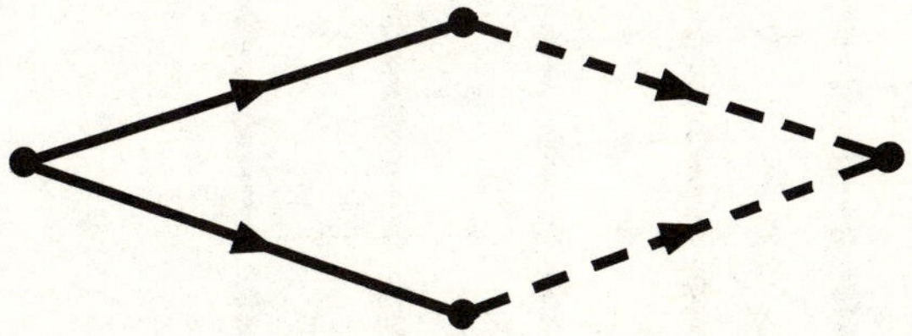

This picture is known to model theorists as amalgamation. Proof theorists think of it as the (Church-Rosser) diamond property (no connection with Jensen's $\Diamond$ of section 5.2 below). It represents a situation where any two developments from the left node are compatible: they can be fitted into a common further development. One should imagine the branches of the tree (3) all being pulled together into a single line.

Omitting types is harder to represent on a page, but I think I have a duty to try. Picture (6) is an attempt. Arbitrarily far up the tube there are threads running, but no thread runs right through the tube. The person who lays the threads in the tube can only fix a finite amount of each thread at a time, so his rival can keep pulling the ends out through the holes in the sides of the tube:

(6)

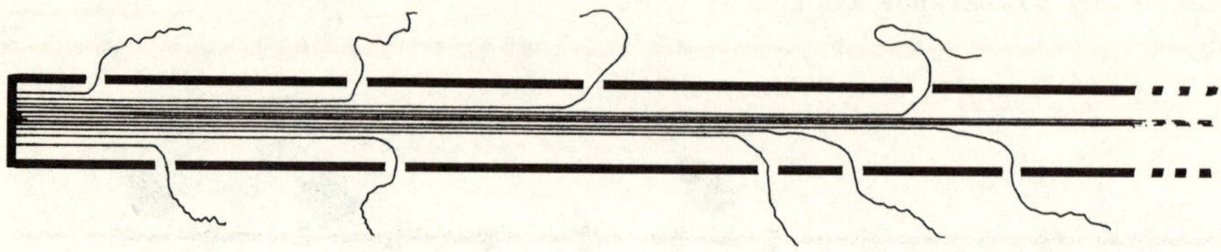

In Chapter 6 we start to meet two-dimensional constructions:

(7)

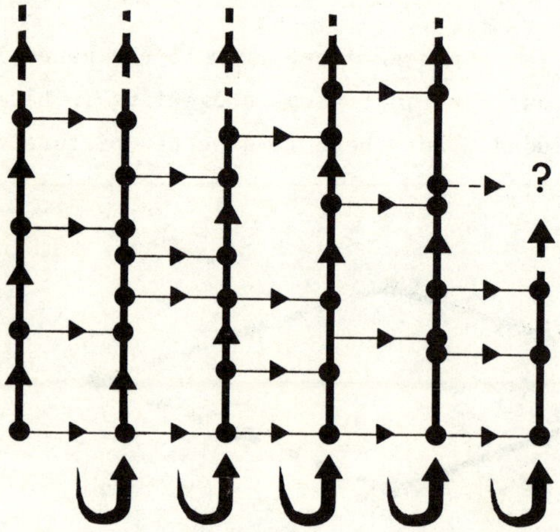

The columns are built up one by one, starting on the left. Note that each advance in the construction is an amalgamation problem of type (5), as the column being constructed is hooked onto the previous one. Although it goes beyond this book, it is perhaps worth mentioning that Shelah has made constructions which extend (7) into higher dimensions. Thus in three dimensions one builds up a pile of slices (like a pack of Kraft cheese). The top slice is built up column by column as in (7). It has to be fitted onto the slice below, so that each advance is now a three-dimensional amalgamation problem.

1.2 MODEL THEORY

Structures, homomorphisms and substructures

A structure A consists of the following ingredients (1)-(5):

(1) a non-empty set called the domain of A, in symbols dom(A); the elements of dom(A) are called the elements of A, and the cardinality of A, $|A|$, is defined to be the cardinality of dom(A);

(2) a set of function symbols, constants and relation symbols; each function symbol and each relation symbol is n-ary (i.e. of arity n) for some unique positive integer n; the set of symbols is called the signature of A (or in some books the similarity type of A);

(3) for each $n \geq 1$ and each n-ary function symbol F in the signature of A, a function F^A: $(\text{dom } A)^n \to \text{dom } A$;

(4) for each constant c in the signature of A, an element c^A of dom(A);

(5) for each $n \geq 1$ and each n-ary relation symbol R in the signature of A, a relation $R^A \subseteq (\text{dom } A)^n$.

If L is the signature of A, we say that A is an L-structure.

EXAMPLE. An ordered abelian group G can be regarded as a structure of signature $\{0,+,-,\leq\}$ where 0 is a constant, + a 2-ary function symbol, - a 1-ary function symbol and $\leq$ a 2-ary relation symbol. The domain of G is the set of group elements, the zero element is 0^G, the group operation is $+^G$, and so on.

We write $\bar{a}$, $\bar{b}$ etc. for sequences of elements of a structure. An n-tuple from A is a sequence of elements of A, of length n. A tuple is an n-tuple for some $n < \omega$. If $\bar{a}$ is $(a_0,\ldots,a_{n-1})$ and f is a map, then $f\bar{a}$ is $(fa_0,\ldots,fa_{n-1})$.

Two structures are said to be similar iff they have the same signature. A homomorphism f: A $\to$ B, where A and B are similar structures with signature L, is a map f: dom(A) $\to$ dom(B) satisfying the following three conditions:

(6) for each $n \geq 1$, each n-ary function symbol F in L and each n-tuple $\bar{a}$ from A, $f(F^A(\bar{a})) = F^B(f\bar{a})$;

(7) for each constant c in L, $f(c^A) = c^B$;

(8) for each $n \geq 1$, each n-ary relation symbol R in L and each n-tuple $\bar{a}$ from A, $\bar{a} \in R^A \Rightarrow f\bar{a} \in R^B$.

An embedding f: A → B is a homomorphism such that (8) holds in the strong form:

(9) f is injective; and for each $n \geq 1$, each n-ary relation symbol R in L and each n-tuple $\bar{a}$ from A, $\bar{a} \in R^A \Leftrightarrow f\bar{a} \in R^B$.

An isomorphism f: A → B is a homomorphism which is bijective from dom(A) to dom(B), and such that the inverse map f^{-1}: dom(B) → dom(A) is also a homomorphism. One can show that an isomorphism is the same thing as a surjective embedding.

An inclusion, or extension map, is an embedding f: A → B such that dom(A) ⊆ dom(B) and f is the identity on dom(A). If there exists an inclusion f: A → B, we say that A is a substructure of B and B is an extension of A, in symbols A ⊆ B.

LEMMA 1.2.1 (Substructure lemma). *Let* B *be a structure of signature* L. *Let* X *be a set of elements of* B, *and let* λ *be an infinite cardinal such that* $|X| \leq \lambda$ *and* L *contains at most* λ *symbols. Suppose either that* $X \neq \emptyset$ *or that* L *has at least one constant. Then there is a unique smallest substructure* A *of* B *such that* X ⊆ dom(A) *(smallest in the sense that if* C *is any substructure of* B *with* X ⊆ dom(C), *then* A ⊆ C*). The domain of* A *is the smallest set which contains* X *and all elements* c^B *(*c *a constant in* L*) and is closed under the functions* F^B *(*F *a function symbol in* L*). Thus* A *has cardinality at most* λ.

Proof. Cf. Cohn (1981) section II.5 or Grätzer (1979) section 1.9. □

The substructure A in the lemma is called the substructure of B generated by X, in symbols $\langle X \rangle_B$. If we had allowed structures to have empty domains, we could have dropped the requirement that $X \neq \emptyset$ or L has a constant.

Languages

Each signature L allows us to build terms and formulas for talking about L-structures. The terms of L are defined inductively: (i) every constant of L and every variable x, y, z, x_1, x_2 etc. is a term of L; (ii) if $n \geqslant 1$, F is an n-ary function symbol of L and $t_1, \ldots, t_n$ are terms of L, then the expression $F(t_1,\ldots,t_n)$ is a term of L. The atomic formulas of L are the expressions of form s=t (where s, t are terms of L) or $R(t_1,\ldots,t_n)$ (where for some $n \geqslant 1$, R is an n-ary relation symbol of L and $t_1, \ldots, t_n$ are terms of L).

Formulas of L can be built up from atomic formulas in various ways. For example first-order formulas are those built up from atomic formulas by any (finite) number of applications of $\neg$ ("not"), $\wedge$ ("and"), $\vee$ ("or"), $\rightarrow$ ("if ... then"), $\leftrightarrow$ ("iff"), $\forall x$ ("for all x"), $\exists x$ ("there is x"). For example the sentence "Every non-zero element has a multiplicative inverse" can be written as a first-order formula:

(10) $$\forall x(\neg x=0 \rightarrow \exists y\ x.y=1).$$

Two important examples of non-first-order ways of building formulas are infinitary conjunction $\bigwedge$ and infinitary disjunction $\bigvee$: if $\{\phi_i : i \in I\}$ is a set of formulas, then $\bigwedge_{i\in I}\phi_i$ is the formula which says that all the formulas ϕ_i ($i \in I$) are true, and $\bigvee_{i\in I}\phi_i$ says that at least one of the formulas ϕ_i is true.

The class of all formulas of L built up in the ways described above is called $L_{\infty\omega}$. The first subscript $_\infty$ means that there is no bound on the size of conjunctions and disjunctions; the second subscript $_\omega$ means that we can only have finitely many variables $x_1, \ldots, x_n$ in a quantifier $\forall x_1 \ldots x_n$ or $\exists x_1 \ldots x_n$. Actually the definitions above only allowed quantifiers $\forall x$ or $\exists x$ which contain a single variable x. But we regard $\forall x_1 \ldots x_n$ as shorthand for $\forall x_1 \forall x_2 \ldots \forall x_n$, and likewise with $\exists$.

On the same principle, $L_{\omega\omega}$ is the set of formulas of $L_{\infty\omega}$ in which we only form conjunctions and disjunctions of finitely many formulas at a time. Since finite conjunctions and disjunctions can be built up with $\wedge$ and $\vee$, $L_{\omega\omega}$ is in practice the set of first-order formulas of L.

More generally, $L_{\kappa\lambda}$ is the set of formulas built up from the symbols of L in the ways described above, allowing conjunctions and disjunctions of fewer than κ formulas at a time, and quantifiers $\forall\bar{x}$ and $\exists\bar{y}$ where $\bar{x}$, $\bar{y}$ are sequences of fewer than λ variables. Thus $L_{\omega_1\omega}$, which we shall meet several times, has finite and countable conjunctions and disjunctions but only finite quantifiers.

If we start from the atomic formulas of a signature L and allow formulas to be built according to some fixed set of rules, the resulting collection of formulas is called a <u>language</u> of signature L. For example $L_{\omega\omega}$, $L_{\omega_1\omega}$ and $L_{\infty\omega}$ are three different languages of signature L. A <u>first-order language</u> is a language of form $L_{\omega\omega}$.

There will be no harm if we use the symbol L for languages as well as signatures. Each language L determines its signature, and hence it makes sense to speak of L-<u>structures</u> when L is a language. Similarly if L is a first-order language, it is clear what language is meant by $L_{\omega_1\omega}$.

We shall freely use standard abbreviations, such as $x \neq y$ for $\neg x=y$. We write $\bar{x}$ for a tuple $(x_0,\ldots,x_{n-1})$ of variables. (But sometimes $\bar{x}$ will mean an infinite sequence of variables.) A formula $\phi(\bar{x})$ is a formula ϕ whose free variables (i.e. variables y not bound by quantifiers $\forall y$ or $\exists y$) all occur in $\bar{x}$, and when nothing is said to the contrary we assume that $\bar{x}$ is a sequence of distinct variables; when a formula $\phi(x_0,\ldots,x_{n-1})$ is introduced, and t_0, ..., t_{n-1} are terms, then $\phi(t_0,\ldots,t_{n-1})$ is the resulting formula when each free x_i in ϕ is replaced by t_i.

A <u>negated atomic</u> formula is a formula $\neg\psi$ where ψ is atomic. A formula is <u>basic</u> iff it is either atomic or negated atomic.

Let L be a language and A an L-structure. A <u>closed term</u> of L is a term of L (i.e. of the signature of L) which has no variables. Each closed term t of L names an element t^A of A. A <u>sentence</u> of L is a formula of L with no free variables. Each sentence ϕ of L makes a statement about A; we write $A \models \phi$ ("ϕ is <u>true</u> in A", "A is a <u>model</u> of ϕ" or "A <u>satisfies</u> ϕ") iff this statement about A is true.

A <u>theory</u> is a set of sentences. A <u>first-order theory</u> is a theory in a first-order language.

If L is a signature and A is an L-structure, then the <u>theory</u>

of A, Th(A), is the set of all sentences of $L_{\omega\omega}$ which are true in A. We say that two L-structures A and B are elementarily equivalent, in symbols $A \equiv B$, iff $Th(A) = Th(B)$.

For any theory T, $A \models T$ ("A is a model of T") means $A \models \phi$ for all $\phi \in T$; when nothing is said to the contrary, we assume that T is in some fixed language L and A is an L-structure, i.e. that A has no irrelevant symbols. For a first-order theory T, $A \models T$ means the same as $T \subseteq Th(A)$. $T \vdash \phi$ ("T entails ϕ", "ϕ is a consequence of T") means that every model of T is a model of ϕ.

We say that $\phi(\bar{x})$ and $\psi(\bar{x})$ are equivalent modulo T iff $T \vdash \forall\bar{x}(\phi \leftrightarrow \psi)$. More generally we write $\Phi(\bar{x})$ for a set Φ of formulas of form $\phi(\bar{x})$, and we say that $\Phi(\bar{x})$ and $\Psi(\bar{x})$ are equivalent modulo T iff $T \vdash \forall\bar{x}(\bigwedge\Phi \leftrightarrow \bigwedge\Psi)$. We say that two theories are equivalent iff they have exactly the same models.

When K is the class of all models of the theory T, we say that T axiomatises K, or is a set of axioms for K. If T axiomatises the class of fields, then we say T is the theory of fields. In a similar spirit we talk of the theory of groups, the theory of lattices etc. This looks dangerous terminology: there are different possible choices of signature for groups, and even in one fixed signature there are infinitely many ways of axiomatising the class of groups. But in practice I think there should be no problem. For example we shall prove a theorem about "strict universal Horn theories" and then apply it to "the theory of groups". If the reader has in his hand a set of axioms for groups which isn't strict universal Horn, he should look for another set which is. There will always be a natural choice which works. Thus a natural signature for fields, or more generally for any rings with 1, consists of the symbols $+, -, \cdot, 0, 1$; we can call this the signature of rings with 1. A natural set of axioms for fields in this signature is given at (11) below.

The following lemma is proved for first-order formulas in elementary courses, but it is true quite generally:

LEMMA 1.2.2 (Lemma on constants). *Let* $\phi(x_1,\ldots,x_n)$ *be a formula of signature* L, T *a theory of signature* L *and* $c_1, \ldots, c_n$ *distinct constants not in* L. *Then* $T \vdash \phi(c_1,\ldots,c_n)$ *iff* $T \vdash \forall x_1\ldots x_n\phi$.

Proof. For the first-order case, cf. Shoenfield (1967) p. 33f. □

Expansions and parameters

Let L and L^+ be signatures with $L \subseteq L^+$. Let B be an L^+-structure. We can convert B into an L-structure by deleting the symbols not in L (without removing any elements of B). Then A is called the L-reduct of B, in symbols $B|L$, and B is called an expansion of A to L^+. (Note the difference between expansions and extensions: an expansion adds new symbols but no new elements, while an extension adds new elements but no new symbols.) If W is the set of symbols which are in L^+ but not in L, then we can refer to L^+ as L(W). The same notation applies if L and L^+ are languages rather than signatures. For example if L is a first-order language and W is a set of constants which are not in L, then L(W) is the first-order language got from L by adding all the constants in W.

Suppose L is a language and A is an L-structure. Maybe not all the elements of A are named by constants of L. However, we can add to L a new constant $\tilde{a}$ for each element a of A. (For example we can choose each constant $\tilde{a}$ to be the element a itself. These new constants will be "symbols" only in a formal sense, but this will be no problem.) The expanded language is called L(A), and A can be expanded to an L(A)-structure D by putting $\tilde{a}^D = a$ for each $a \in \mathrm{dom}(A)$. We shall very often write $A \models \phi(a_1,\ldots,a_n)$ ("the elements $a_1, \ldots, a_n$ satisfy $\phi(x_1,\ldots,x_n)$ in A") for $D \models \phi(\tilde{a}_1,\ldots,\tilde{a}_n)$. This abbreviated notation doesn't show what are the constants naming the elements. This is deliberate - the choice of constants is never important.

Likewise if L is a language, A an L-structure, $\phi(\bar{x})$ a formula of L and $\bar{a}$ a tuple from A, we write $\phi(\bar{a})$ for the formula which results if we replace each variable x_i by a constant to name a_i. If $\Phi(\bar{x})$ is a set of formulas then $\Phi(\bar{a})$ is $\{\phi(\bar{a}) : \phi(\bar{x}) \in \Phi\}$.

When we add new constants to a language in order to name elements of a structure, the added constants and the elements named by them are known as parameters. If $\bar{a}$ is a sequence $(a_i : i < \gamma)$ of elements of A, then we write $(A,\bar{a})$ for the structure A expanded so that the elements a_i are named by new constants. Sometimes we shall compare two structures $(A,\bar{a})$ and $(B,\bar{b})$; in such cases we always understand that A and B are similar structures, $\bar{a}$ and $\bar{b}$ are sequences of the same length γ, and for each $i < \gamma$, the same new constant c_i is used to name a_i in $(A,\bar{a})$ and b_i in $(B,\bar{b})$.

Let A be an L-structure, $\phi(x,\bar{y})$ a formula of L and $\bar{a}$ a tuple from A. Then we write $\phi(A,\bar{a})$ for the set of elements b of A such that $A \models \phi(b,\bar{a})$. Sets of the form $\phi(A,\bar{a})$, where ϕ is first-order, are said to be first-order definable with parameters. Sets of the form $\phi(A)$, where $\phi(x)$ is first-order, are said to be first-order definable without parameters.

Let D be an expansion of A to L(A), so that every element a of A has a name $\tilde{a}$ in D. The diagram of A, diag(A), is the set of all basic sentences of L(A) which are true in D. The positive diagram of A, $\mathrm{diag}^+(A)$, is the set of all atomic sentences of L(A) which are true in D. The elementary (or complete) diagram of A, eldiag(A), is the set of all first-order sentences of L(A) which are true in D.

We say that a homomorphism $f\colon A \to B$ preserves the formula $\phi(\bar{x})$ iff for all tuples $\bar{a}$ from A (of the same length as $\bar{x}$ - we usually assume this without saying it), $A \models \phi(\bar{a})$ implies $B \models \phi(f\bar{a})$. A homomorphism which preserves all first-order formulas is called an elementary embedding.

LEMMA 1.2.3 (Diagram lemma). *Let* A *be an* L*-structure and* B *an* L(A)*-structure. Then the following are equivalent:*

(a) $B \models \mathrm{diag}^+(A)$ *(resp.* diag(A)*,* eldiag(A)*).*

(b) There is a unique homomorphism (resp. embedding, elementary embedding) $f\colon A \to B|L$ *such that* $f(a) = \tilde{a}^B$ *for each* $a \in \mathrm{dom}(A)$.

Proof. Cf. Shoenfield (1967) section 5.2, or Chang & Keisler (1973) section 2.1. □

Quantifier prefixes

A quantifier-free formula is one in which no quantifiers occur. A formula is said to be universal, or a $\forall_1$ formula, iff it has the form $\forall\bar{x}\psi$ for some quantifier-free ψ; we allow $\bar{x}$ to be empty, so that every quantifier-free formula counts as a $\forall_1$ formula. A $\forall_1$ theory is a theory consisting of $\forall_1$ formulas. For example an equational theory is a theory consisting of sentences of form $\forall\bar{x}(s=t)$ where s and t are terms. Any equational theory is a first-order $\forall_1$ theory.

The class of all models of an equational theory is what universal algebraists know as a variety. The classes of groups, abelian groups, rings, boolean algebras, lattices, left modules over a fixed ring are examples of varieties (provided each class is put into a suitable signature).

Dually, we say a formula is existential, or an $\exists_1$ formula, iff it has the form $\exists\bar{x}\psi$ for some quantifier-free ψ. The negation of a universal formula is equivalent to an existential formula, and *vice versa*.

A formula is said to be a $\forall_2$ formula iff it is of form $\forall\bar{x}\exists\bar{y}\psi$ for some quantifier-free formula ψ. As before, $\bar{x}$ and $\bar{y}$ are allowed to be empty, so that every universal formula and every existential formula is $\forall_2$. An example of a $\forall_2$ theory is the theory of fields:

(11)
$$\forall xyz\ x+(y+z) = (x+y)+z.$$
$$\forall xy\ \ x+y = y+x.$$
$$\forall x\ \ x+0 = x.$$
$$\forall x\ \ x+(-x) = 0.$$
$$\forall xyz\ x.(y.z) = (x.y).z.$$
$$\forall xy\ \ x.y = y.x.$$
$$\forall x\ \ x.1 = x.$$
$$\forall xyz\ x.(y+z) = x.y + x.z.$$
$$0 \neq 1.$$
$$\forall x\exists y(x\neq 0 \rightarrow x.y = 1).$$

Generalising the definitions above, we say that a formula is a $\forall_n$ (resp. $\exists_n$) formula iff it can be written as $\forall\bar{x}_1\exists\bar{x}_2\forall\bar{x}_3\ldots\bar{x}_n\psi$ (resp. $\exists\bar{x}_1\forall\bar{x}_2\exists\bar{x}_3\ldots\bar{x}_n\psi$) with n alternating quantifier blocks and ψ quantifier-free. Again any of the blocks $\forall\bar{x}_i$, $\exists\bar{x}_i$ can be empty.

A chain of L-structures is a sequence $(A_i : i < \gamma)$ of L-structures such that for all $i < j < \gamma$, $A_i \subseteq A_j$. There is a unique L-structure B such that $\mathrm{dom}(B) = \bigcup_{i<\gamma}\mathrm{dom}(A_i)$ and $A_i \subseteq B$ for each $i < \gamma$; this L-structure B is called the union of the chain, in symbols $\bigcup_{i<\gamma}A_i$.

LEMMA 1.2.4 (Lemma on preservation). *Let* A *be an* L-*structure,* $\bar{a}$ *a tuple of elements of* A *and* $\phi(\bar{x})$ *a formula of* L. *Then:*

(a) If ϕ *is universal,* $B \subseteq A$, $\bar{a}$ *are elements of* B *and* $A \models \phi(\bar{a})$ *then* $B \models \phi(\bar{a})$. *In particular if* A *is a model of a* $\forall_1$ *theory* T *and* $B \subseteq A$, *then* B *is also a model of* T.

(b) If ϕ *is existential,* $f\colon A \rightarrow C$ *is an embedding and* $A \models \phi(\bar{a})$ *then* $C \models \phi(f\bar{a})$. *("Embeddings preserve* $\exists_1$ *formulas.")*

(c) *If* $(A_i : i < \gamma)$ *is a chain of* L*-structures with* $A = A_0$*, and* ϕ *is a* $\forall_2$ *first-order formula such that* $A_i \models \phi(\bar{a})$ *for all* $i < \gamma$*, then* $\bigcup_{i<\gamma} A_i \models \phi(\bar{a})$*. Thus the union of a chain of models of a* $\forall_2$ *first-order theory* T *is again a model of* T.

Proof. Cf. Shoenfield (1967) section 5.2. □

Elementary substructures

Let A and B be L-structures. We say that A is an elementary substructure of B, and B is an elementary extension of A, in symbols $A \preccurlyeq B$, iff $A \subseteq B$ and the inclusion map from A to B is an elementary embedding. The definition implies at once that if $A \preccurlyeq B$ and $B \preccurlyeq C$ then $A \preccurlyeq C$; and that if $A \preccurlyeq C$, $B \preccurlyeq C$ and $A \subseteq B$ then $A \preccurlyeq B$.

LEMMA 1.2.5 (Tarski-Vaught criterion for elementary substructures). *Suppose* L *is a first-order language,* A *and* B *are* L*-structures and* $A \subseteq B$*. Then the following are equivalent:*

(a) $A \preccurlyeq B$.

(b) *For every formula* $\phi(x,\bar{y})$ *of* L *and every tuple* $\bar{a}$ *from* A*, if* $B \models \exists x\phi(x,\bar{a})$ *then there is* $b \in dom(A)$ *such that* $B \models \phi(b,\bar{a})$.

Proof. Cf. Chang & Keisler (1973) Proposition 3.1.2. □

LEMMA 1.2.6 (Downward Löwenheim-Skolem theorem). *Let* B *be a structure of signature* L*. Let* X *be a set of elements of* B*, and let* λ *be an infinite cardinal such that* $|X| \leqslant \lambda$ *and* L *contains at most* λ *symbols. Then there is an* L*-structure* A *such that* $A \preccurlyeq B$*,* $X \subseteq dom(A)$ *and* A *has cardinality* $\leqslant \lambda$.

Proof. Cf. Chang & Keisler (1973) Theorem 3.1.6. □

Let L be a first-order language. In view of Lemma 1.2.6, we define the cardinality of L, $|L|$, to be the least infinite cardinal $\geqslant$ the number of symbols in the signature of L. In fact this cardinal is the same as the number of essentially different formulas of L, where we count two formulas as not-essentially-different iff the second comes from the first by a one-to-one change of variables.

An elementary chain is a chain $(A_i : i < \gamma)$ in which $A_i \preccurlyeq A_j$ whenever $i < j < \gamma$.

LEMMA 1.2.7 (Elementary chain lemma). *Let* $(A_i : i < \gamma)$ *be an elementary chain. Then* $\bigcup_{i<\gamma} A_i$ *is an elementary extension of every* A_j $(j < \gamma)$.

Proof. Cf. Chang & Keisler (1973) Theorem 3.1.13. □

Fragments

When dealing with a language of form $L_{\omega_1\omega}$, we sometimes need to pick out countable sets of formulas which are better behaved than the language $L_{\omega_1\omega}$ as a whole. We proceed as follows.

Let L be a countable first-order language, whose variables are the countable set $\{x_n : n < \omega\}$. A fragment of $L_{\omega_1\omega}$ is a countable set Φ of formulas of $L_{\omega_1\omega}$ such that (i) every atomic formula of L is in Φ, (ii) if ϕ, ψ are in Φ then so are $\neg\phi$, $\phi\wedge\psi$, $\phi\vee\psi$, $\phi\to\psi$, $\phi\leftrightarrow\psi$, $\forall x_n\phi$, $\exists x_n\phi$ (for all $n < \omega$), (iii) if $\phi(x) \in \Phi$ and t is a term of L then $\phi(t) \in \Phi$, (iv) if ϕ is in Φ then every subformula of ϕ is in Φ, (v) if $\neg\bigwedge\Psi$ (resp. $\neg\bigvee\Psi$) is in Φ then so is $\bigvee\{\neg\psi : \psi \in \Psi\}$ (resp. $\bigwedge\{\neg\psi : \psi \in \Psi\}$), (vi) each formula in Φ has at most finitely many free variables. If A and B are L-structures, we define $A \preccurlyeq_\Phi B$ to mean that $A \subseteq B$ and for every formula $\phi(\bar{x}) \in \Phi$ and every tuple $\bar{a}$ from A, $A \models \phi(\bar{a})$ iff $B \models \phi(\bar{a})$.

For example L itself is a fragment of $L_{\omega_1\omega}$, and $\preccurlyeq_L$ coincides with $\preccurlyeq$.

LEMMA 1.2.8. *(a) For every formula* ϕ *of* $L_{\omega_1\omega}$ *with just finitely many free variables, there is a fragment of* $L_{\omega_1\omega}$ *containing* ϕ.

(b) Let B *be an* L*-structure,* X *a finite or countable set of elements of* B*, and* Φ *a fragment of* $L_{\omega_1\omega}$. *Then there is an* L*-structure* A *of cardinality at most* ω*, such that* $A \preccurlyeq_\Phi B$ *and* $X \subseteq \mathrm{dom}(A)$.

(c) Let Φ *be a fragment of* $L_{\omega_1\omega}$ *and* $(A_i : i < \gamma)$ *a chain of* L*-structures such that* $A_i \preccurlyeq_\Phi A_j$ *whenever* $i < j < \gamma$. *Then for each* $j < \gamma$, $A_j \preccurlyeq_\Phi \bigcup_{i<\gamma} A_i$.

Proof. Cf. Keisler (1971) Chapters 4, 21. □

Other ideas from model theory will be explained as and when they are needed. For the present it remains only to note some pieces of set-theoretic symbolism.

Set theory

Ordinals are written α, β, γ etc., or i, j, k etc. Usually δ is a limit ordinal. 0 is not a limit ordinal. Natural numbers are finite ordinals, and are written m, n etc. Ordinals are von Neumann ordinals, so that each ordinal is the set of its predecessors. A cardinal λ is identified with the first ordinal of cardinality λ. Thus the infinite cardinals are ω_0 (= ω), ω_1, ω_2 etc. Cardinals are written κ, λ, μ etc.

A sequence $(a_i : i < \gamma)$ is said to be non-repeating iff $a_i \neq a_j$ whenever $i \neq j$. The sequence $(X_i : i < \gamma)$ is an (increasing) chain iff $X_i \subseteq X_j$ whenever $i < j < \gamma$; we say the chain is continuous iff for every limit ordinal $\delta < \gamma$, $X_\delta = \bigcup_{i<\delta} X_i$.

Exercises for 1.2.

1. Let A and B be L-structures with dom(A) $\subseteq$ dom(B). Show that $A \subseteq B$ iff: (i) for each $n \geqslant 1$ and n-ary function symbol F of L, F^A is the restriction of F^B to dom(A); (ii) for each constant c of L, $c^A = c^B$; and (iii) for each $n \geqslant 1$ and n-ary relation symbol R of L, $R^A = R^B \cap (\text{dom } A)^n$.

2. (a) In the signature with function symbols $\cdot$, $^{-1}$ and constant 1, write down a first-order set of axioms for the class of groups. (b) Do the same for torsion-free abelian groups, in the signature whose symbols are +, − and 0. [In this case you will need infinitely many axioms.] (c) Do the same for left modules over a ring R with 1, using the signature whose symbols are +, −, 0, and 1-ary function symbols $\underline{r}$ for each $r \in R$ (where $\underline{r}(x)$ means x multiplied on the left by r). (d) Do the same for lattices, in the signature whose symbols are $\wedge$ and $\vee$. (e) Write a sentence of $L_{\omega_1\omega}$ (any signature) which says "There are just finitely many elements in the domain".

3. (a) Using the same signature as in Exercise 2(a), write a $\forall_1$ first-order formula $\phi(x)$ such that for any group G, $\phi(G)$ is the centre Z(G) of G. (b) Use (a) and Lemma 1.2.4(a) to deduce that if H is a subgroup of G then $H \cap Z(G) \subseteq Z(H)$. (c) Extend (a), (b) to all terms Z_n $(n < \omega)$ of the upper central series (cf. Kargapolov & Merzljakov (1979) section 16).

4. Show that if f: A → B is an elementary embedding, then f is an embedding.

5. Let $\phi(\bar{x})$, $\psi(\bar{x})$ be formulas of a first-order language L and T a theory in L. Show that ϕ and ψ are equivalent modulo T if and only if for every

model A of T and every tuple $\bar{a}$ from A, $A \models \phi(\bar{a}) \Leftrightarrow A \models \psi(\bar{a})$.

6. Let A and B be L-structures such that $A \subseteq B$, and let $\bar{a}$ be a sequence listing all the elements of A. Show that $A \preccurlyeq B$ iff $(A,\bar{a}) \equiv (B,\bar{a})$.

7. Show that if $A \preccurlyeq C$, $B \preccurlyeq C$ and $A \subseteq B$ then $A \preccurlyeq B$.

REFERENCES FOR CHAPTER 1

1.1: Kratylos, who is said to have influenced Plato, used to communicate by wiggling his finger. (According to Aristotle, he did this because he had noticed that things keep changing, and he was afraid that if he said anything, it would turn false by the time he had finished saying it.) Picture (3) has been very influential in mathematical logic. Perhaps it comes from Cantor (1872). The "shovel-everything-in" idea in (4) is due to Fraissé (1953), who also pointed out the importance of amalgamation in this context. Picture (5) is well known throughout mathematics, for example in pushouts. The version of Church & Rosser (1936) seems to be one of its first appearances. The present popularity of (7) in model theory is due to Keisler (1970). For the higher-dimensional versions see Shelah (1983 b,c); they are foreshadowed in Shelah (1975 b). I think none of these authors printed his pictures on the page.

1.2: Chang & Keisler (1973) is the authoritative text on first-order model theory up to about 1970. But it is rather too encyclopedic for a quick introduction. For that I recommend the opening chapters of either Bell & Slomson (1969) or Sacks (1972) for those who know some logic; readers whose logical background is weaker might start with Bridge (1977) or Malitz (1979). Keisler (1977) is highly recommended as a survey, and it now has the advantage of being available in Russian too. Speaking of other languages, German readers should find Potthoff (1981) helpful. In French I very much hope that Poizat (198-) will be published soon.

2 GAMES AND FORCING

Bricks are made chiefly from clay and shale. Clay, a plastic earth, is constituted largely of sand and alumina and may contain various quantities of chalk, iron, manganese dioxide, etc. Shale is a laminated deposit of clay rock which is capable of being reduced to a plastic condition when broken up and ground to a fine state of division.

W. B. McKay, Building Construction (1970)

What are models made of? Often an algebraic structure is built out of other structures by forming products, unions of chains and so forth. In this chapter we shall build a structure out of raw materials, by induction up the ordinals. Then we shall start to analyse what can be got by this method of construction. The central idea is that many different tasks have to be performed by the end of the construction. We think of each task as being assigned to a separate builder. Each builder can regard the other builders as rivals who keep interfering in his attempts to carry out his task. Is there a strategy which will enable him to finish his job regardless of what the rest do? Of course this will depend on the task, and on the properties of the building materials.

2.1 A WAY OF BUILDING MODELS

COMPACTNESS THEOREM FOR FIRST-ORDER LOGIC. *Let* T *be a first-order theory such that each finite subset of* T *has a model. Then* T *has a model.*

The reader probably knows this theorem already. Nevertheless I shall prove it, because the proof which I shall give is the basis for all the constructions in this book. The aim is to construct a model of T, and we do it by building up inductively a theory which contains T and has a "canonical" model. This type of construction is known as a Henkin construction; the name means nothing very precise, but it recalls two papers (1949; 1963) of Leon Henkin.

Let T be a set of atomic sentences of signature L. We say that T is =-closed in L iff (i) for every closed term t of L, the sentence t=t is in T, and (ii) for every atomic formula $\phi(x)$ of L and all closed terms s, t of L, if T contains $\phi(s)$ and at least one of the sentences s=t, t=s, then T also contains $\phi(t)$.

THEOREM 2.1.1 (Canonical model theorem). *Let* L *be a signature with at least one constant. Let* T *be an =-closed set of atomic sentences of signature* L. *Then there is an* L*-structure* A *such that:*

(a) *every element of* A *is of form* t^A *for some closed term* t *of* L*;*

(b) *for every atomic sentence* ϕ *of* L*,* $A \models \phi$ *iff* $\phi \in T$.

This L*-structure* A *is unique up to isomorphism.*

Proof. I give the barest sketch of the proof. Details can be found for example in Shoenfield (1967) section 4.2, Chang & Keisler (1973) Lemma 2.1.2 or Malitz (1979) Lemma 3.7.7.

We define a relation $\sim$ on the set of closed terms of L by: $s \sim t$ iff $(s=t) \in T$. By =-closure of T, this is an equivalence relation. Writing $t^\sim$ for the equivalence class of t, we take dom(A) to be the set of all equivalence classes $t^\sim$ of closed terms t of L; this set is non-empty because L has a constant. For each constant c of L we put $c^A = c^\sim$. For each $n \geqslant 1$ and n-ary function symbol F of L, we put $F^A(t_1^\sim,\ldots,t_n^\sim) = (F(t_1,\ldots,t_n))^\sim$; this is well-defined since T is =-closed. For each $n \geqslant 1$ and n-ary relation symbol R of L, we put $(t_1^\sim,\ldots,t_n^\sim) \in R^A$ iff $R(t_1,\ldots,t_n) \in T$; the definition is sound since T is =-closed. An induction on complexity shows that for every closed term t of L, $t^A = t^\sim$. (a) and (b) then follow easily. For the uniqueness, see Exercise 1 below.

□

The model of T constructed in Theorem 2.1.1 is called the canonical model of T.

For the next step in the proof of the compactness theorem, we axiomatise the notion of a finitely consistent set. Intuitively one should think of "consistent" as meaning "could be true in some structure".

For example if p is a set of sentences true in a structure A, and $\phi \wedge \psi$ is in p, then by the usual inductive definition of "true", ϕ and ψ must also be true in A, and so $p \cup \{\phi,\psi\}$ is again a set of sentences true in A. Hence if each finite subset of a set p is "consistent" and $\phi \wedge \psi \in p$, then each finite subset of $p \cup \{\phi,\psi\}$ must be "consistent" too. This is what clause (5) below says. The reader should try to justify clause (6) along the same lines. (At this point I usually get the class to write down the remaining clauses of (1)-(12) out of their own heads.) The cardinality restriction in (13) is for a technical reason: we don't want a "consistent" set to use up all the available closed terms.

Incidentally clauses (4)-(13) are irrelevant for the algebraic applications in Chapters 3 and 4. For simplicity I have left out the four clauses for $\rightarrow$ and $\leftrightarrow$; one can take $\phi \rightarrow \psi$, $\phi \leftrightarrow \psi$ to be shorthand for $\neg\phi \vee \psi$, $(\phi \wedge \psi) \vee (\neg\phi \wedge \neg\psi)$ respectively.

Let L be a first-order language of cardinality λ, and W a set of closed terms of L. By a <u>notion of consistency</u> (for L) we mean a set N of sets of sentences of L such that the following thirteen clauses hold for every set $p \in N$:

(1) for every closed term t of L, $p \cup \{t=t\} \in N$;

(2) for every atomic formula $\phi(x)$ of L and all closed terms s, t of L, if $\phi(s) \in p$ and p contains at least one of s=t, t=s, then $p \cup \{\phi(t)\} \in N$;

(3) for every atomic sentence ϕ of L, ϕ and $\neg\phi$ are not both in p;

(4) for every sentence ϕ of L, if $\neg\neg\phi \in p$ then $p \cup \{\phi\} \in N$;

for all sentences ϕ, ψ of L,

(5) if $\phi \wedge \psi \in p$ then $p \cup \{\phi,\psi\} \in N$;

(6) if $\neg(\phi \wedge \psi) \in p$ then either $p \cup \{\neg\phi\} \in N$ or $p \cup \{\neg\psi\} \in N$;

(7) if $\phi \vee \psi \in p$ then either $p \cup \{\phi\} \in N$ or $p \cup \{\psi\} \in N$;

(8) if $\neg(\phi \vee \psi) \in p$ then $p \cup \{\neg\phi,\neg\psi\} \in N$;

for all variables x and formulas $\psi(x)$ of L,

(9) if $\forall x\psi \in p$ then for every closed term t of L, $p \cup \{\psi(t)\} \in N$;

(10) if $\neg\forall x\psi \in p$ then for some closed term t of L, $p \cup \{\neg\psi(t)\} \in N$;

(11) if $\exists x\psi \in p$ then for some closed term t of L, $p \cup \{\psi(t)\} \in N$;

(12) if $\neg\exists x\psi \in p$ then for every closed term t of L, $p \cup \{\neg\psi(t)\} \in N$;

and finally

(13) if $\alpha < \lambda$, $(p_i)_{i<\alpha}$ is an increasing chain of sets $\in N$ and $|\bigcup_{i<\alpha} p_i \smallsetminus p_0| < \lambda$, then $\bigcup_{i<\alpha} p_i \in N$.

The elements p of a notion of consistency N are known as <u>conditions</u>.

THEOREM 2.1.2. *If* N *is a notion of consistency and* $p \in N$, *then* p *has a model.*

<u>Proof</u>. We shall construct by induction an increasing chain $(p_i)_{i<\lambda}$ of conditions $\in N$, so that $|p_{i+1} \smallsetminus p_i|$ is finite for each $i < \lambda$. At the end we shall have constructed a set $\bigcup_{i<\lambda} p_i$, or $\bigcup \bar{p}$ for short. We begin by putting $p_0 = p$. At each limit ordinal $\delta < \lambda$ we put $p_\delta = \bigcup_{i<\delta} p_i$; this is allowed by (13) and the assumption that each $|p_{i+1} \smallsetminus p_i|$ is finite.

We set ourselves λ tasks to perform at the successor stages. These tasks correspond to the instances of (1), (2), (4)-(12) above. For example the task that corresponds to clause (7) with a particular pair of formulas ϕ, ψ is:

(14) If $\phi \vee \psi \in \bigcup \bar{p}$, then put at least one of ϕ, ψ into $\bigcup \bar{p}$.

Likewise the task that corresponds to clause (9) with a particular formula $\psi(x)$ and a particular closed term t is:

(15) If $\forall x\psi \in \bigcup \bar{p}$, then put $\psi(t)$ into $\bigcup \bar{p}$.

It should be clear from these examples what the rest of the tasks are. List all these tasks as τ_i $(i < \lambda)$, so that each task occurs λ times in the list. When p_i has been chosen, choose p_{i+1} so that task τ_i is performed if possible. For example suppose τ_i is (15). If $\forall x\psi \notin p_i$, there is nothing to do, and p_{i+1} can be p_i. But if $\forall x\psi \in p_i$, then by (9), $p_i \cup \{\psi(t)\} \in N$, and we can put $p_{i+1} = p_i \cup \{\psi(t)\}$. Since each task is listed λ times, it will eventually be completed.

At the end of the construction, let U be the set of atomic sentences in $\bigcup \bar{p}$. By the tasks corresponding to (1) and (2), U is =-closed in L, where L is the language for which N is a notion of consistency. Hence U has a canonical model A by Theorem 2.1.1. We assert that for every sentence ϕ of L,

$$(16) \qquad \phi \in \bigcup \bar{p} \Rightarrow A \models \phi, \qquad \text{and} \qquad \neg\phi \in \bigcup \bar{p} \Rightarrow A \models \neg\phi.$$

(16) is proved by induction on the complexity of ϕ. I consider two cases and leave the rest to the reader.

Suppose ϕ is atomic. If $\phi \in \bigcup \bar{p}$ then $\phi \in U$ so that $A \models \phi$ by the choice of A. If $\neg\phi \in \bigcup \bar{p}$ then $\phi \notin \bigcup \bar{p}$ by (3) and hence $\phi \notin U$, so that $A \models \neg\phi$ by the choice of A.

Suppose ϕ is $\exists x\psi$. If $\exists x\psi \in \bigcup \bar{p}$, then by the task corresponding to (11) there is a closed term t such that $\psi(t) \in \bigcup \bar{p}$, and so $A \models \psi(t)$ by induction hypothesis. Therefore t^A satisfies $\psi(x)$ in A, and so $A \models \exists x\psi$. If $\neg\exists x\psi \in \bigcup \bar{p}$, then by the tasks corresponding to (12), if t is any closed term of L, $\neg\psi(t) \in \bigcup \bar{p}$. Hence by induction hypothesis (the second half of (16)), $A \models \neg\psi(t)$. By the choice of A, every element of A is of form t^A for some closed term t. Hence $A \models \neg\exists x\psi$.

Thus (16) is proved. Since $p \subseteq \bigcup \bar{p}$, it follows that $A \models p$ and hence p has a model. □

Now we can prove the compactness theorem. Let T be a theory in a first-order language L of cardinality λ. Let W be a set $\{c_i : i < \lambda\}$ of new constants. Let N be the following set of sets of sentences of L(W): $p \in N$ iff (i) fewer than λ constants from W occur in sentences in p p, and (ii) every finite subset of p has a model. It is easy to verify that N is a notion of consistency for L(W), and that $T \in N$. So by Theorem 2.1.2, some L(W)-structure is a model of T.

□ Compactness theorem

Let us analyse the proof of Theorem 2.1.2. There were λ tasks to perform during the construction of $(p_i)_{i<\lambda}$. Each task could be completed, provided that λ sets p_{i+1} were assigned to it.

Now imagine that each task τ belongs to some model theorist;

call him τ^*. There are λ model theorists; imagine that they share out the work between them, so that each model theorist is responsible for choosing some sets p_{i+1} when the preceding sets p_j ($j \leqslant i$) have been chosen. Every model theorist τ^* can think of himself as playing a game against all the other model theorists: he wins his game iff task τ has been completed by the time the chain $(p_i)_{i<\lambda}$ is finished. Then each model theorist has a winning strategy for his own game, provided that he is able to choose λ of the sets p_{i+1}.

This raises the question: what other tasks can a model theorist guarantee to carry out, provided he is allowed to choose λ of the sets p_{i+1}? Read on for the answer.

Exercises for 2.1.

1. Supply the details of the proof of Theorem 2.1.1. To show the uniqueness of A up to isomorphism, prove that A also has the following property (c): For every L-structure B which is a model of T, there is a unique homomorphism $f: A \to B$.

2. If the symbols $\to$, $\leftrightarrow$ are not regarded as shorthand, what should be the clauses for $\phi\to\psi$, $\neg(\phi\to\psi)$, $\phi\leftrightarrow\psi$ and $\neg(\phi\leftrightarrow\psi)$ in the definition of notion of consistency?

3. Complete the proof of (16).

4. Let N be a notion of consistency for L. Let M be the set of all subsets of sets $\in$ N. Show that M is also a notion of consistency for L.

5. Let N be a notion of consistency for L. Suppose $p \in N$, and write N/p for the set of all $q \in N$ such that $p \subseteq q$. Show that N/p is a notion of consistency for L.

6. Deduce the following from the compactness theorem for first-order logic: (a) Let ϕ be a sentence and T a theory, both first-order, and suppose that ϕ is equivalent to T (i.e. ϕ and T have exactly the same models). Then ϕ is equivalent to the conjunction of a finite number of sentences from T. (b) Let ϕ be a sentence and Ψ a set of sentences, both in a first-order language L, and suppose that for any L-structure A, the question whether or not A is a model of ϕ depends only on which sentences in Ψ are true in A. Then ϕ is equivalent to a finite boolean combination of sentences in Ψ (i.e. a combination by means of $\neg$, $\wedge$ and $\vee$).

7. An <u>inverse system of sets</u>, of length γ, is defined to consist of sets X_α $(\alpha < \gamma)$ and maps $f_{\alpha\beta}: X_\beta \to X_\alpha$ $(\alpha \leqslant \beta < \gamma)$ such that for all $\alpha \leqslant \beta \leqslant \zeta < \gamma$, $f_{\alpha\beta}.f_{\beta\zeta} = f_{\alpha\zeta}$. Show that if X_α and $f_{\alpha\beta}$ $(\alpha, \beta < \gamma)$ form an inverse system of sets, and each X_α is a finite non-empty set, then there is a family of elements $(x_\alpha : \alpha < \gamma)$ such that for all $\alpha \leqslant \beta < \gamma$, $x_\alpha \in X_\alpha$ and $f_{\alpha\beta}(x_\beta) = x_\alpha$.

Model theorists often use the compactness theorem in order to construct elementary extensions, and also to prove that certain sets are not first-order definable. The next four exercises illustrate these applications.

8. (a) (Upward Löwenheim-Skolem theorem) Let A be an infinite L-structure and λ a cardinal, and suppose $|A| \leqslant \lambda$ and $|L| \leqslant \lambda$. Show that there exists an elementary extension B of A such that $|B| = \lambda$. [Use the diagram lemma, Lemma 1.2.3. Take the L-reduct of eldiag(A) $\cup$ $\{c_i \neq c_j : i < j < \lambda\}$ where $\{c_i : i < \lambda\}$ are new constants.] (b) Under the hypotheses of (a), suppose $\phi(x)$ is a first-order formula of L and $\phi(A)$ is infinite. Show that B can be chosen so that $|\phi(B)| = \lambda$.

9. Let A be an L-structure, and for every $i \leqslant \omega$ let $\phi_i(x, \bar{y}_i)$ be a first-order formula of L and $\bar{a}_i$ a tuple of elements of A, such that if $i < j \leqslant \omega$ then $\phi_i(A, \bar{a}_i) \subset \phi_j(A, \bar{a}_j)$. Show that there is an elementary extension B of A such that $\bigcup_{i<\omega} \phi_i(B, \bar{a}_i) \subset \phi_\omega(B, \bar{a}_\omega)$. [Consider the theory U = eldiag(A) $\cup$ $\{\neg\phi_i(c, \bar{a}_i) : i < \omega\} \cup \{\phi_\omega(c, \bar{a}_\omega)\}$, where c is a new constant. Use compactness to show that U has a model.]

10. Let G be a group with an element g of infinite order. Show that G has an elementary extension H in which the set of all powers of g is not first-order definable with parameters. [If the set of powers of g is $\phi_\omega(G, \bar{a}_\omega)$, apply Exercise 9 with suitable $\phi_i(x, \bar{a}_i)$ to construct an elementary extension H of G such that $\phi_\omega(H, \bar{a}_\omega)$ is not the set of powers of g. Repeat if necessary, forming an elementary chain.]

11. Show that there exists a group whose commutator subgroup is not first-order definable with parameters. [Cf. proof of Exercise 10.]

2.2 *GAMES*

Before we start to apply games, it will be as well if I say a few words about what games are.

All our games will be between two players, $\forall$ and $\exists$. I shall make $\forall$ male and $\exists$ female. (One can read $\forall$ as Abelard and $\exists$ as Eloise -

Peter Abelard was a twelfth-century Parisian logician who used to play games with Eloise, the niece of a canon of Notre Dame.)

Each game G will have a fixed length, which is an ordinal γ. The ordinal γ, regarded as the set of all smaller ordinals, is partitioned into two sets $M_\forall$ and $M_\exists$, respectively the set of $\forall$'s moves and the set of $\exists$'s moves. The players play G by choosing one by one a sequence of objects x_i $(i < \gamma)$. Player $\forall$ decides what x_i shall be if $i \in M_\forall$, while player $\exists$ has the choice of x_i if $i \in M_\exists$. At each stage the player who makes the choice is allowed to know all the previous choices of both players. When they have finished playing, the result is a sequence $(x_i)_{i<\gamma}$, $\bar{x}$ for short, which we call a play of G.

The game will also have some rules that restrict the possible choices. In particular all the objects chosen must come from some fixed set. There may be other restrictions: for example in virtually all the games in this book, the play $\bar{x}$ must form an increasing chain of sets.

Finally G says who wins each play. Putting it more formally, the set of all possible plays of G is partitioned into two sets; the plays in one set are called wins for $\forall$, and the plays in the other are called wins for $\exists$. There are no draws.

Suppose two players play the game G up to but not including the β-th move, so that they pick $(x_i)_{i<\beta}$. Then $(x_i)_{i<\beta}$ is called a position of length β. The β-th position of the play $\bar{x}$, in symbols $\bar{x}|\beta$, is the initial segment $(x_i)_{i<\beta}$. There is just one initial position, before any choices have been made, namely the empty sequence $<>$.

A strategy for a player in a game G is a set of instructions which tell this player how to move. For example a strategy for player $\exists$ can be thought of as a family σ of functions σ_i $(i \in M_\exists)$, so that each σ_i is defined on the set of all positions of length i. Player $\exists$ is said to use σ in the play $\bar{x}$ iff for each $i \in M_\exists$, $x_i = \sigma_i(\bar{x}|i)$.

If a player uses a strategy σ, then all his choices are determined by σ and the previous choices of the other player. So we can equally well think of a strategy as a family of functions σ_i which are functions of the previous moves of the other player. We shall use this style when it is convenient.

We say that a strategy σ for a player in a game G is winning iff that player wins every time he or she uses σ. So at most one player can have a winning strategy. A game where at least one player has a winning strategy is said to be *determined*. There are some games which are not determined - see Exercises 4 and 5.

All the definitions given so far are symmetrical between the two players. But in practice model theorists tend to take sides and support the player ∃. This helps to decide which player is labelled when we set up a game: if the important question is whether a certain player has a winning strategy, then that player is usually called ∃.

Let $\bar{y}$ be a position in a game G. We can define the notion of a *strategy from position* $\bar{y}$ *onwards*. Such a strategy tells a player how to move in the rest of the game, assuming that the players have just reached position $\bar{y}$. Technically such a strategy will be a family of functions depending on the moves after $\bar{y}$, but we won't need the details. A strategy σ for a player from position $\bar{y}$ onwards is said to be *winning* iff that player wins whenever he or she uses σ for the rest of the play after $\bar{y}$. A position $\bar{y}$ is said to be *winning* for a player (and the player is said to be in *winning position* at $\bar{y}$) iff that player has a winning strategy from position $\bar{y}$ onwards. So for example a player has a winning strategy for G iff the initial position is winning for that player.

Exercises for 2.2.

1. Let $\bar{y}$ be a position in a game G, before the play has finished. Prove: (a) If $\bar{y}$ is winning for a player, then that player can guarantee to be still in winning position after the next move. (b) If $\bar{y}$ is not winning for a player, then the other player can guarantee that the former player is still not in winning position after the next move.

2. Let σ be a strategy for player ∃ in the game G, thought of as a family of functions σ_i $(i \in M_\exists)$ defined on positions in G. Show that σ can also be represented as a family of functions σ^i $(i \in M_\exists)$ which define player ∃'s choice at the i-th stage as a function of player ∀'s choices at previous stages. Define the family of functions σ^i in terms of the functions σ_i, and *vice versa*.

3. Show that every game of finite length is determined. [Use Exercise 1(b).]

For topologists: the games to be studied in this book nearly all have the following topological form. (I give this as background information - I am not convinced that it is helpful for understanding anything later.)

4. Write ${}^{\omega}2$ for the set of all maps $f: \omega \to \{0,1\}$. A condition is a map $p: Y \to \{0,1\}$ where Y is a finite subset of ω; write $M(p)$ for the set of maps $f \in {}^{\omega}2$ which extend p. ${}^{\omega}2$ is topologised by taking as basic open sets the sets $M(p)$ for conditions p. Given a non-empty closed set $F \subseteq {}^{\omega}2$ and a set $P \subseteq F$, players $\forall$ and $\exists$ play the game $G(F,P)$ of length ω as follows. The players choose an increasing sequence $p_0 \subseteq p_1 \subseteq \ldots$ of conditions so that $F \cap M(p_i) \neq \emptyset$ for each $i < \omega$; player $\forall$ chooses p_i iff i is even. Player $\exists$ wins iff $P \cap \bigcap_{i<\omega} M(p_i) \neq \emptyset$. (These games are a version of the so-called Banach-Mazur games.) Show: (a) Player $\exists$ has a winning strategy for $G(F,P)$ iff P is comeagre in F. (b) Player $\forall$ has a winning strategy for $G(F,P)$ iff there is a condition p such that $M(p) \setminus P$ is comeagre in $M(p)$. (c) If P_n $(n < \omega)$ are subsets of F, then player $\exists$ has a winning strategy for $G(F, \bigwedge_{n<\omega} P_n)$ iff player $\exists$ has a winning strategy for each game $G(F,P_n)$ $(n < \omega)$. (d) If P is a Borel set then $G(F,P)$ is determined.

5. Consider games $G({}^{\omega}2,-)$ as in Exercise 4. (a) Show that if τ is a strategy for one player in such a game, then there is a family $\bar{p}^{\alpha}$ $(\alpha < 2^{\omega})$ of plays in which that player uses τ, such that if $\alpha < \beta < 2^{\omega}$ then for some $n < \omega$, $M(p^{\alpha}_n) \cap M(p^{\beta}_n) = \emptyset$. (b) Show that there is a set $P \subseteq {}^{\omega}2$ such that the game $G(F,P)$ is not determined. [List the possible strategies of either player as τ^{α} $(\alpha < 2^{\omega})$. Inductively build up P and ${}^{\omega}2 \setminus P$ so that for each $\alpha < 2^{\omega}$ there is a play $\bar{p}$ in which one player uses τ^{α} but $\bar{p}$ is in the wrong one of P, ${}^{\omega}2 \setminus P$; use part (a).]

6. For a fixed non-empty closed set $F \subseteq {}^{\omega}2$, let C be the set $\{p : p$ is a condition such that $F \cap M(p) \neq \emptyset\}$. We call a set $D \subseteq C$ pre-dense iff for every condition $p \in C$ there is a condition $q \in D$ such that $F \cap M(p) \cap M(q) \neq \emptyset$; we call D dense iff for every condition $p \in C$ there is a condition $q \in D$ such that $q \supseteq p$. Given any set Q of subsets of C, we say that an element f of F is Q-generic iff for every $D \in Q$ there is $p \in D$ such that $f \in M(p)$. (a) Show that for every set $D \subseteq C$, D is pre-dense iff the set $F \setminus \bigcup\{M(p) : p \in D\}$ is nowhere dense in F. (b) Let Q be a countable family of pre-dense subsets of C, and let P be the set of all Q-generic elements of F. Show that player $\exists$ has a winning strategy for $G(F,P)$. (c) Show that every dense subset of C is pre-dense.

2.3 *FORCING*

For the rest of this book we shall be building structures by games, using notions of consistency to define the allowed moves. The notions of consistency will come in various shapes and sizes, but they will nearly all have certain crucial features in common. I don't want to overburden the opening definitions, so for the moment I pick out just two of these features, which control how the elements of the structure will be named. A notion of consistency with these features will be called a notion of forcing. The first step in every application will be to find the right notion of forcing. The exercises below explore three other features (6)-(8) which notions of forcing usually have.

Until further notice (in fact until the middle of Chapter 5), L is a fixed countable first-order language. (In section 5.3 and Chapter 8 we drop the assumption of countability.) $W = \{c_i : i < \omega\}$ is a set of distinct new constants, which are known as <u>witnesses</u>. L(W) is the first-order language got from L by adding these constants.

By a <u>notion of forcing</u> (for L(W)) we mean a notion of consistency N for L(W) which satisfies the two extra clauses:

(1) if $p \in N$, t is a closed term of L(W) and c is a witness which occurs nowhere in p or t, then $p \cup \{t=c\} \in N$;

(2) at most finitely many witnesses occur in any one $p \in N$.

(Note that (13) of section 2.1 is irrelevant here. The cardinality λ of the language is ω, and trivially N is closed under unions of finite chains.)

Elements of N are called N-<u>conditions</u>, or simply <u>conditions</u> when the context allows. An N-<u>construction sequence</u> is a chain $\bar{p} = (p_i)_{i<\omega}$ of N-conditions. We write $\bigcup\bar{p}$ for $\bigcup_{i<\omega} p_i$. There is a least =-closed set U in L(W) which contains all the atomic sentences $\in \bigcup\bar{p}$; we write $A^+(\bar{p})$ for the canonical model of U, and $A(\bar{p})$ for the L-reduct of $A^+(\bar{p})$. When $\bar{p}$ is fixed by the context, we can abbreviate $A^+(\bar{p})$ and $A(\bar{p})$ to A^+ and A respectively. Indiscriminately we call A^+ or A the <u>structure compiled by</u> $\bar{p}$, or more briefly the <u>compiled structure</u>.

Let P be a property which the set $\bigcup\bar{p}$ can have or fail to have; since $\bigcup\bar{p}$ determines the compiled structure $A^+(\bar{p})$, this includes properties of $A^+(\bar{p})$. Let X be a subset of ω. Imagine the following game $G_N(P;X)$ (written $G(P;X)$ when N is fixed by the context). Players $\forall$ and $\exists$ choose between them the stages p_i of an N-construction sequence $\bar{p}$; player $\exists$ has the choice of p_i when $i \in X$, and otherwise player $\forall$ chooses p_i. Player $\exists$ wins iff at the end of the game, $\bigcup\bar{p}$ has property P.

We shall say that the game $G_N(P;X)$ is <u>standard</u> iff player $\forall$ has the choice of p_0 and both players have infinitely many moves; in other words, X is an infinite and coinfinite subset of $\omega \setminus \{0\}$.

For example the game $G_N(P;\text{odds})$ is standard, where 'odds' is the set of odd positive integers. In fact every standard game $G_N(P;X)$ is equivalent to $G_N(P;\text{odds})$, in the sense that each player has a winning strategy for $G_N(P;X)$ if and only if he has one for $G_N(P;\text{odds})$. I leave the formal proof of this to the reader, but the reason is as follows. If a player controls the choice of $p_i, p_{i+1}, \ldots, p_{i+k}$, then this player loses nothing if he makes $p_i, \ldots, p_{i+k-1}$ all equal to his preferred choice of p_{i+k}, and the other player loses nothing if he assumes that the first player has done this. Hence in any block of moves of the same player, only the last move is significant.

We say that a property P is N-<u>enforceable</u> (or simply <u>enforceable</u> when N is clear) iff player $\exists$ has a winning strategy for some (or equivalently, for all) standard games $G_N(P;X)$.

The next lemma is a refinement of Theorem 2.1.2.

LEMMA 2.3.1. *Let* N *be a notion of forcing. Then the following property* P *is* N-*enforceable: "The compiled structure* $A^+(\bar{p})$ *is a model of* $\bigcup\bar{p}$, *and each element of* $A^+(\bar{p})$ *is of form* $c^{A^+(\bar{p})}$ *for infinitely many witnesses* c".

<u>Proof</u>. The language L(W) is countable, and in any standard game player $\exists$ has countably many moves. Let her use them to carry out the countably many tasks described in the proof of Theorem 2.1.2, together with the following countably many tasks:

(3) (If t is a closed term of L(W) and $n < \omega$) put $t=c_i$ into $\bigcup\bar{p}$ for some witness c_i with $i \geqslant n$.

She can do this because N satisfies (1), (2) above. Then exactly as in the proof of (16) in section 2.1, $A^+(\bar{p})$ will be a model of $\bigcup\bar{p}$. If the tasks (3) are performed, then every element of $A^+(\bar{p})$ will be named by infinitely many witnesses. □

Many of the arguments in the chapters below will aim to show that certain properties are enforceable. It will be helpful to relativise this notion and think of properties which are "enforceable as soon as certain sentences have been put into $\bigcup\bar{p}$". The following definition makes this precise.

Let N be a notion of forcing, q an N-condition and P a property. We say that q <u>forces</u> P, in symbols $q \Vdash_N P$ or $q \Vdash P$, iff: for any position $(p_0,\ldots,p_k)$ in a standard game $G_N(P;X)$, if $q \subseteq p_k$ then the position is winning for player ∃. In other words, as soon as q has got into $\bigcup\bar{p}$, player ∃ can be sure of winning. The next lemma puts this definition into the form which we shall usually use in calculations:

LEMMA 2.3.2. *Let* N *be a notion of forcing,* q *an* N*-condition and* P *a property. Then the following are equivalent:*

(a) q *forces* P.

(b) *In* $G_N(P;\text{odds})$, *if player* ∀ *chooses* $p_0 \supseteq q$, *then he puts player* ∃ *into winning position.*

<u>Proof</u>. (a) immediately implies (b). For the converse, assume (b). Let $(p_0,\ldots,p_k)$ be a position in a standard game $G_N(P;X)$ with $q \subseteq p_k$; we must show that player ∃ can guarantee to win from this position. We can assume that the choice of p_{k+1} is hers (otherwise let her sit back and wait till her next move). In her own mind let her imagine that the choices of $p_0, \ldots, p_{k-1}$ were simply a warming-up, so that the game begins in earnest at p_k. Thus she imagines she is playing $G(P;Y)$ where $Y = \{n - k : n \in X, n \geqslant k\}$. Now $G(P;Y)$ is a standard game in which player ∀ has just made his first choice $\supseteq q$, so by (b), he has put player ∃ into winning position. If she plays to win $G(P;Y)$, she will win $G(P;X)$ too. □

LEMMA 2.3.3. *Let* N *be a notion of forcing,* q *an* N*-condition and* P *a property.*

(a) q *forces* P *iff* P *is* (N/q)*-enforceable, where* N/q *is the notion of forcing consisting of all* N*-conditions* p *such that* $q \subseteq p$.

(b) P *is N-enforceable iff every N-condition forces* P.

(c) *If* q *forces* P *and* $p \supseteq q$ *then* p *forces* P.

(d) *If for every* $p \supseteq q$ *there is* $r \supseteq p$ *such that* r *forces* P, *then* q *forces* P.

(e) (Conjunction lemma) *Suppose* P *is the conjunction of the properties* P_i $(i < \omega)$. *Then* q *forces* P *iff* q *forces each* P_i $(i < \omega)$.

Proof. (a)-(c) are immediate from the definitions.

To prove (d), assume that for every $p \supseteq q$ there is $r \supseteq p$ which forces P. In a play of $G_N(P;\text{odds})$, suppose player $\forall$ chooses p_0 with $q \subseteq p_0$. Then by assumption some $r \supseteq p_0$ forces P. By choosing p_1 to be r, player $\exists$ puts herself in winning position; so p_0 was already a winning position for her, and thus q forces P. (We are using Lemma 2.3.2.)

In (e), left to right is trivial. We prove right to left. Partition the odd natural numbers into countably many infinite sets X_i $(i < \omega)$. In a play of $G_N(P;\text{odds})$, suppose player $\forall$ chooses $p_0 \supseteq q$. By assumption player $\exists$ has a winning strategy σ_i for $G_N(P_i;X_i)$ from position (p_0) onwards. Let her play the rest of $G_N(P;\text{odds})$ by using σ_i to choose p_j whenever $j \in X_i$. Let $\bar{p}$ be the resulting play. Then for each odd i, $\bar{p}$ is also a play of $G_N(P_i;X_i)$ in which player $\exists$ uses σ_i and wins. So each property P_i holds. □

Let ϕ be a sentence of the signature of L(W), and let P be the property "$A^+(\bar{p}) \models \phi$". Then we write $G_N(\phi;X)$ for $G_N(P;X)$; we say ϕ is N-enforceable iff P is N-enforceable, and we say q forces ϕ iff q forces P.

Here ϕ need not be first-order. If ϕ is in $L(W)_{\omega_1\omega}$ (the language got by adding countable conjunctions and disjunctions to the operations of L(W)), then we can characterise those conditions which force ϕ; so says the following theorem.

THEOREM 2.3.4. *Let* N *be a notion of forcing and* q *an* N-*condition.*

(a) q *forces every logically true sentence.*

(b) *If* q *forces* ϕ *and* $\phi \vdash \psi$, *then* q *forces* ψ.

(c) *Let* ϕ *be an atomic sentence of* L(W). *Then* q *forces* ϕ *iff for every* N-*condition* $p \supseteq q$ *there is an* N-*condition* $r \supseteq p$

with $\phi \in r$.

(d) *Let* ϕ *be a conjunction* $\bigwedge_{i<\omega}\phi_i$. *Then* q *forces* ϕ *iff for each* $i < \omega$, q *forces* ϕ_i.

(e) *Let* $\psi(x_1,\ldots,x_n)$ *be a formula. Then* q *forces* $\forall x_1\ldots x_n\psi$ *iff for every* n*-tuple* $\bar{c}$ *of witnesses,* q *forces* $\psi(\bar{c})$.

(f) *Let* ϕ *be a sentence of* $L(W)_{\omega_1\omega}$. *Then* q *forces* $\neg\phi$ *iff there is no* N*-condition* $p \supseteq q$ *which forces* ϕ.

Proof. (a) and (b) are immediate from the definition of forcing, and (d) is a special case of the conjunction lemma, Lemma 2.3.3(e).

We prove (e). Left to right is a special case of (b). For right to left, it suffices (by Lemma 2.3.3(a)) to show that if all the sentences $\psi(\bar{c})$ are enforceable, then so is $\forall\bar{x}\psi$. Suppose all the sentences $\psi(\bar{c})$ are enforceable. Then by Lemma 2.3.1 and the conjunction lemma, so is the property "All the sentences $\psi(\bar{c})$ are true in $A^+(\bar{p})$, and every element of $A^+(\bar{p})$ is named by a witness". But if this property holds, then clearly $A^+(\bar{p}) \models \forall\bar{x}\psi$.

There remain (c) and (f). As a lemma for both, we claim

(4) if ϕ is an atomic sentence and p is an N-condition, then p forces $\neg\phi$ iff no condition $\supseteq$ p contains ϕ.

For suppose first that some condition $r \supseteq p$ contains ϕ. Then by playing $p_0 = r$ in $G_N(\neg\phi;\text{odds})$, player $\forall$ puts ϕ into $\bigcup\bar{p}$ and hence prevents player $\exists$ from making $\neg\phi$ true; so p doesn't force $\neg\phi$. Second, suppose no condition $\supseteq$ p contains ϕ. Then let player $\exists$ play $G_N(\neg\phi;\text{odds})$ so that $\bigcup\bar{p}$ is =-closed (she can do this by Lemma 2.3.1), and hence so that $A^+(\bar{p}) \models \phi$ iff $\phi \in \bigcup\bar{p}$ (by the definition of $A^+(\bar{p})$ as canonical model). If player $\forall$ began with $p_0 \supseteq p$, then by supposition $\phi \notin \bigcup\bar{p}$ and so player $\exists$ wins. This proves (4).

Now we prove (c). If $\phi \in r$ then certainly r forces ϕ, as in the proof of (4). So if for every $p \supseteq q$ there is an $r \supseteq p$ which contains ϕ, then for every $p \supseteq q$ there is an $r \supseteq p$ which forces ϕ; and so q forces ϕ by Lemma 2.3.3(d). Conversely if there is $p \supseteq q$ such that no $r \supseteq p$ contains ϕ, then by (4), p forces $\neg\phi$. Let player $\forall$ in $G_N(\phi;\text{odds})$ choose $p_0 = p$, and for his remaining infinitely many moves let him borrow the habits of player $\exists$ and play to make $A^+(\bar{p}) \models \neg\phi$. Then player $\exists$ loses the game, and so q doesn't force ϕ.

To prove (f), we first show that for every N-condition q and every sentence ϕ of $L(W)_{\omega_1\omega}$,

(5) either some $p \supseteq q$ forces ϕ, or some $p \supseteq q$ forces $\neg\phi$.

(5) is proved by induction on the complexity of ϕ. (i) If ϕ is atomic suppose that no $p \supseteq q$ forces $\neg\phi$; then by (4), for every $p \supseteq q$ there is $r \supseteq p$ which contains ϕ, and so by (c), p already forces ϕ. (ii) If (5) holds for a sentence ϕ, then it holds for $\neg\phi$ too, since by (b), p forces $\neg\neg\phi$ iff p forces ϕ. (iii) Suppose that ϕ is $\bigwedge_{i<\omega}\phi_i$, and suppose that no $p \supseteq q$ forces $\neg\phi$. By (b) it follows that for each $i < \omega$, no $p \supseteq q$ forces $\neg\phi_i$. So by induction hypothesis, if $p \supseteq q$ then there is some $r_i \supseteq p$ which forces ϕ_i. By Lemma 2.3.3(d), it follows that q forces each ϕ_i, and so q forces ϕ by (d). (iv) If ϕ is of form $\forall\bar{x}\psi$, then (e) reduces us to Case (iii). Every sentence of $L(W)_{\omega_1\omega}$ is equivalent to one built up from atomic formulas by $\neg$, $\bigwedge$ and $\forall$; so by (b) the proof of (5) is complete.

Finally we prove (f). For left to right, suppose there is some $p \supseteq q$ which forces ϕ. Then let player $\forall$ put $p_0 = p$ in $G_N(\neg\phi;\text{odds})$, and play his remaining moves to ensure that $A^+(\bar{p}) \models \phi$; he wins, and so q doesn't force $\neg\phi$. For right to left, suppose no condition $p \supseteq q$ forces ϕ. If p is any condition $\supseteq q$, then no $r \supseteq p$ forces ϕ, and so by (5), some $r \supseteq p$ forces $\neg\phi$. Therefore by Lemma 2.3.3(d), q forces $\neg\phi$. □

Let me close by stating three more properties which notions of forcing often have. I leave their consequences to the exercises.

If π is a permutation of the set W of witnesses, we write $\pi\phi$ for the formula which results if we replace each witness c in ϕ by πc; if p is a set of formulas then πp is $\{\pi\phi : \phi \in p\}$. Notions of forcing N often have the properties:

(6) For each permutation π of the set of witnesses, if p is an N-condition then so is πp.

(7) N has a least element 0_N.

(8) For every N-condition p and every atomic sentence ϕ of $L(W)$, at least one of $p \cup \{\phi\}$ and $p \cup \{\neg\phi\}$ is an N-condition.

Exercises for 2.3.

1. Show that if N is a notion of forcing with property (6), then for every N-condition p, every sentence ϕ and every permutation π of the set of witnesses, p forces ϕ iff πp forces $\pi\phi$.

2. Let N be a notion of forcing with property (6). Let $\psi(x_1,\ldots,x_n)$ be a formula of $L(W)_{\omega_1\omega}$ in which at most finitely many witnesses occur, and let q be an N-condition. Show that q forces $\forall x_1\ldots x_n\psi$ iff for some sequence $(c_1,\ldots,c_n)$ of distinct witnesses not occurring in q or in ψ, q forces $\psi(c_1,\ldots,c_n)$. [For right to left let $d_1, \ldots, d_n$ be any witnesses. It suffices, by Theorem 2.3.3(d), to prove that if $p \supseteq q$ then some $r \supseteq p$ forces $\psi(\bar{d})$. Choose distinct witnesses $c'_1, \ldots, c'_n$ not in p or $\psi(\bar{d})$, and use (6) to show that $p \cup \{c'_1=d_1,\ldots,c'_n=d_n\}$ forces $\psi(\bar{d})$.]

3. Show that if N is a notion of forcing with property (7), then a property P is N-enforceable iff 0_N forces P.

4. Show that if N is a notion of forcing with property (8), then it is enforceable that $\mathrm{diag}(A(\bar{p})) \subseteq \bigcup\bar{p}$.

5. Let N be a notion of forcing and P a property. Show that (a)-(c) are equivalent: (a) For every N-condition q there is an N-condition $p \supseteq q$ such that either p forces P or p forces the property not-P. (b) For every N-condition q, q forces not-P iff there is no N-condition $p \supseteq q$ which forces P. (c) Every position in any play of a standard game $G_N(P;X)$ is winning for one of the players.

6. Let M and N be notions of forcing for L(W). We say that M is <u>dense in</u> N iff $M \subseteq N$ and for every condition $p \in N$ there is a condition $q \in M$ such that $q \supseteq p$. Show that if M is dense in N, then for every property P, P is N-enforceable iff P is M-enforceable.

7. Let N be a notion of forcing for L(W). Define "$p \Vdash^s \phi$" (p <u>strongly forces</u> ϕ) for conditions $p \in N$ and first-order sentences ϕ of L(W) as follows, by induction on the complexity of ϕ. (i) For atomic ϕ, $p \Vdash^s \phi$ iff $p \vdash \phi$. (ii) $p \Vdash^s \phi\wedge\psi$ iff $p \Vdash^s \phi$ and $p \Vdash^s \psi$. (iii) $p \Vdash^s \phi\vee\psi$ iff either $p \Vdash^s \phi$ or $p \Vdash^s \psi$. (iv) $p \Vdash^s \neg\phi$ iff there is no condition $q \supseteq p$ such that $q \Vdash^s \phi$. (v) $p \Vdash^s \exists x\psi(x)$ iff for some closed term t of L(W), $p \Vdash^s \psi(t)$. We take $\rightarrow$, $\leftrightarrow$ to be defined in terms of $\wedge$, $\vee$, $\neg$, and $\forall x\psi$ to be $\neg\exists x\neg\psi$. Define "$p \Vdash^w \phi$" (p <u>weakly forces</u> ϕ) to mean $p \Vdash^s \neg\neg\phi$. Show:
(a) If $p \Vdash^s \phi$ and $p \subseteq q$ then $q \Vdash^s \phi$. (b) If $p \Vdash^s \phi$ then $p \Vdash^w \phi$.

(c) $p \Vdash^w \phi$ iff for all $q \supseteq p$ there is $r \supseteq q$ such that $r \Vdash^w \phi$. (d) $p \Vdash^w \neg\phi$ iff there is no $q \supseteq p$ such that $q \Vdash^w \phi$. (e) $p \Vdash^w \phi \wedge \psi$ iff $p \Vdash^w \phi$ and $p \Vdash^w \psi$. (f) $p \Vdash^w \forall x \psi(x)$ iff for every closed term t, $p \Vdash^w \psi(t)$. (g) Every condition weakly forces every logically true first-order sentence of $L(W)$. (h) If $p \Vdash^w \phi$ and $\phi \vdash \psi$ then $p \Vdash^w \psi$. (j) It can happen that for some atomic sentence ϕ and some condition p, p doesn't strongly force $\phi \vee \neg\phi$. [For (g) and (h), choose a particular proof-calculus for first-order logic and go by induction on the lengths of proofs.]

REFERENCES FOR CHAPTER 2

2.1: The compactness theorem for countable first-order languages is due to Gödel (1930). Without the cardinality restriction it is due to Mal'tsev (1936) and Henkin (1949). The proof given above, emphasising notions of consistency, is based on Smullyan (1963).

Exercise 11: for more on this theme see van den Dries et al. (1982); they construct elementarily equivalent groups whose commutator subgroups are not elementarily equivalent.

2.2: Exercise 4(a,b) was conjectured by S. Mazur and proved by S. Banach, in about 1930. Cf. Oxtoby (1971) Chapter 6.

2.3: Types of forcing tend to be grouped into two species, strong and weak, according to the rules for forcing $\phi \vee \psi$ and $\exists x \psi$; cf. Exercise 7. In weak forcing, all logically true sentences are everywhere forced; in strong forcing only intuitionistic laws are forced by all conditions. (Our forcing is weak.) The motivations behind the two kinds of forcing are quite different. Strong forcing is simply the classical-style truth definition for intuitionist logic, and the defining clauses correspond to the meanings of the intuitionist connectives, cf. Kripke (1965). Weak forcing is a way of looking at dense subsets of a partial order (cf. Exercise 6 of section 2.2). It's a historical accident that Cohen (1963), in his brilliant work on independence in set theory, discovered weak forcing via strong forcing and the $\neg\neg$-interpretation of classical in intuitionist logic. Abraham Robinson (Barwise & Robinson 1970) adjusted Cohen's definitions to give a general way of constructing models. Keisler's survey (1973) shows how Robinson's weak forcing generalises omitting-types constructions, cf. Chapter 5 below. Ziegler (1980) finally removed the red herring of strong forcing by defining Robinson's weak forcing in terms of Banach-Mazur games.

3 EXISTENTIAL CLOSURE

Tant qu'on a de nouveaux éléments à introduire, on doit craindre d'avoir à recommencer tout son travail; or il n'arrivera jamais qu'on n'ait plus de nouveaux éléments à introduire ... Et alors des points qui n'étaient pas définissables deviendront susceptibles d'être définis; d'autres qui l'étaient cesseront de l'être.

Poincaré, La logique de l'infini (1913)

Poincaré's essay, published in 1913, was part of the metaphysical hubbub that followed the discovery of the foundational paradoxes. It is a pity that nobody read his words mathematically. In fact he raised two excellent questions. First, when and how can elements be added to a structure so that the result is another structure of the same kind? And second, if new elements are added, what does this do to definability of elements or sets of elements within the structure? To the best of my knowledge, the earliest frontal attack on the first question was by B. H. Neumann in his paper (1943) on 'Adjunction of elements to groups'; in principle he gave a general solution for all varieties. Progress on the second question was more gradual.

Why is this relevant to forcing? The answer is that in a construction by forcing we are continually adding new elements - if not to a structure, then at least to a partial description of a structure. When a structure is compiled by games, we can enforce that it contains "all possible kinds of element".

This chapter and the next are about a particularly simple kind of forcing invented by Abraham Robinson. He called it finite forcing to distinguish it from the cruder infinite forcing that we shall meet in Chapter 5. In Robinson's (finite) forcing it is always enforceable that the compiled structure is existentially closed. In section 3.2 I say what this means, after laying the groundwork on adjunction of elements in section 3.1. Section 3.3 takes existentially closed groups as a concrete example; the high point is Martin Ziegler's beautiful theorem on the form of resultants in groups. Section 3.4 develops Robinson's machinery.

3.1 ADJUNCTION OF ELEMENTS

Let L be a first-order language and T a theory in L. The extension problem for T runs as follows. Given a model A of T, an existential formula $\phi(\bar{y})$ of L and a tuple $\bar{a}$ of elements of A, when does there exist a model B of T such that $A \subseteq B$ and $B \models \phi(\bar{a})$?

For example if A is a group with elements a and b, when is there a group $B \supseteq A$ in which a and b are conjugate? This is a case of the extension problem for groups: take T to be the theory of groups (cf. section 1.2), and take $\phi(y,z)$ to be the existential formula $\exists x(x^{-1}yx = z)$. The answer is known. There is such an extension B iff a and b have the same order, in other words, iff $A \models \bigwedge_{n<\omega}(a^n = 1 \leftrightarrow b^n = 1)$. Cf. Higman et al. (1949).

Perhaps this is the place to remark that in this book, if a group H is an extension of G, that means only that $H \supseteq G$, i.e. that G is a subgroup of H. Group theorists usually reserve the word "extension" for the case where G is a normal subgroup of G.

Another well-known example of an extension problem is where T is the theory of fields. Let A be a field and $p(\bar{a},x)$, $q(\bar{b},x)$ two polynomials with coefficients $\bar{a}$, $\bar{b}$ from A. When do $p(\bar{a},x)$ and $q(\bar{b},x)$ have a common root in some field extending A? The formula $\phi(\bar{y},\bar{z})$ here is $\exists x(p(\bar{y},x) = 0 \wedge q(\bar{z},x) = 0)$; I leave it to the reader to check that this can be written out as an $\exists_1$ first-order formula in the signature of rings with 1. Algebra texts (Cohn (1974 b) section 7.4, Jacobson (1974) section 5.4) give a criterion which solves this extension problem in terms of a determinant in $\bar{a}$ and $\bar{b}$, called the resultant.

There is a general solution to the extension problem. For each $\exists_1$ formula $\phi(\bar{x})$ of L, let $Res_\phi(\bar{x})$ be the set of all $\forall_1$ formulas $\psi(\bar{x})$ of L such that $T \vdash \forall\bar{x}(\phi \rightarrow \psi)$. Res_ϕ is called the resultant of ϕ. We have:

THEOREM 3.1.1. *Let* T *be a theory in a first-order language* L, *let* A *be an* L*-structure,* $\bar{a}$ *a tuple of elements of* A *and* $\phi(\bar{x})$ *an* $\exists_1$ *formula of* L. *Then the following are equivalent:*

(a) *For some model* B *of* T, $A \subseteq B$ *and* $B \models \phi(\bar{a})$.

(b) $A \models \bigwedge \mathrm{Res}_\phi(\bar{a})$.

Proof. (a) ⇒ (b): Suppose there is such a model B. For each formula $\psi(\bar{x})$ in Res_ϕ, $B \models \forall\bar{x}(\phi \to \psi)$ since B is a model of T. But also $B \models \phi(\bar{a})$, and hence $B \models \psi(\bar{a})$. Then since ψ is a $\forall_1$ formula, $A \models \psi(\bar{a})$ by the lemma on preservation (Lemma 1.2.4(a)).

(b) ⇒ (a): Assume $A \models \bigwedge \mathrm{Res}_\phi(\bar{a})$. Expand A by introducing a tuple $\bar{d}$ of distinct new constants to name the elements $\bar{a}$, so that an element in $\bar{a}$ is named by a different constant each time it occurs in $\bar{a}$; write D for the expanded structure. We can write ϕ as $\exists\bar{y}\chi(\bar{x},\bar{y})$ where χ is quantifier-free. By the diagram lemma (Lemma 1.2.3), to find B as in (a) it suffices to show that $T \cup \mathrm{diag}(D) \cup \{\chi(\bar{d},\bar{c})\}$ has a model, where $\bar{c}$ is a tuple of distinct new constants. If this theory has no model, then by the compactness theorem (section 2.1) there is a finite conjunction $\theta(\bar{d},\bar{b})$ of sentences in diag(D) such that $T \cup \{\chi(\bar{d},\bar{c})\} \vdash \neg\theta(\bar{d},\bar{b})$; we have listed as $\bar{b}$ the distinct constants which occur in $\theta(\bar{d},\bar{b})$ but not in $\bar{d}$. By the lemma on constants (Lemma 1.2.2) and a little elementary logic we deduce in turn that $T \cup \{\chi(\bar{d},\bar{c})\} \vdash \forall\bar{z}\neg\theta(\bar{d},\bar{z})$, then that $T \vdash \exists\bar{y}\chi(\bar{d},\bar{y}) \to \forall\bar{z}\neg\theta(\bar{d},\bar{z})$, and finally that $T \vdash \forall\bar{x}(\phi \to \forall\bar{z}\neg\theta(\bar{x},\bar{z}))$. (This last step is why we needed to replace $\bar{a}$ by distinct constants $\bar{d}$; $\bar{a}$ might have contained a repetition.) Therefore $\forall\bar{z}\neg\theta(\bar{x},\bar{z}) \in \mathrm{Res}_\phi$ and so $A \models \forall\bar{z}\neg\theta(\bar{a},\bar{z})$ by assumption. This implies $D \models \forall\bar{z}\neg\theta(\bar{d},\bar{z})$ and hence $D \models \neg\theta(\bar{d},\bar{b})$, contradicting the choice of $\theta(\bar{d},\bar{b})$ as a sentence true in D. □

Note that A in the theorem was not required to be a model of T. That allows us to draw a useful corollary. Write $T_\forall$ for the set of all $\forall_1$ sentences ψ of L such that $T \vdash \psi$.

COROLLARY 3.1.2. *An* L*-structure* A *is a model of* $T_\forall$ *if and only if there is a model* B *of* T *such that* $A \subseteq B$.

Proof. This is the special case of the theorem when ϕ is a trivially true sentence, such as $\exists x\, x=x$. □

For example there is a $\forall_1$ first-order theory T such that for any ring A with 1, A is embeddable in a skew field if and only if $A \models T$. P. M. Cohn (1974 a) showed that such a theory T must be infinite.

Theorem 3.1.1 is perhaps not too helpful as it stands - the definition of Res_ϕ is too ineffective. If T and L are recursively enumerable (r.e. for short), then each set Res_ϕ is r.e. and hence equivalent to a recursive set (cf. Exercise 3 below). But this is small beer. We want a much sharper description of Res_ϕ, and preferably a description that an algebraist can use.

Unfortunately (fortunately?) in algebra there is no substitute for hard work. But general theory can give some useful pointers. Let me illustrate this with an example from commutative rings, together with a discussion that generalises the example.

Question: Given a commutative ring A (with 1 - we always assume this) and an element a of A, when is there a commutative ring $B \supseteq A$ containing an element b such that $a^2b = a$?

The first step towards an answer is to form the polynomial ring A[x] with indeterminate x. Let I be the ideal in A[x] generated by the polynomial $a^2x - a$. Then in A[x]/I, the element x+I satisfies the equation $(a+I)^2(x+I) = a+I$. If the map $g: d \mapsto d+I$ embeds A into A[x]/I, then we have our required element b. Now g is an embedding if and only if

(1) $$A \cap I = \{0\}.$$

Conversely if there is a commutative ring $B \supseteq A$ with an element b such that $a^2b = a$, then we can map $f: A[x] \to B$ homomorphically by taking f(d) = d for all d in A, and f(x) = b. Then $I \subseteq \ker(f)$ and $A \cap \ker(f) = \{0\}$, so that (1) holds. So the question reduces to determining when $A \cap I = \{0\}$.

Suppose (1) fails. This means that for some non-zero element d of A and some polynomial $p(x) = \Sigma_{i\leqslant k} b_i x^i \in A[x]$ we have $d = p(x).(a^2x - a)$. Identifying coefficients of x, we have:

(2) $$a^2b_k = 0,\quad a^2b_{k-1} = ab_k,\ \ldots,\ a^2b_0 = ab_1,\quad ab_0 + d = 0.$$

In other words, (1) fails if and only if for some $k \geqslant 0$ the following holds:

(3) $$A \models \exists y_0 \ldots y_k z(a^2y_k = 0 \wedge a^2y_{k-1} = ay_k \wedge \ldots \wedge a^2y_0 = ay_1 \wedge ay_0 + z = 0 \wedge z \neq 0).$$

The sentence in (3) simplifies to: $\exists y(a^{k+2}y = 0 \wedge ay \neq 0)$. So we have the solution to our extension problem. A commutative ring $B \supseteq A$ exists with an element b such that $a^2b = a$, if and only if

(4) $$A \models \forall y(a^k y = 0 \rightarrow ay = 0) \qquad \text{for all } k \geqslant 2.$$

A short calculation shows that it suffices to take $k = 2$, so that if T is the theory of commutative rings and $\phi(w)$ is $\exists x\ w^2x = w$, then Res_ϕ is equivalent modulo T to the single formula $\forall y(w^2y = 0 \rightarrow wy = 0)$.

A formula is said to be <u>strict universal Horn</u> iff it has the form

(5) $$\forall \bar{x}(\chi_1 \wedge \ldots \wedge \chi_k \rightarrow \psi)$$

with $\chi_1, \ldots, \chi_k, \psi$ all atomic. We allow $\bar{x}$ to be empty. Also we allow $k = 0$, in which case (5) reduces to $\forall \bar{x}\psi$. A formula is <u>universal Horn</u> iff it is either strict universal Horn or of form

(6) $$\forall \bar{x}\neg(\chi_1 \wedge \ldots \wedge \chi_k),$$

where again $\chi_1, \ldots, \chi_k$ are atomic. Note that the sentences in (4) are strict universal Horn. This is not an accident. The next theorem explains why.

A <u>strict universal Horn theory</u> is a first-order theory consisting of strict universal Horn sentences. Every equational theory is of this form. An example of a strict universal Horn theory which is not equational is the theory of torsion-free abelian groups. This consists of sentences axiomatising the class of abelian groups (these can be chosen equational), together with:

(7) $$\forall x(nx = 0 \rightarrow x = 0) \qquad \text{for all } n \geqslant 1$$

where nx means x+...+x (n times).

A primitive formula is a formula of form

(8) $\exists\bar{x}(\psi_1 \wedge \ldots \wedge \psi_k \wedge \neg\sigma_1 \wedge \ldots \wedge \neg\sigma_m)$

where $\psi_1, \ldots, \psi_k, \sigma_1, \ldots, \sigma_m$ are atomic. We call the formula (8) positive primitive (often abbreviated to p.p. in the literature) iff $m = 0$. If the language has no relation symbols, then a positive primitive formula says that a certain finite system of equations has a simultaneous solution. There is a theorem of elementary logic which says that every $\exists_1$ first-order formula is logically equivalent to a disjunction of primitive formulas.

THEOREM 3.1.3. *Let* T *be a strict universal Horn theory in a first-order language* L.

(a) *If* $\phi(\bar{x})$ *is a positive primitive formula of* L, *then* Res_ϕ *is equivalent modulo* T *to a set of strict universal Horn formulas of* L.

(b) *If* $\phi(\bar{x})$ *is a primitive formula of* L, *then* Res_ϕ *is equivalent modulo* T *to a set of universal Horn formulas of* L.

Proof. First let $\phi(\bar{x})$ be a positive primitive formula $\exists\bar{y}\psi(\bar{x},\bar{y})$, where ψ is a conjunction of atomic formulas. Let $\Phi(\bar{x})$ be the set of strict universal Horn formulas in Res_ϕ. It will be enough to show that if A is a model of T and $A \models \bigwedge\Phi(\bar{a})$, then there is a model B of T, $B \supseteq A$, such that $B \models \phi(\bar{a})$. (For then if $\psi(\bar{x}) \in \mathrm{Res}_\phi$ we have $B \models \psi(\bar{a})$ and so $A \models \psi(\bar{a})$.) Suppose that $A \models T$ and $A \models \bigwedge\Phi(\bar{a})$ but there is no such model $B \supseteq A$. We shall derive a contradiction.

The first step is to add a "free" solution of $\psi(\bar{a},\bar{y})$ to A. For this we introduce new constants $\bar{c}$ to name the solution, and we define C^+ to be the canonical model of the set

(9) $\{\theta : \theta \text{ is an atomic sentence such that } T \cup \mathrm{diag}^+(A) \cup \{\psi(\bar{a},\bar{c})\} \vdash \theta\}$.

Let C be the L-reduct of C^+. Since $C^+ \models \mathrm{diag}^+(A)$, there is a homomorphism $g: A \to C$ (as in the diagram lemma, Lemma 1.2.3). Also C is a model of T. I omit the proof of this (which uses the assumption that T is strict universal Horn), and instead I point to some examples.

If T is the theory of groups and $\psi(\bar{a},\bar{y})$ is a conjunction of equations $s_i(\bar{a},\bar{y})=1$ $(i \in I)$, let $F_{\bar{c}}$ be the free group generated by the elements $\bar{c}$, and let N be the normal subgroup of the free product $A*F_{\bar{c}}$ which is generated by the elements $s_i(\bar{a},\bar{c})$. Then C will be $(A*F_{\bar{c}})/N$. It is clear what the homomorphism g is; in fact g is an embedding iff $A \cap N = \{1\}$.

If T is the theory of abelian groups and $\psi(\bar{a},\bar{y})$ a conjunction of equations $s_i(\bar{a},\bar{y})=0$, then let $F_{\bar{c}}$ be the free abelian group generated by $\bar{c}$, and let H be the subgroup of $A \oplus F_{\bar{c}}$ generated by the elements $s_i(\bar{a},\bar{c})$. Then C will be $(A \oplus F_{\bar{c}})/H$, and again we have an embedding iff $A \cap H = \{0\}$. The same description works for modules.

In the case of commutative rings we form the polynomial ring $A[\bar{c}]$ and then factor out a suitable ideal I to get $C = A[\bar{c}]/I$, as in the example above.

We return to the proof. By construction, $C \models \psi(g\bar{a},\bar{c})$. If g is an embedding, then we can identify A with its image under g, and C will be an extension of A containing a solution of $\psi(\bar{a},\bar{y})$. But we assumed A has no such extension. Therefore g is not an embedding. This means that g fails to preserve some negated atomic formula. So there is an atomic sentence $\theta(\bar{a},\bar{d})$ such that $A \models \neg\theta(\bar{a},\bar{d})$ but $C \models \theta(g\bar{a},\bar{d})$. Since C^+ was the canonical model of (9), it follows by the compactness theorem that there is a finite conjunction $\chi(\bar{a},\bar{d},\bar{e})$ of sentences in $\mathrm{diag}^+(A)$ such that $T \vdash \chi(\bar{a},\bar{d},\bar{e}) \wedge \psi(\bar{a},\bar{c}) \rightarrow \theta(\bar{a},\bar{d})$. Using the lemma on constants (Lemma 1.2.2) three times, this yields that $T \vdash \forall\bar{x}(\exists\bar{y}\psi(\bar{x},\bar{y}) \rightarrow \forall\bar{z}\bar{w}(\chi(\bar{x},\bar{z},\bar{w}) \rightarrow \theta(\bar{x},\bar{z})))$. Now $\forall\bar{z}\bar{w}(\chi(\bar{x},\bar{z},\bar{w}) \rightarrow \theta(\bar{x},\bar{z}))$ is strict universal Horn, so we have proved that it lies in Φ. But $A \models \neg\forall\bar{z}\bar{w}(\chi(\bar{a},\bar{z},\bar{w}) \rightarrow \theta(\bar{x},\bar{z}))$, since $\chi(\bar{a},\bar{d},\bar{e})$ and $\neg\theta(\bar{a},\bar{d})$ are both true in A. This contradicts our assumption that $A \models \bigwedge\Phi(\bar{a})$. Thus part (a) is proved.

For part (b) let ϕ be $\exists\bar{y}(\psi(\bar{x},\bar{y}) \wedge \neg\sigma_1(\bar{x},\bar{y}) \wedge \ldots \wedge \neg\sigma_m(\bar{x},\bar{y}))$ where ψ is as before and $\sigma_1, \ldots, \sigma_m$ are atomic. Let $\Phi(\bar{x})$ now be the set of all universal Horn formulas in Res_ϕ. Again it suffices to assume that $A \models T$ and $A \models \bigwedge\Phi(\bar{a})$ but there is no model B of T such that $A \subseteq B$ and $B \models \phi(\bar{a})$, and deduce a contradiction. We begin by constructing C just as before, by adding a free solution of $\psi(\bar{a},\bar{y})$ to A. By the argument of part (a), the homomorphism $g: A \rightarrow C$ is an embedding; without loss we can suppose $A \subseteq C$. Now since C is not an extension of A satisfying $\phi(\bar{a})$, we must have $C \models \sigma_i(\bar{a},\bar{c})$ for some i $(1 \leqslant i \leqslant m)$. But then by the same argument as for part (a), there is a conjunction $\chi(\bar{a},\bar{e})$ of finitely many sentences from $diag^+(A)$, such that $T \vdash \chi(\bar{a},\bar{e}) \wedge \psi(\bar{a},\bar{c}) \rightarrow \sigma_i(\bar{a},\bar{c})$. By the lemma on constants, $T \vdash \forall\bar{x}(\exists\bar{y}(\psi(\bar{x},\bar{y}) \wedge \neg\sigma_i(\bar{x},\bar{y})) \rightarrow \forall\bar{z}\neg\chi(\bar{x},\bar{z}))$. So $\forall\bar{z}\neg\chi(\bar{x},\bar{z})$ is in Φ, and again we have a contradiction to $A \models \bigwedge\Phi(\bar{a})$. □

The general theory can usefully be pushed one step further. Let us say that the theory T in language L has the amalgamation property iff the following holds:

(10) If A, B, C are L-structures such that $A \subseteq B$ and $A \subseteq C$, and all three are models of T, then there are a model D of T such that $B \subseteq D$, and an embedding $g: C \rightarrow D$ which is the identity on A.

For example T is known to have the amalgamation property if T is any of the following theories: groups, abelian groups, left modules over a fixed ring, lattices, distributive lattices, skew fields.

A (strict) quantifier-free Horn formula is the same as a (strict) universal Horn formula, except that it is quantifier-free.

THEOREM 3.1.4. *Let* T *be a strict universal Horn theory in a first-order language* L *whose signature has at least one constant. Suppose* T *has the amalgamation property. Then:*

(a) If $\phi(\bar{x})$ *is a positive primitive formula of* L, *then* Res_ϕ *is equivalent modulo* T *to a set of strict quantifier-free Horn formulas of* L.

(b) If $\phi(\bar{x})$ *is a primitive formula of* L, *then* Res_ϕ *is equivalent modulo* T *to a set of quantifier-free Horn formulas of* L.

Proof. Copy the proof of Theorem 3.1.3, but instead of extending A to C, extend $\langle\bar{a}\rangle_A$ to C. (The assumption that L has a constant is needed to make sure that $\langle\bar{a}\rangle_A$ is well-defined when $\bar{a}$ is empty.) By the lemma on preservation (Lemma 1.2.4(a)), $\langle\bar{a}\rangle_A$ is a model of T. By (10), since $\langle\bar{a}\rangle_A \subseteq A$ and $\langle\bar{a}\rangle_A \subseteq C$, there is a model D of T such that $D \supseteq A$ and C can be embedded in D over $\langle\bar{a}\rangle_A$; so if $C \models \phi(\bar{a})$ then $D \models \phi(\bar{a})$ by the lemma on preservation (Lemma 1.2.4(b)). The effect of working with $\langle\bar{a}\rangle_A$ in place of A is that any sentences from $diag^+\langle\bar{a}\rangle_A$ can be written as $\chi(\bar{a})$ with no further constants added. This disposes of the variables $\bar{z}$ and $\bar{w}$ in the previous proof. □

In the proof of Theorem 3.1.3 we formed C by adding a free solution of $\psi(\bar{a},\bar{y})$ to A. Speaking practically, this is often the best way to start looking for resultants. The first step is to construct the free algebra $F_{\bar{c}}$ in the appropriate class, and then to form its free product with A, $A*F_{\bar{c}}$. The feasibility of the next step depends on whether we have a good algebraic description of $A*F_{\bar{c}}$. In commutative rings, as we saw, $A*F_{\bar{c}}$ is just the polynomial ring $A[\bar{c}]$, and this makes calculations rather easy. Abelian groups and modules are even easier. Nilpotent groups of class 3 are about at the limit of what one can handle by brute force; in higher nilpotency classes we don't know what free products look like. The exercises below and section 4.4 will illustrate all this.

Not all interesting theories are strict universal Horn, of course. (Cf. Exercise 5(a).) But finding resultants for other theories seems to be more a matter of insight and good luck.

One such theory which has been studied very thoroughly is the theory T of fields. Here every primitive formula is equivalent to a positive primitive one, since an inequation $s \neq t$ can be rewritten as $\exists w\ (s-t)w = 1$. (This is known in some contexts as Rabinowitsch's trick.) One of the achievements of nineteenth century mathematics was to find a way of computing, for every positive primitive formula ϕ of the language of rings, a single quantifier-free formula equivalent to Res_{ϕ} modulo T. Van der Waerden (1950) Chapter XI describes the method in detail, remarking that "In practice the required calculations are very often too complicated to be carried out effectively". General model-theoretic arguments show quite easily that the resultants for T can be be brought to this form (cf. Exercises 13, 20 in section 3.2 below), but they give no useful guidance on how to compute the formulas in question.

Exercises for 3.1.

1. Let L be the first-order language of rings with 1. (a) Show that if $p(\bar{x})$ is a polynomial in $\bar{x}$ with integer coefficients, then the polynomial equation $p(\bar{x})=0$ can be written as an atomic formula $\phi(\bar{x})$ of L, assuming the L-structures under discussion are rings. (b) Show the same when $p(\bar{x})$ has rational coefficients, assuming the L-structures under discussion are rings of characteristic 0.

2. Show that each of the following classes is axiomatised by a $\forall_2$ first-order theory T, and describe the class of models of $T_{\forall}$ in each case:

(a) The class of algebraically closed fields, in the signature with +, −, ·, 0, 1. (b) The class of commutative von Neumann regular rings (i.e. commutative rings such that for each element x there is y with xyx = x) in the same signature as (a). (c) The class of boolean algebras, in the signature with ∧, ∨, 0, 1. [(a) integral domains, (b) commutative rings with no non-zero nilpotent elements, (c) distributive lattices with 0 and 1.]

A language is said to be recursive iff its terms and formulas are encoded as natural numbers so that all the basic syntactic operations, such as substitution for a free variable or conjunction of two formulas, go over into recursive functions. Virtually every countable first-order language that one meets in practice can be encoded as a recursive language.

3. Let L be a recursive first-order language and T an r.e. theory in L. (a) Show that the set $U = \{\theta : \theta$ is a sentence of L and $T \vdash \theta\}$ is r.e. (b) (Craig's trick) Show that the theory T is equivalent to a recursive theory in L. [Replace θ by $\theta\wedge\ldots\wedge\theta$ (k times) where k is the Gödel number of a computation putting θ into T.]

The definition of (cartesian) product of groups generalises at once to products $\Pi_{i\in I}A_i$ of arbitrary L*-structures; cf. Chang & Keisler (1973) p. 177 or Cohn (1981) p. 49. We say that a theory* T *is closed under non-trivial products iff for every non-empty family* A_i (i ∈ I) *of models of* T, $\Pi_{i\in I}A_i$ *is also a model of* T. *We say that* T *is closed under products iff it is closed under non-trivial products and the one-element structure (whose element satisfies all atomic formulas) is a model of* T.

4. Let T be a theory which is closed under non-trivial products. Show that if ψ_i (i ∈ I) are a non-empty family of atomic sentences and $T \vdash \bigvee_{i\in I}\psi_i$, then $T \vdash \psi_i$ for some i ∈ I.

5. (a) Show that every strict universal Horn theory is closed under products, and every universal Horn theory is closed under non-trivial products. (b) Show that Theorem 3.1.3(a) remains true if we replace "strict universal Horn theory" by "theory closed under products".

6. Let T be a theory in a first-order language L. We say the universal sentence problem for T is solvable iff there is an algorithm which determines, given any $\forall_1$ formula θ of L, whether or not $T \vdash \theta$. (a) Show that if the universal sentence problem for T is solvable, then there is an algorithm for determining, given any $\exists_1$ formula $\phi(\bar{x})$ and $\forall_1$ formula $\psi(\bar{x})$

of L, whether or not $\psi \in \mathrm{Res}_\phi$. (b) Assuming that T is closed under non-trivial products, show that if there is an algorithm which determines, for any $\exists_1$ formula $\phi(\bar{x})$ and any atomic formula $\psi(\bar{x})$ of L, whether or not $\psi \in \mathrm{Res}_\phi$, then the universal sentence problem for T is solvable.

7. Let T be a theory in a first-order language L with at least one constant. Show that if $T_\forall$ has the amalgamation property, then for every $\exists_1$ formula $\phi(\bar{x})$ of L, Res_ϕ is equivalent modulo T to a set of quantifier-free formulas of L.

8. Let T be a strict universal Horn theory in a first-order language L with at least one constant, and let U be the set of atomic sentences θ of L such that $T \vdash \theta$. Show that U is =-closed in L, and that the canonical model of U is a model of T. Use this to complete the proof of Theorem 3.1.3.

9. Suppose T is a strict universal Horn theory satisfying the amalgamation property. Suppose that in Theorem 3.1.1, a group G of automorphisms of A is given. Show that in (a) of the theorem we can add "and the automorphisms in G extend so that G acts as a group of automorphisms on B".

The remaining exercises calculate resultants in various theories. Some need specialist knowledge.

10. Let R be a ring with 1 and T the theory of left R-modules (cf. Exercise 2(c) of section 1.2). (a) Show that if $\phi(\bar{x})$ is $\exists\bar{y}\bigwedge_i(\Sigma_j\mu_{ij}y_j = \Sigma_j\nu_{ij}x_j)$ where μ_{ij}, ν_{ij} are ring elements, then Res_ϕ is equivalent modulo T to the set of all equations $\Sigma_{i,j}\lambda_i\nu_{ij}x_j$ such that the family (λ_i) of ring elements satisfies $\Sigma_i\lambda_i\mu_{ij} = 0$ for each j. (b) Show that if $\phi(\bar{x})$ is $\exists\bar{y}(\bigwedge_i(\Sigma_j\mu_{ij}y_j = \Sigma_j\nu_{ij}x_j) \wedge \bigwedge_h(\Sigma_j\mu'_{hj}y_j \neq \Sigma_j\nu'_{hj}x_j))$, then Res_ϕ is equivalent modulo T to the set of equations given in (a) together with, for all h, the set of all inequations $\Sigma_j(\nu'_{hj} - \Sigma_i\sigma_i\nu_{ij})x_j \neq 0$ such that the family (σ_i) of ring elements satisfies $\mu'_{hj} = \Sigma_i\sigma_i\mu_{ij}$ for all j. (c) The ring R is said to be <u>left coherent</u> iff for every finite n and R-linear $\alpha: R^n \to R$, ker α is finitely generated. Show that if R is left coherent then the resultants in both (a) and (b) are equivalent modulo T to finite sets of formulas of the forms shown. [If R is left coherent, then for all finite n and m, every R-linear $\alpha: R^n \to R^m$ has finite kernel, cf. Chase (1960) Theorem 2.1.]

11. Let L be the first-order language of rings, with symbols +, −, ·, 0, 1, and in L let T be the theory of commutative rings. Show the following: (a) If $\phi(x,y)$ is $x|y$ (i.e. $\exists z\ xz = y$), then Res_ϕ is equivalent modulo T to the single formula $\forall t(ta = 0 \to tb = 0)$. (b) If $n \geqslant 1$ and $\phi(x)$ is $\exists uw(x=u+w \wedge u^2|u \wedge uw=0 \wedge w^n=0)$, then Res_ϕ is equivalent modulo T to $\forall t(ta^{n+1}=0 \to ta^n=0)$. (c) If $n \geqslant 1$ and $\phi(x)$ is $\exists y(y^2=y\neq 0 \wedge xy=y \wedge (x-y)^n\neq 0)$, then Res_ϕ is equivalent modulo T to $x^n\neq 0 \wedge \forall t\ t(1-x)\neq 0$. (d) If $\phi(x)$ is $\exists y(y^2=y\neq 0 \wedge x|y)$ then Res_ϕ is equivalent modulo T to the set of formulas $x^n \neq 0$ $(n \geqslant 1)$. (e) If A is a commutative ring with 1, a an element of A, a is not nilpotent and 1−a is not invertible, then there is a commutative ring $B \supseteq A$ containing an element $b \neq 0$ such that $ab = b^2 = b$ and a−b is not nilpotent. [Use (c) and compactness.]

12. Let L be as in Exercise 11. A <u>local ring</u> will mean a commutative ring with 1 which has exactly one maximal ideal. (a) Show that there is a $\forall_2$ theory in L which axiomatises the class of local rings. (b) Let T be a theory as in (a), and for some $k \geqslant 0$ let $\phi(\bar{y})$ be the formula $\exists x\ \Sigma_{i\leqslant k} y_i x^i = 0$. Show that Res_ϕ is equivalent modulo T to the set of all formulas of the following form:

$$\forall w_0 \ldots w_m u_0 \ldots u_{k+m} (\bigwedge_{1\leqslant i\leqslant k+m} \Sigma_{p+q=i} (y_p w_q) = u_i y_0 w_0 \to y_0 w_0 = 0).$$

13. Let L be as in Exercise 11 but without the symbol 1. In L, let T be the theory of rings, not necessarily with a unit element. Let $\phi(x,y)$ be $\exists uvw(xu = uv \wedge wuv = xwu \wedge wyu \neq ywu)$. Show that Res_ϕ is equivalent modulo T to the set of formulas $\forall zt(y \neq nx + zx + xt)$ (n an integer). (Note that Res_ϕ expresses that y is not in the subring generated by x; in rings with 1 this is expressible by a single formula.)

14. Let R be a commutative ring with 1, A an R-algebra and $\{(\psi_i : A \to A) : i < \mu\}$ a family of maps. Show that the following are equivalent: (a) Each ψ_i is R-linear. (b) There is an R-algebra $B \supseteq A$ containing elements b_i, c_i $(i < \mu)$ such that for all $i < \mu$ and a in A, $\psi_i(a) = b_i a c_i$.

15. Let K be an ordered field, $\bar{x}$ a tuple of indeterminates and $p_0, \ldots, p_{k-1}$ polynomials $\in K[\bar{x}]$. Show that the following are equivalent:
(a) There is a common root of $p_0, \ldots, p_{k-1}$ in some ordered field $J \supseteq K$.
(b) For all $n < \omega$ and all polynomials $q_0, \ldots, q_{k-1}, r_0, \ldots, r_{n-1} \in K[\bar{x}]$, and all elements $\lambda_0, \ldots, \lambda_{n-1} \geqslant 0$ in K, $\Sigma_{i<k} q_i p_i \neq 1 + \Sigma_{j<n} \lambda_j r_j^2$.

3.2 *EXISTENTIALLY CLOSED MODELS*

A structure A in a class of structures is said to be existentially closed iff every finite system of equations and inequations over A which has a solution in some extension of A within the class already has a solution inside A. Actually this definition is only correct if the signature of the structure contains no relation symbols. A formally correct definition is as follows.

Let L be a first-order language and K a class of L-structures. Let A be a structure in K. We say that A is existentially closed (e.c. for short) in K iff:

(1) For every $\exists_1$ formula $\phi(\bar{x})$ of L and every tuple $\bar{a}$ of elements of A, if there is a structure B in K such that $B \supseteq A$ and $B \models \phi(\bar{a})$, then already $A \models \phi(\bar{a})$.

Every $\exists_1$ formula of L is logically equivalent to a disjunction of primitive formulas, and so in (1) we can replace "$\exists_1$" by "primitive". In a signature with no relation symbols, a primitive formula says exactly that some finite system of equations and inequations has a solution, so that (1) reduces to our original definition.

An existentially closed group is a group which is e.c. in the class of groups; likewise an e.c. ring, an e.c. Lie algebra, and so on. If T is a theory in L and K is the class of all models of T, then an e.c. structure in K is called an e.c. model of T.

EXAMPLE. One form of Hilbert's Nullstellensatz says that if A is an algebraically closed field and E is a finite system of equations and inequations with coefficients from A, such that some field extending A contains a solution of E, then A already contains a solution of E. (Cf. Jacobson (1980) p. 425.) It follows that an e.c. field is the same thing as an algebraically closed field.

The usual definition of "algebraically closed field" is essentially the same as the definition of "existentially closed" in the first paragraph above, except that instead of "finite system of equations and inequations" we have just "equation in one variable". For most classes of structures this gives a very much weaker notion than e.c. There is a halfway notion: we say A is algebraically closed (a.c. for short) in K iff every finite system of equations over A with a solution in some extension of A within K already has a solution in A. Overlooking relation symbols, this is equivalent to (1) with "$\exists_1$" replaced by "positive primitive". A.c. is a weaker property than e.c., and since our methods will always give e.c. models as easily as a.c. ones, there is not much point in dwelling on the weaker notion. (But see Exercise 10 of section 3.3 for a little information.) Jacobson (1980) remarks that existential closure "perhaps gives the real meaning of the algebraic closedness property of a field".

Model theorists studied e.c. models intensively in the early 1970s, and a great deal is known about them. In this section I shall confine myself to Four Memorable Theorems and some of their corollaries.

The first memorable theorem says that many e.c. structures exist. Both the theorem and its proof are immediate generalisations of Steinitz's proof that every field can be extended to an algebraically closed field.

A class K of structures is said to be inductive iff for every chain of structures in K, the union of the chain is also in K.

THEOREM 3.2.1. *Let* L *be a first-order language and* K *an inductive class of* L*-structures. Then for every structure* A *in* K *there is an e.c. structure* B *in* K *which is an extension of* A.

Proof. First we show that for every structure A in K there is a structure A^* in K, $A^* \supseteq A$, such that for every $\exists_1$ formula $\phi(\bar{x})$ of L and all tuples $\bar{a}$ from A, if there is $C \supseteq A^*$ in K such that $C \models \phi(\bar{a})$, then already $A^* \models \phi(\bar{a})$. (NB the tuples $\bar{a}$ are from A, not from A^*.) For this we well-order as $(\phi_i, \bar{a}_i)$ $(i < \lambda)$ the set of all pairs $(\phi, \bar{a})$ such that ϕ is an $\exists_1$ formula of L and $\bar{a}$ is a tuple from A. We define a chain $(A_i)_{i \leqslant \lambda}$ of structures in K by:

(2)
$$A_0 = A.$$
$$A_{i+1} = \begin{cases} \text{some } C \text{ in K such that } C \supseteq A_i \text{ and } C \models \phi_i(\bar{a}_i), \text{ if such a } C \text{ exists;} \\ A_i \text{ otherwise.} \end{cases}$$
$$A_\delta = \bigcup_{i<\delta} A_i \qquad \text{when } \delta \text{ is a limit ordinal.}$$

We put $A^* = A_\lambda$. To show that A^* has the required property, let $\phi(\bar{x})$ be an $\exists_1$ formula of L and $\bar{a}$ a tuple from A, and suppose that $C \models \phi(\bar{a})$ for some C in K, $C \supseteq A^*$. There is some $i < \lambda$ such that $(\phi,\bar{a})$ is $(\phi_i,\bar{a}_i)$. Since $C \supseteq A^* \supseteq A_i$, (2) implies that $A_{i+1} \models \phi(\bar{a})$. But $A_{i+1} \subseteq A^*$, and so $A^* \models \phi(\bar{a})$ by the lemma on preservation (Lemma 1.2.4(b)).

Now let A be any structure in K. We define $A^{(i)}$ inductively for each $i \leqslant \omega$, by: $A^{(0)} = A$, $A^{(i+1)} = A^{(i)*}$, $A^{(\omega)} = \bigcup_{i<\omega} A^{(i)}$. We put $B = A^{(\omega)}$. To show that B is e.c., let $\phi(\bar{x})$ be an $\exists_1$ formula of L and $\bar{b}$ a tuple from B, and suppose that $C \models \phi(\bar{b})$ for some C in K, $C \supseteq B$. Since $\bar{b}$ is finite, it already lies inside some $A^{(i)}$ with $i < \omega$. Since $A^{(i)} \subseteq C$, the choice of $A^{(i+1)}$ gives at once that $A^{(i+1)} \models \phi(\bar{b})$, and hence $B \models \phi(\bar{b})$ by the lemma on preservation again. □

Recall from section 1.2 that a $\forall_2$ first-order formula is one of form $\forall\bar{x}\exists\bar{y}\psi(\bar{x},\bar{y},\bar{z})$ with ψ quantifier-free. A $\forall_2$ theory is a set of $\forall_2$ sentences. We allow either or both of $\bar{x}$, $\bar{y}$ to be empty, so that every $\forall_1$ or $\exists_1$ formula is $\forall_2$. (In particular every variety is axiomatised by a $\forall_2$ theory. We showed in section 1.2 that the class of fields has a set of $\forall_2$ axioms.) By the lemma on preservation (Lemma 1.2.4(c)), the class of models of a $\forall_2$ theory is inductive. So the theorem immediately implies:

COROLLARY 3.2.2. *If* T *is a* $\forall_2$ *first-order theory then every model of* T *can be extended to an e.c. model of* T. □

A set-theoretic aside: if A is a field and B is an algebraically closed field extending A, then without using the axiom of choice we can find the algebraic closure of A inside B. But there are models of Zermelo-Fraenkel set theory without choice in which some fields have no algebraic closure. Hence some form of the axiom of choice is needed in order to prove Theorem 3.2.1, even in cases where B can be chosen to sit symmetrically over A.

The second memorable result is a characterisation of the e.c. models of $\forall_2$ theories. Recall that $T_\forall$ is the set of all $\forall_1$ first-order consequences of T.

THEOREM 3.2.3. *Let* T *be a* $\forall_2$ *theory in a first-order language* L *and let* A *be an* L-*structure. Then the following are equivalent:*

(a) A *is an e.c. model of* T.

(b) A *is a model of* $T_\forall$, *and for every* $\exists_1$ *formula* $\phi(\bar{x})$ *of* L,

$A \models \forall\bar{x}(\bigwedge Res_\phi(\bar{x}) \to \phi(\bar{x}))$.

(c) A *is an e.c. model of* $T_\forall$.

Proof. (a) ⇒ (b): Assume (a). Then certainly A is a model of $T_\forall$. Let $\phi(\bar{x})$ be an $\exists_1$ formula of L and $\bar{a}$ a tuple from A such that $A \models \bigwedge Res_\phi(\bar{a})$. We must show that $A \models \phi(\bar{a})$. By Theorem 3.1.1 there is a model B of T such that $A \subseteq B$ and $B \models \phi(\bar{a})$. But A is an e.c. model of T, so it follows that $A \models \phi(\bar{a})$ as required.

(b) ⇒ (c): Assume (b). Then A is a model of $T_\forall$. Suppose $B \supseteq A$, B is a model of $T_\forall$, $\phi(\bar{x})$ is an $\exists_1$ formula of L and $\bar{a}$ is a tuple from A such that $B \models \phi(\bar{a})$. We must show that $A \models \phi(\bar{a})$. By Corollary 3.1.2 there is a model C of T such that $B \subseteq C$, and $C \models \phi(\bar{a})$ by the lemma on preservation (Lemma 1.2.4(b)). Hence $A \models \bigwedge Res_\phi(\bar{a})$ by Theorem 3.1.1, and so $A \models \phi(\bar{a})$ by (b).

(c) ⇒ (a): Assume (c). To prove (a) it suffices to show that A is a model of T (cf. Exercise 3). Now (c) asserts that A is a model of $T_\forall$, and so by Corollary 3.1.2 there is a model B of T such that $A \subseteq B$. Since T is a $\forall_2$ theory, a typical sentence in T is of form $\forall\bar{x}\exists\bar{y}\chi(\bar{x},\bar{y})$, where χ is quantifier-free. For any tuple $\bar{a}$ from A, $B \models \exists\bar{y}\chi(\bar{a},\bar{y})$ since B is a model of T. But A is an e.c. model of $T_\forall$ and B is a model of $T_\forall$; so it follows that $A \models \exists\bar{y}\chi(\bar{a},\bar{y})$. Since $\bar{a}$ was arbitrary, $A \models \forall\bar{x}\exists\bar{y}\chi(\bar{x},\bar{y})$. Therefore A is a model of T as required. □

COROLLARY 3.2.4. *Let* T *be a* $\forall_2$ *theory in a first-order language* L *and let* A *be an* L*-structure. Then* A *is an e.c. model of* T *iff:*

A *is a model of* $T_\forall$, *and for every* $\exists_1$ *formula* $\phi(\bar{x})$ *of* L *and all tuples* $\bar{a}$ *from* A, *if* $A \models \neg\phi(\bar{a})$ *then there is some* $\exists_1$ *formula* $\chi(\bar{x})$ *of* L *such that* $A \models \chi(\bar{a})$ *and* $T \vdash \forall\bar{x}(\chi \to \neg\phi)$.

Proof. The condition paraphrases (b) in the theorem. □

COROLLARY 3.2.5. *Let* T *be a* $\forall_2$ *theory in a first-order language* L, *and let* A, B *be e.c. models of* T *such that* $A \subseteq B$. *Then for every* $\forall_3$ *formula* $\phi(\bar{x})$ *of* L *(i.e. formula of form* $\forall\bar{y}\exists\bar{z}\forall\bar{w}\psi(\bar{x},\bar{y},\bar{z},\bar{w})$ *with* ψ *quantifier-free), if* $\bar{a}$ *is a tuple from* A *and* $B \models \phi(\bar{a})$ *then* $A \models \phi(\bar{a})$.

Proof. Let $\phi(\bar{x})$ be as shown, and $\bar{a}$ a tuple from A such that $B \models \phi(\bar{a})$. We must prove that for every tuple $\bar{b}$ from A,

$A \models \exists\bar{z}\forall\bar{w}\psi(\bar{a},\bar{b},\bar{z},\bar{w})$. Now by assumption $B \models \exists\bar{z}\forall\bar{w}\psi(\bar{a},\bar{b},\bar{z},\bar{w})$, so for some tuple $\bar{c}$ from B, $B \models \forall\bar{w}\psi(\bar{a},\bar{b},\bar{c},\bar{w})$. Since $\forall\bar{w}$ can be paraphrased as $\neg\exists\bar{w}\neg$, and B is an e.c. model of T, by Corollary 3.2.4 there is an $\exists_1$ formula $\chi(\bar{x},\bar{y},\bar{z})$ of L such that $B \models \chi(\bar{a},\bar{b},\bar{c})$ and $T \vdash \forall\bar{x}\bar{y}\bar{z}(\chi \rightarrow \forall\bar{w}\psi(\bar{x},\bar{y},\bar{z},\bar{w}))$. But A is also an e.c. model of T, and so $A \models \exists\bar{z}\chi(\bar{a},\bar{b},\bar{z})$. Choose $\bar{d}$ in A so that $A \models \chi(\bar{a},\bar{b},\bar{d})$. Then $A \models \forall\bar{w}\psi(\bar{a},\bar{b},\bar{d},\bar{w})$, and hence $A \models \exists\bar{z}\forall\bar{w}\psi(\bar{a},\bar{b},\bar{z},\bar{w})$, as required. □

We say that a theory T has the joint embedding property iff for every two models A, B of T there is a third model C of T into which both A and B can be embedded. For example the theory of groups has the joint embedding property: take C to be A×B. The theory of fields of a fixed characteristic has it too. But the theory of fields doesn't; two fields of different characteristics can never be embedded together in a third field.

COROLLARY 3.2.6. *Let* T *be a* $\forall_2$ *theory in a first-order language* L, *and suppose* T *has the joint embedding property. Then if* A *and* B *are any two e.c. models of* T, *exactly the same* $\forall_2$ *sentences of* L *are true in* A *as in* B.

Proof. Embed both A and B into a third model C of T, and use Theorem 3.2.1 to extend C to an e.c. model D of T. Since A and B are both isomorphic to substructures of D, the result follows from Corollary 3.2.5. □

While Theorem 3.2.3 is still fresh in our minds, let me note that by (b) in that theorem, if A is an e.c. model of the $\forall_2$ theory T, then for every $\exists_1$ formula $\phi(\bar{x})$,

$$(3) \qquad A \models \forall\bar{x}(\bigwedge \mathrm{Res}_\phi(\bar{x}) \leftrightarrow \phi(\bar{x})).$$

Right to left in (3) is by the definition of Res_ϕ. It follows that any property of form $\mathrm{Res}_\phi(\bar{x})$ is actually first-order definable in A. For example we saw at the beginning of section 3.1 that if T is the theory of groups and ϕ is $\exists x(x^{-1}yx = z)$, then Res_ϕ is a set of formulas which say "y and z are elements of the same order". So this latter property is first-order expressible in e.c. groups. It is most emphatically not expressible first-order in groups in general. This is a rather dramatic illustration of Poincaré's observation at the opening of this chapter, that new things may become definable when we add elements. Section 3.3 will develop this theme as a way of studying the algebraic properties of an e.c. model.

Let K be a class of structures and A a structure in K. We say that A is a strong amalgamation base in K iff the following holds:

(4) If B, C are structures in K and $A \subseteq B$ and $A \subseteq C$, then there exist a structure D in K such that $B \subseteq D$, and an embedding $g: C \to D$ which is the identity on A, such that $(\text{dom } B) \cap g(\text{dom } C) = \text{dom } A$.

In short, any two extensions of A in K can be amalgamated in K over A with A as the overlap. For example every group is a strong amalgamation base in the class of groups, because of free products with amalgamation (cf. Rotman (1973) Theorem 11.35). Hence the next result will interest group theorists less than some other people:

THEOREM 3.2.7. *Let* T *be a* $\forall_2$ *theory in a first-order language* L, *and let* A *be an e.c. model of* T. *Then* A *is a strong amalgamation base in the class of models of* T.

Proof. I shall derive the theorem from a slightly stronger lemma which has some independent interest. Roughly speaking, it says that the defining property for A to be e.c. can be lifted to formulas containing parameters which are not in A.

LEMMA 3.2.8. *Under the hypotheses of the theorem, suppose* B, C *are models of* T *such that* $A \subseteq B$ *and* $A \subseteq C$. *Then there exist a model* D *of* T *such that* $B \subseteq D$, *and an embedding* $g: C \to D$ *such that* g *is the identity on* A, *and for every* $\exists_1$ *formula* $\psi(\bar{x},\bar{y})$ *of* L, *if* $\bar{b}$, $\bar{c}$ *are tuples from* B, C *respectively such that* $D \models \psi(\bar{b},g\bar{c})$, *then there is already a tuple* $\bar{a}$ *in* A *such that* $C \models \psi(\bar{a},\bar{c})$.

Proof of lemma. Taking an isomorphic copy of C if necessary, we can suppose that $(\text{dom } B) \cap (\text{dom } C) = \text{dom } A$. This allows us to use the elements of these structures without ambiguity as constants naming themselves. Forming diagrams with these constants, consider the theory:

(5) $T \cup \text{diag}(B) \cup \text{diag}(C) \cup \{\neg\psi(\bar{b},\bar{c}) : \bar{b}, \bar{c}$ are tuples from B, C and $\psi(\bar{x},\bar{y})$ is an $\exists_1$ formula of L such that $C \models \neg\psi(\bar{a},\bar{c})$ for all tuples $\bar{a}$ from A$\}$.

We claim that (5) has a model.

For if not, then by the compactness theorem there are a finite conjunction $\chi(\bar{b},\bar{d})$ of sentences in diag(B) and a set of sentences $\psi_i(\bar{b},\bar{c},\bar{d})$ $(i < n)$ such that

(6) $T \cup \text{diag}(C) \vdash \chi(\bar{b},\bar{d}) \to \psi_0(\bar{b},\bar{c},\bar{d}) \vee \ldots \vee \psi_{n-1}(\bar{b},\bar{c},\bar{d})$, and

(7) for each $i < n$, there is no $\bar{a}$ in A such that $C \models \psi_i(\bar{a},\bar{c},\bar{d})$,

where all constants not in L are shown, and $\bar{b}$, $\bar{c}$, $\bar{d}$ are tuples of distinct elements from (dom B) $\smallsetminus$ (dom A), (dom C) $\smallsetminus$ (dom A) and dom A respectively. Since $\chi(\bar{b},\bar{d})$ comes from diag(B), we have $B \models \exists\bar{x}\chi(\bar{x},\bar{d})$, and hence $A \models \exists\bar{x}\chi(\bar{x},\bar{d})$ since A is e.c. Then there is $\bar{a}$ in A such that $A \models \chi(\bar{a},\bar{d})$, and so $C \models \chi(\bar{a},\bar{d})$ since χ is quantifier-free and $A \subseteq C$. But by the lemma on constants (Lemma 1.2.2), (6) yields

(8) $T \cup \text{diag}(C) \vdash \forall\bar{x}(\chi(\bar{x},\bar{d}) \to \psi_0(\bar{x},\bar{c},\bar{d}) \vee \ldots \vee \psi_{n-1}(\bar{x},\bar{c},\bar{d}))$,

and hence there is $i < n$ such that $C \models \psi_i(\bar{a},\bar{c},\bar{d})$. This contradicts (7), proving the claim.

Let D be a model of (5). Since $D \models \text{diag}(B)$, by the diagram lemma (Lemma 1.2.3) there is an embedding of B into $D|L$ taking each element b to b^D. Identifying b with b^D, we can assume that $B \subseteq D|L$. Since $D \models \text{diag}(C)$, the diagram lemma gives an embedding $g: C \to D|L$ such that $gc = c^D$ for each c in C. Since $a^D = a$ for each element a of A, g is the identity on A. The final clause of the lemma holds because of the last part of (5). □ Lemma

The theorem follows from the lemma by taking $\psi(x,y)$ to be $x=y$.

□

Let me mention one place where Theorem 3.2.7 gave welcome new information. By a result of Grätzer et al. (1973), if K is a variety of modular lattices which contains a non-distributive lattice, then the amalgamation property fails in K. Efforts to construct strong amalgamation bases explicitly in such varieties ran into formidable complexities. Theorem 3.2.7 supplies such strong amalgamation bases at once, though non-effectively.

A theory T in a first-order language L is said to be model-complete iff whenever A and B are models of T and $A \subseteq B$, then $A \preccurlyeq B$. (We assume here, as always in such contexts, that A and B are L-structures.)

This notion is useful for the following reason. In model theory the natural maps to study are the elementary embeddings. But the maps that tend to appear in algebra are usually not elementary embeddings, and the elementary embeddings are often hard to recognise when they do occur. This has been and still is a real obstacle in the way of applying model-theoretic notions to algebra. Thus any result of the form "All maps with such-and-such algebraic properties are elementary" can serve as a messenger between model theorists and algebraists. Our fourth memorable theorem, Theorem 3.2.10 below, has turned out to be a powerful device for finding model-complete theories.

First we should paraphrase the notion of model-completeness.

LEMMA 3.2.9. *Let* L *be a first-order language and* T *a theory in* L. *Then the following are equivalent:*

(a) T *is model-complete.*

(b) *Every model of* T *is e.c.*

(c) *Modulo* T, *every* $\exists_1$ *formula of* L *is equivalent to a* $\forall_1$ *formula of* L.

(d) *Modulo* T, *every formula of* L *is equivalent to a* $\forall_1$ *formula of* L.

Proof sketch. (a) ⇒ (b) is immediate. (b) implies that if $\phi(\bar{x})$ is an $\exists_1$ formula of L and $\bar{c}$ are parameters, then for every substructure A of any model of $T \cup \{\phi(\bar{c})\}$, if $A \models T$ then $A \models \phi(\bar{c})$. A variant of the proof of Theorem 3.1.1 now shows that $\phi(\bar{c})$ is equivalent modulo T to a set of $\forall_1$ sentences. Compactness reduces the set to a single sentence $\theta(\bar{c})$, and then $\phi(\bar{x})$ is equivalent modulo T to $\theta(\bar{x})$ by the lemma on constants. Thus (b) implies (c). Using $\neg$, (c) implies also that modulo T, every $\forall_1$ formula of L is equivalent to an $\exists_1$ formula of L. Using this and (c), we can change blocks of existential quantifiers to blocks of universal, and vice versa, until any formula is brought to $\forall_1$ form; this gives (d). Finally (d) implies (a) by the lemma on preservation (Lemma 1.2.4(a)). □

We say that a theory T is λ-categorical iff up to isomorphism, T has exactly one model of cardinality λ. For example a well-known back-and-forth proof shows that the theory of dense linear orderings without endpoints is ω-categorical. A very similar argument shows that the theory of atomless boolean algebras is ω-categorical (cf. Exercise 1 of section 6.3 below). Both these theories can be written as $\forall_2$ theories. By a theorem of Steinitz, the theory of algebraically closed fields of a fixed characteristic is λ-categorical for every uncountable cardinal λ; this theory is also $\forall_2$.

THEOREM 3.2.10. *Let* T *be a* $\forall_2$ *theory in a first-order language* L. *Suppose that* T *has no finite models, and for some cardinal* $\lambda \geqslant |L|$, T *is* λ-*categorical. Then* T *is model-complete.*

Proof. Suppose the conclusion fails. Then by (b) in the lemma, there are models $A \subseteq B$ of T, an $\exists_1$ formula $\phi(\bar{x})$ of L and a tuple $\bar{a}$ from A such that $B \models \phi(\bar{a})$ but $A \models \neg\phi(\bar{a})$. Let L' be L with an added 1-ary relation symbol P, and let B' be the expansion of B got by putting $P^{B'} = \mathrm{dom}(A)$. Since T has no finite models, A is infinite. So by a combination of the downward Löwenheim-Skolem theorem (Lemma 1.2.6) and Exercise 8(b) of section 2.1, there is a structure D' which is elementarily equivalent to B' and has $|P^{D'}| = \lambda$. By considering what can be said about A in Th(B'), we find a substructure C of $D'|L$ such that $\mathrm{dom}(C) = P^{D'}$, $C \models T$ and for some tuple $\bar{c}$ from C, $D'|L \models \phi(\bar{c})$ but $C \models \neg\phi(\bar{c})$. It follows that C is not an e.c. model of T. But by Theorem 3.2.1 and Exercise 1 below, T has an e.c. model of cardinality λ. Thus T has two models of cardinality which are not isomorphic, contradicting that T is λ-categorical. □

Exercises for 3.2.

1. Suppose that K in Theorem 3.2.1 is the class of all models of some $\forall_2$ theory in L. Show that B in the theorem can be chosen to be of cardinality at most $|A| + |L|$. [Use the downward Löwenheim-Skolem theorem.]

2. Show that the statement "Every group can be extended to an e.c. group" can be proved in Zermelo-Fraenkel set theory without the axiom of choice. [Amalgamate free solutions, cf. the proof of Theorem 3.1.3.]

3. Show that if $T \subseteq T'$ are first-order theories and A is a model of T' which is an e.c. model of T, then A is an e.c. model of T'.

4. Let T be a $\forall_2$ theory in a first-order language L. Show the following. (a) If $(A_i)_{i<\gamma}$ is a chain of e.c. models of T, then $\bigcup_{i<\gamma} A_i$ is an e.c. model of T. (b) If B is an e.c. model of T and A a substructure of B such that for every $\exists_1$ formula $\phi(\bar{x})$ of L and every tuple $\bar{a}$ from A, $B \models \phi(\bar{a})$ implies $A \models \phi(\bar{a})$, then A is also an e.c. model of T.

5. Let T be a theory in a first-order language L, and let A be a model of T. Show that the following are equivalent: (a) A is an e.c. model of T. (b) For every $C \models T$, if $A \subseteq C$ then there are $D \succcurlyeq A$ and an embedding $g: C \to D$ such that g is the identity on A.

6. Show that if T is a $\forall_2$ theory in a countable first-order language L, then the class of e.c. models of T is axiomatised by a sentence of the infinitary language $L_{\omega_1\omega}$. [Cf. Corollary 3.2.4.]

7. Let T be an r.e. theory in a recursive first-order language L. Let L' be the language got from L by adding countably many new constants c_0, c_1, ... The set of sentences of L' can be Gödel numbered. Let E be the set of all subsets X of ω such that for some e.c. model A of T with diagram written in L', $X = \{n < \omega$: the sentence with Gödel number n is in diag(A)$\}$. Show that E is a Π^0_2 set.

8. Let T be a theory in a first-order language L. Show that T has the joint embedding property iff for all $\forall_1$ sentences ϕ, ψ of L, if $T \vdash \phi \vee \psi$ then either $T \vdash \phi$ or $T \vdash \psi$.

9. Let T be a $\forall_2$ theory in a first-order language L. We write T^o for the set of all $\forall_2$ sentences ϕ of L such that $T_\forall = (T \cup \{\phi\})_\forall$. T^o is called the <u>Kaiser hull</u> of T. (a) Show that T^o is the set of all $\forall_2$ sentences of L which are true in every e.c. model of T. (b) Show that if T has the joint embedding property, then for every e.c. model A of T, T^o is exactly the set of $\forall_2$ sentences of L which are true in A.

10. Show that every model-complete theory is equivalent to a $\forall_2$ first-order theory.

11. Let T be a theory in a first-order language L. Show that the following are equivalent: (a) T is model-complete. (b) For every model A of T and every first-order sentence ϕ which uses only symbols from T and diag(A), either $T \cup \text{diag}(A) \vdash \phi$ or $T \cup \text{diag}(A) \vdash \neg\phi$. (c) If A, B and C are models of T with $A \subseteq B$ and $A \subseteq C$, then there are an elementary extension D of B and an embedding $g: C \to D$ which is the identity on A.

If T *is a theory in a first-order language* L, *a* <u>*model-companion*</u> *of* T *is a theory* U *in* L *such that* $T_\forall = U_\forall$ *and* U *is model-complete. A theory is said to be* <u>*companionable*</u> *iff it has a model-companion.*

12. Let T be a $\forall_2$ first-order theory. Show that the following are equivalent: (a) T is companionable. (b) The class of e.c. models of T is the class of models of some first-order theory. (c) Every model of T^o (cf. Exercise 9) is an e.c. model of T. (d) T^o is the model-companion of T, unique up to equivalence of theories.

13. Let T be a $\forall_2$ theory in a first-order language L with at least one constant. (a) Show that if, for every $\exists_1$ formula $\phi(\bar{x})$ of L, Res_ϕ is equivalent modulo T to a single $\forall_1$ formula of L, then T is companionable. (b) Show that if T is companionable and $T_\forall$ has the amalgamation property, then for every $\exists_1$ formula $\phi(\bar{x})$ of L, Res_ϕ is equivalent modulo T to a single quantifier-free formula $\psi(\bar{x})$. (c) Show that the converse of (a) fails. [For (b) use Exercise 7 of section 3.1 and compactness. For (c) consider the set of sentences $\forall xy\neg(P_0x \wedge P_iy)$ $(0 < i < \omega)$, and let ϕ be the $\exists_1$ sentence $\exists x P_0 x$.]

A theory T *in a first-order language* L *is said to* <u>*have quantifier elimination*</u> *iff for every formula* $\phi(\bar{x})$ *of* L *there is a quantifier-free formula* $\psi(\bar{x})$ *of* L *which is equivalent to* ϕ *modulo* T.

14. Let T be a theory in a first-order language L with at least one constant. Show that the following are equivalent: (a) T has quantifier elimination. (b) For every $\exists_1$ formula $\phi(\bar{x})$ of L there is a quantifier-free formula $\psi(\bar{x})$ of L which is equivalent to ϕ modulo T. (c) T is model-complete and $T_\forall$ has the amalgamation property.

15. Let T be a $\forall_2$ theory in a first-order language L. Let $\phi(\bar{x})$ be an $\exists_1$ formula of L. Suppose Res_ϕ is equivalent modulo T to a set $\{\psi_j : j < \omega\}$, and there is an e.c. model A of T such that for each $i < \omega$, A contains a tuple $\bar{a}_i$ satisfying $\bigwedge_{j<i}\psi_j(\bar{x}) \wedge \neg\psi_i(\bar{x})$. Deduce that T is not companionable.

The remaining exercises consider particular theories. Some need specialist knowledge.

16. Let T be the theory of commutative rings with 1, as in Exercise 11 of section 3.1. (a) Show that T is not companionable. [Use Exercise 15 together with (d) of Exercise 11 of section 3.1.] (b) Show that if A is an e.c. model of T and I is the nilradical of A, then every nonconstant monic polynomial in (A/I)[x] has a root in A/I, and that A/I has no minimal idempotents.

17. Let T be the theory of local rings, as in Exercise 12 of section 3.1. Show that T is not companionable, and that every e.c. model A of T has the following properties: (a) Writing M for the maximal ideal of A, A/M is an algebraically closed field. (b) Every nonconstant monic polynomial in A[x] has a root in A. (c) A is henselian. (d) Every prime ideal of A is infinitely generated. (e) Writing ann for annihilator, if $a_1, \ldots, a_n \in M$ then $(a_1,\ldots,a_n) = \text{ann ann } (a_1,\ldots,a_n)$. (f) Every finite intersection of non-zero ideals of A is non-zero.

18. Let c be a fixed prime or 0, and $1 \leqslant n < \omega$. Write L(n,c) for the class of all groups embeddable in linear groups of degree n over fields of characteristic c. Let L be the first-order language of groups. (a) Show that for all n and c, L(n,c) is axiomatised by a set of $\forall_1$ sentences of L. (b) Show that a group A in L(1,c) is e.c. in this class iff $A \cong \bigoplus_{p \text{ prime} \neq c} Z(p^\infty) \oplus Q^{(\lambda)}$ for some cardinal λ. (c) Show that if n > 1, then a group A in L(n,c) is e.c. in this class iff A = G.SL(n,F) for some group G which is e.c. in L(1,c) and some algebraically closed field F. (d) Show that the class of groups isomorphic to SL(n,F), as F ranges over all algebraically closed fields of characteristic c, is axiomatised by a $\forall_2$ first-order theory which is λ-categorical for all uncountable λ.

19. Let R be a ring with 1 and T the theory of left R-modules. (a) Show that T is companionable iff R is left coherent. (b) Show that R is von Neumann regular iff for every model A of T, Th(A) is model-complete. [For (a) use Exercise 13 together with Exercise 10 of section 3.1.]

20. Let T be the theory of fields. Use Theorem 3.2.10 to show that T is companionable, and then use Exercise 13 to show that for T, every resultant is equivalent to a single quantifier-free formula. [The model-companion is the theory of algebraically closed fields.]

21. A <u>graph</u> is a set X on which is defined an irreflexive symmetric

relation R; an edge is a pair {a,b} such that aRb. The graph is 2-colourable iff it contains no cycles of odd length. (There is a theorem justifying this name!) Graphs can be thought of as L-structures where L is the first-order language whose signature contains only the 2-ary relation symbol R. (a) Show that the class of 2-colourable graphs is axiomatised by a universal Horn theory T in L. (b) Describe the e.c. models of T. Show that up to isomorphism there is just one countable e.c. model of T, and its automorphism group has just two orbits. (c) Show that T is companionable. (d) Show that T does not have the amalgamation property.

3.3 E.C. GROUPS

Groups and logic mix well. In 1961 Graham Higman showed that group theorists had been in a position to prove theorems about recursively enumerable sets for decades without realising it. (Cf. Fact 3.3.1 below.) This revelation might have been expected to persuade group theorists that logicians are completely redundant. In fact it marked the beginning of an era of cooperation. In 1972 Angus Macintyre renewed the momentum by using model theory to find new connections between groups and recursion theory; his paper (1972 b) is one of the paradigms of applied model theory.

A high proportion of the research on e.c. groups has been aimed at the question: How different from each other can they be? The model-theoretic methods described later in this book have played a large part in providing the answers. In this section I collect together some things which are true of every e.c. group. They are part algebraic, part logical and part somewhere between. That's the way with e.c. groups.

We need a page or so of definitions. (Group theorists will know them, but they may be interested to see how a model theorist expresses them.)

Throughout the section, L will be the first-order language of groups, with function symbols $\cdot$ and $^{-1}$, and constant 1. Then L is easily encoded as a recursive language (cf. Exercise 3 of section 3.1), and we can list its terms recursively as $t_0, t_1, \ldots$. To avoid pother about the number of variables in a tuple, when $t_i(x_0,\ldots,x_{n-1})$ is applied to $(a_0,\ldots,a_{m-1})$ with $m < n$, we assume that $x_m, \ldots, x_{n-1}$ are replaced by 1. T shall be a set of axioms in L for the class of groups.

A <u>presentation</u> is an ordered pair $\langle \bar{c};\Phi(\bar{c})\rangle$ where $\bar{c}$ is a sequence of distinct constants not in L, and $\Phi(\bar{x})$ is a set of equations of L. We call $\bar{c}$ the <u>generators</u> of the presentation and $\Phi(\bar{c})$ its <u>relations</u>. Writing $L(\bar{c})$ for L with the constants $\bar{c}$ added, let U be the set of all atomic sentences θ of $L(\bar{c})$ such that $T \cup \Phi(\bar{c}) \vdash \theta$. Then U is =-closed in $L(\bar{c})$, so that it has a canonical model G (cf. Theorem 2.1.1). One can quickly show that G is a group. We call G, or more loosely the reduct $G|L$, the <u>group presented by</u> $\langle \bar{c};\Phi(\bar{c})\rangle$. We say that a group H <u>has</u> the presentation $\langle \bar{c};\Phi(\bar{c})\rangle$ iff $H \cong G|L$.

To give an example familiar from elementary algebra: the dihedral group D_n has the presentation $\langle c,d;c^n=1,d^2=1,cd=dc^{n-1}\rangle$.

In fact any group has a presentation. Let H be a group and $\bar{c}$ a sequence which lists a set of generators of H, and let $\Phi(\bar{x})$ be the set of all equations $\phi(\bar{x})$ such that $H \models \phi(\bar{c})$. Then H has the presentation $\langle \bar{c};\Phi(\bar{c})\rangle$. When $\bar{c}$ here is finite, the set $\{i < \omega : (t_i(\bar{c})=1) \in \Phi\}$ is called the <u>word problem</u> of H, and we say that the <u>word problem of</u> H <u>is solvable</u> iff this set is recursive. This doesn't depend on the choice of generators $\bar{c}$. For suppose we choose a different finite sequence $\bar{d}$ of generators of H, and we put $\Psi(\bar{x}) = \{\text{equation } \psi : H \models \psi(\bar{d})\}$. Then each d_j is of form $s_j(\bar{c})$ for some term $s_j(\bar{x})$, and so $(t_i(\bar{x})=1) \in \Psi$ iff $t_i(s_0(\bar{x}),s_1(\bar{x}),\ldots)=1 \in \Phi$. Hence Ψ is 1-1 reducible to Φ. Thus the sets $\{i : (t_i=1) \in \Phi\}$ and $\{i : (t_i=1) \in \Psi\}$ are recursively isomorphic (cf. Rogers (1967) p. 85 for the relevant theorem of Myhill), so that one is recursive if and only if the other is.

We shall use free products G*H of groups. G*H is the group with presentation $\langle \bar{c}\bar{d};\ \Phi(\bar{c}) \cup \Psi(\bar{d})\rangle$ where $\langle \bar{c};\Phi(\bar{c})\rangle$ and $\langle \bar{d};\Psi(\bar{d})\rangle$ are presentations of G and H respectively. Both G and H are canonically embedded in G*H.

A group is said to be <u>finitely presented</u> iff it has a presentation $\langle \bar{c};\Phi(\bar{c})\rangle$ in which both $\bar{c}$ and Φ are finite. Likewise it is <u>recursively presented</u> iff it has a presentation in which both the set of generators and the set of relations are recursive.

FACT 3.3.1 (Higman's embedding theorem). *Let* G *be a finitely generated group. Then* G *is recursively presented if and only if* G *is embeddable in a finitely presented group.*

<u>Proof</u>. A muscular application of HNN extensions; cf. Lyndon

& Schupp (1977) section IV.7. □

FACT 3.3.2. *Let* G *be a group with subgroups* H, K *and let* α: H $\to$ K *be an isomorphism. Then there is a group* G' $\supseteq$ G *containing an element* g *such that for all* h $\in$ H, $g^{-1}hg = \alpha h$.

Proof. Cf. Lyndon & Schupp (1977) section IV.2. □

We shall write h^g for $g^{-1}hg$. Also we shall write [a,b] for the commutator $a^{-1}b^{-1}ab$.

FACT 3.3.3. *Let* G *be a group and suppose* g_i $(i < \omega)$ *are elements of* G. *Then there is a group* G' $\supseteq$ G *containing elements* a, b *such that for each* $i < \omega$, $g_i = [[a,b^{2i+1}],a]$.

Proof. Neumann & Neumann (1959). □

Now we go to business. From the previous two sections we already know the following:

THEOREM 3.3.4. *(a) Let* G *be an e.c. group. Then* G $\neq$ {1}, *and for every* $\exists_1$ *formula* $\phi(\bar{x})$ *of* L *and every tuple* $\bar{a}$ *from* G, G $\models \phi(\bar{a})$ *iff* G $\models \bigwedge Res_\phi(\bar{a})$.

(b) If ϕ *is a positive primitive formula of* L, *then* Res_ϕ *is equivalent modulo* T *to an r.e. set of strict quantifier-free Horn formulas.*

Proof. (a) G has an extension satisfying $\exists x\, x \neq 1$, so G itself satisfies this sentence since G is e.c. The rest of (a) is from Theorem 3.2.3(b) and the definition of Res_ϕ.

(b) The theory T of groups has the amalgamation property (cf. Rotman (1973) p. 269f.), and so by Theorem 3.1.4(a), Res_ϕ is equivalent modulo T to a set of strict quantifier-free Horn formula. Hence it must be equivalent to the set Φ of strict quantifier-free Horn formulas in Res_ϕ. Since Res_ϕ is r.e. (cf. Exercise 3 of section 3.1), so is Φ. □

We shall prove that a converse to Theorem 3.3.4(b) holds. This will throw a flood of light on e.c. groups. To simplify statements below, I shall say that a set $\Phi(\bar{x})$ of formulas is a resultant iff Φ is equivalent modulo T to the resultant of some $\exists_1$ formula $\phi(\bar{x})$ of L.

Likewise a formula ψ <u>is a resultant</u> iff $\{\psi\}$ is a resultant. Occasionally we shall need to know what the formula ϕ is, but only occasionally.

LEMMA 3.3.5. *(a) For every* $n < \omega$, *the formula* $x_0=1 \wedge \ldots \wedge x_{n-1}=1 \rightarrow y=1$ *is a resultant.*

(b) For every $n < \omega$, *if* $\bar{x}$, $\bar{y}$ *are respectively* $(x_0,\ldots,x_{n-1})$, $(y_0,\ldots,y_{n-1})$, *then the set* $\{t_i(\bar{x})=1 \rightarrow t_i(\bar{y})=1 : i < \omega\}$ *is a resultant.*

<u>Proof</u>. We first show that

(1) $x=1 \rightarrow y=1$ is (equivalent to) the resultant of $\exists uvw\ y = x^u.x^v.x^w$.

Clearly if $y = x^u.x^v.x^w$ and $x = 1$, then (given that we are in a group) $y = 1$. In the other direction, let G be a group with elements a, b such that $a=1 \rightarrow b=1$. In other words, either $b = 1$ or $a \neq 1 \neq b$. We have to show that there is a group $H \supseteq G$ containing elements g, h, k such that $b = a^g.a^h.a^k$. There are two cases.

<u>Case 1</u>: $b = 1$. Let C, D be two copies of $\langle a\rangle_G$ generated by elements c, d respectively. In the group $G\times C\times D$, all the elements a, c, d, cd have the same order. So by Fact 3.3.2 applied three times, there is a group $H \supseteq G\times C\times D$ containing elements u, v, w such that $a^u = c$, $a^v = d$, $(cd)^w = a^{-1}$. Then in H, $b = 1 = a^1.a^{uw}.a^{vw}$.

<u>Case 2</u>: $a \neq 1 \neq b$. Then let C be a copy of $\langle a\rangle_G$ generated by an element c, and consider the free product $G*C$. By properties of free products (cf. Rotman (1973) Chapter 11), both ca and $c^{-1}b$ have infinite order. So by Fact 3.3.2 again, there is a group $H \supseteq G*C$ containing elements u, v such that $a^u = c$ and $(ca)^v = c^{-1}b$. Then in H, $b = a^u.a^{uv}.a^v$.

This proves (1). Considering the ways in which the formula $x_0=1 \wedge \ldots \wedge x_{n-1}=1 \rightarrow y=1$ can be true, (1) shows that this formula is the resultant of the positive primitive formula

(2) $$\exists u_0v_0w_0\ldots u_{n-1}v_{n-1}w_{n-1}\ (y = \Pi_{i<n}(x_i^{u_i}.x_i^{v_i}.x_i^{w_i})).$$

(b) We claim that the set of all formulas $t_i(\bar{x})=1 \rightarrow t_i(\bar{y})=1$ $(i < \omega)$ is the resultant of the formula

(3) $$\exists wz\ (\bigwedge_{i,j<n}[x_i^w,y_j]=1 \wedge \bigwedge_{i<n} x_i^z=x_i^w.y_i).$$

To take the easy direction first, suppose that the elements a_i^g all commute with the elements b_j, and that each a_i^h is $a_i^g.b_i$. Then for any term $t(\bar{x})$ we compute:

$$(4)\qquad t(\bar{a})^h = t(\bar{a}^h) = t(\bar{a}^g).t(\bar{b}) = t(\bar{a})^g.t(\bar{b}).$$

If $t(\bar{a}) = 1$ then $t(\bar{a})^h = t(\bar{a})^g = 1$ and so $t(\bar{b}) = 1$ too.

In the other direction, suppose G is a group with tuples $\bar{a}$, $\bar{b}$ of elements such that $t(\bar{a}) = 1 \rightarrow t(\bar{b}) = 1$ for all terms $t(\bar{x})$. This means that there is a homomorphism α: $\langle\bar{a}\rangle_G \rightarrow \langle\bar{b}\rangle_G$ taking each a_i to b_i. Now let $\bar{C}$ be a copy of $\langle\bar{a}\rangle_G$, with generators $\bar{c}$ corresponding to $\bar{a}$. Then by Fact 3.3.2 there is a group $H \supseteq G\times C$ containing an element g such that $a_i^g = c_i$ for each $i < n$. Each element c_i commutes with all elements of G, and with the elements b_j in particular. Also there is an embedding β: $\langle\bar{a}\rangle_G \rightarrow H$ such that $a_i = c_i.b_i$ for each i; here we are using the existence of α. Hence by Fact 3.3.2 there is a group $G' \supseteq H$ containing an element h such that $a_i^h = c_i.b_i = a_i^g.b_i$ for each $i < n$. So

$$(5)\qquad G' \models \exists wz\,(\bigwedge_{i,j<n}[a_i^w,b_i]=1 \wedge \bigwedge_{i<n} a_i^z=a_i^w.b_i)$$

as required. □

We write $\mathrm{Hom}_n(\bar{x},\bar{y})$ for the formula (3). By Theorem 3.3.4(a), if G is an e.c. group and $\bar{a}$, $\bar{b}$ are tuples from G, then $G \models \mathrm{Hom}_n(\bar{a},\bar{b})$ if and only if there is a homomorphism from $\langle\bar{a}\rangle_G$ to $\langle\bar{b}\rangle_G$ taking each a_i to b_i.

Let me pause for a moment to show that Lemma 3.3.5 and its proof give algebraic information.

THEOREM 3.3.6. *Let* G *be an e.c. group. Then:*

(a) G *is simple.*

(b) If H *and* K *are finitely generated subgroups of* G, *then* H×K *is embeddable in* G.

Proof. (a) Suppose a and b are any two elements of G, both $\neq 1$. Then by (1) and Theorem 3.3.4(a), b is the product of three conjugates of a, and so b is in the normal subgroup generated by a. Hence G is simple.

(b) Let H, K be generated by an n-tuple $\bar{a}$ and an m-tuple $\bar{b}$ respectively. Let C, D be copies of H, K generated by $\bar{c}$, $\bar{d}$ corresponding to $\bar{a}$, $\bar{b}$ respectively. By Fact 3.3.2 there is a group $G' \supseteq G\times C\times D$ containing elements g, h such that $\bar{a}^g = \bar{c}$ and $\bar{b}^h = \bar{d}$. By Theorem 3.2.1 we can assume that G' is an e.c. group. Now we have

(6) $$G' \models \exists uv(\bigwedge_{\substack{i<n,\\ j<m}} [a_i^u, b_j^v]=1 \wedge \mathrm{Hom}_{n+m}(\bar{a}^u\bar{b}^v, \bar{a}^u 1\ldots 1))$$

and so the same holds in G since G is an e.c. group. Thus G contains elements u, v such that (i) every element of $\langle\bar{a}^u\rangle_G$ commutes with every element of $\langle\bar{b}^v\rangle_G$, and (ii) there is a homomorphism from $\langle\bar{a}^u\bar{b}^v\rangle_G$ which is the identity on $\langle\bar{a}^u\rangle_G$ and takes $\langle\bar{b}^v\rangle_G$ to 1. Thus $\langle\bar{a}^u\bar{b}^v\rangle_G$ is $\langle\bar{a}^u\rangle_G\times\langle\bar{b}^v\rangle_G$, which in turn is isomorphic to H×K. □

The definitive fact about resultants in groups is the following theorem, due to Ziegler:

THEOREM 3.3.7. *Let $\bar{x}$ be a tuple of variables and $\Phi(\bar{x})$ a set of formulas of* L. *The following are equivalent:*

(a) Φ is equivalent (modulo the theory T *of groups) to the resultant of a positive primitive formula $\phi(\bar{x})$ of* L.

(b) Φ is equivalent (modulo T*) to an r.e. set of strict quantifier-free Horn formulas of* L.

<u>Proof</u>. We already have (a) ⇒ (b) by Theorem 3.3.4(b). The startling news is the converse, which we prove as follows.

Suppose $\Phi(\bar{x})$ is an r.e. set $\{\phi_i : i < \omega\}$ of strict quantifier-free Horn formulas. After a trivial rearrangement, every formula ϕ_i has the form shown in Lemma 3.3.5(a), so it is the resultant of a formula $\exists\bar{u}_i\psi_i(\bar{x},\bar{u}_i)$ of form (2). We can take the map $\phi_i \mapsto \psi_i$ to be recursive, and we can assume too that the variables making up the tuples $\bar{u}_i$ are v_n $(n < \omega)$, and that if $i \neq j$ then no variable v_n occurs both in $\bar{u}_i$ and in $\bar{u}_j$.

Introducing two new variables y and z, replace v_n everywhere in the formulas $\psi_i(\bar{x},\bar{u}_i)$ by $[[y,z^{2n+1}],y]$, for each $n < \omega$. This turns each ψ_i into a formula $\chi_i(\bar{x},y,z)$, and again the map is recursive.

Now each formula χ_i is a conjunction of equations in $\bar{x}$, y and z. Let $\Psi(\bar{x},y,z)$ be the set of all equations which occur as conjuncts of formulas χ_i. Then Ψ is an r.e. set; so by a version of Craig's trick (Exercise 3 of section 3.1 - e.g. multiply the left-hand sides by 1^n for suitable n) Ψ is equivalent modulo T to a recursive set $\Psi'(\bar{x},y,z)$ of equations.

The presentation $\langle\bar{x}yz;\Psi'\rangle$ is recursive and presents a finitely generated group H. By Higman's embedding theorem (Fact 3.3.1) H is embeddable in a finitely presented group K. So K has a finite presentation. Adding a finite number of equations if necessary, we can assume that this presentation is of form $\langle\bar{x}yz\bar{w};\Theta\rangle$ where $\Theta(\bar{x},y,z,\bar{w})$ is a finite set of equations. Let $\phi(\bar{x})$ be the formula

$$(7) \qquad \exists yz\bar{x}'y'z'\bar{w}'(\bigwedge\Theta(\bar{x}',y',z',\bar{w}') \wedge \mathrm{Hom}_n(\bar{x}'y'z',\bar{x}yz))$$

where n is the length of the tuple $\bar{x}yz$.

We assert that ϕ is the resultant of Φ (modulo T). In one direction, suppose $G \models \phi(\bar{a})$. Then by choice of ϕ and Θ, there are elements b, c of G and elements $\bar{a}'$, b', c' such that $G \models \bigwedge\Psi'(\bar{a}',b',c')$ and $\langle\bar{a}bc\rangle_G$ is a homomorphic image of $\langle\bar{a}'b'c'\rangle_G$ by a map taking $\bar{a}'$, b', c' to $\bar{a}$, b, c respectively. It follows that $G \models \bigwedge\Psi'(\bar{a},b,c)$ and hence $G \models \bigwedge\Psi(\bar{a},b,c)$. So for each $i < \omega$, $G \models \exists\bar{u}\psi_i(\bar{a},\bar{u})$. This at once gives $G \models \bigwedge\Phi(\bar{a})$.

In the other direction we have to make repeated extensions. Suppose $G \models \bigwedge\Phi(\bar{a})$. Then G has an extension G_0 such that $G_0 \models \exists\bar{u}_0\psi_0(\bar{a},\bar{u}_0)$. In turn G_0 has an extension G_1 such that $G_1 \models \exists\bar{u}_1\psi_1(\bar{a},\bar{u}_1)$. And so on for ω steps. If G_ω is $\bigcup_{i<\omega}G_i$, then for each $i < \omega$, G_ω contains elements $\bar{d}_i$ such that $G_\omega \models \psi_i(\bar{a},\bar{d}_i)$. By Fact 3.3.3 and the choice of the formulas χ_i, there is a group $G' \supseteq G_\omega$ containing elements b, c such that $G' \models \chi_i(\bar{a},b,c)$ for each $i < \omega$. Then $G' \models \bigwedge\Psi'(\bar{a},b,c)$, and so $\langle\bar{a},b,c\rangle_{G'}$ is a homomorphic image of the group H. It follows that $G'' \models \phi(\bar{a})$ for some group $G'' \supseteq G'\times K$, as required. □

Theorem 3.3.7 is rich in corollaries. But because space is short, I confine myself to some facts about finitely generated subgroups of an e.c. group.

COROLLARY 3.3.8. *Let G be an e.c. group. Then:*

(a) Every finitely generated group with solvable word

problem is embeddable in G.

(b) G *has a finitely generated subgroup with unsolvable word problem.*

Proof. (a) Here we use for the first time the fact, noted in Theorem 3.3.4(a), that G contains an element $g \neq 1$. In G any inequation $s \neq t$ is equivalent to $s=t \to g=1$. More generally, any quantifier-free Horn formula is equivalent to a strict quantifier-free Horn formula with g as a parameter.

Now let H be any finitely generated group with solvable word problem. Let $\bar{a}$ generate H. Then the sets $X^+ = \{i < \omega : H \models t_i(\bar{a})=1\}$ and $X^- = \omega \setminus X^+$ are both recursive. Let $\Phi(\bar{x},g)$ be the set of formulas

(8) $$\{t_i(\bar{x})=1 : i \in X^+\} \cup \{t_i(\bar{x})=1 \to g=1 : i \in X^-\},$$

with g as parameter. Since (8) is a recursive set of strict quantifier-free Horn formulas, by Theorem 3.3.7 there is a positive primitive formula $\phi(\bar{x},g)$ such that $\Phi(\bar{x},y)$ is the resultant of $\phi(\bar{x},y)$. Now $G\times H \models \bigwedge \Phi(\bar{a},g)$, and so there is a group $G' \supseteq G\times H$ such that $G' \models \exists\bar{x}\phi(\bar{x},g)$. Since G is an e.c. group, there is $\bar{b}$ in G such that $G \models \phi(\bar{b},g)$ and hence $G \models \bigwedge \Phi(\bar{b},g)$. But $\Phi(\bar{b},g)$ describes $\langle\bar{b}\rangle_G$ completely; in fact it says that $\langle\bar{b}\rangle_G$ is isomorphic to H. So H is embedded in G.

(b) We need a touch of recursion theory. Two disjoint sets $X, Y \subseteq \omega$ are said to be <u>recursively inseparable</u> iff there is no recursive set Z such that $X \subseteq Z$ and $Z \cap Y = \emptyset$. There exist recursively inseparable pairs of r.e. sets (cf. Rogers (1967) section 7.7). Let X, Y be such a pair, and with $g \neq 1$ as before, let $\Phi(x,y,g)$ be the set of formulas

(9) $$\{[[x,y^{2i+1}],x]=1 : i \in X\} \cup \{[[x,y^{2i+1}],x]=1 \to g=1 : i \in Y\}.$$

With the aid of Fact 3.3.3, the same argument as for (a) shows that G contains elements a, b such that $G \models \bigwedge \Phi(a,b,g)$. If the word problem of $\langle a,b\rangle_G$ was solvable, then the set $\{i < \omega : [[a,b^{2i+1}],a] = 1\}$ would be recursive, contradicting the recursive inseparability of X and Y. □

The next corollary uses the recursion-theoretic notion of an arithmetical relation, i.e. a relation on the natural numbers which is definable by a first-order formula of arithmetic.

COROLLARY 3.3.9. *Let* H *be a finitely generated group with arithmetical word problem. Then there is a sentence* χ_H *of* L *such that for every e.c. group* G, $G \models \chi_H$ *if and only if* H *is embeddable in* G.

Proof. We construct the sentence χ_H as if for a particular e.c. group G, but the sentence constructed will be independent of G. Since the additive group of integers has solvable word problem, by Corollary 3.3.8(a) G contains some element g of infinite order. We claim that we can interpret the arithmetic of the natural numbers in G, using g^n to represent the natural number n.

First, there is a positive primitive formula $\phi(x,y,g)$ which expresses that $y \neq x^n$ for all $n \geqslant 0$. For this, use Theorem 3.3.7 and the recursive set of formulas $y=x^n \rightarrow g=1$ $(n \geqslant 0)$. Hence, negating ϕ, there is a $\forall_1$ formula $N(y,g)$ which expresses that $y = g^n$ for some $n \geqslant 0$. Thus $N(y,g)$ represents the statement "y is a natural number".

Next, let R be any r.e. n-ary relation on ω. Then the set of formulas

(10) $$y_1=x^{m_1} \wedge \ldots \wedge y_n=x^{m_n} \rightarrow g=1 \qquad \text{(where } R(m_1,\ldots,m_n) \text{ holds)}$$

is an r.e. set of strict quantifier-free Horn formulas. So by Theorem 3.3.7 again, there is a $\forall_1$ formula $\theta_R(y_1,\ldots,y_n,g)$ such that for all natural numbers $m_1, \ldots, m_n$,

(11) $$R(m_1,\ldots,m_n) \text{ holds} \quad \text{iff} \quad G \models \theta_R(g^{m_1},\ldots,g^{m_n},g).$$

Using $N(y,g)$ and the formulas θ_R, we can at once represent any arithmetical relation. In fact for every arithmetical relation R there is a first-order formula θ_R satisfying (11). This establishes the claim.

Now let H be a finitely generated group with arithmetical word problem. Then for some tuple $\bar{a}$ generating H, the set $R = \{i < \omega : t_i(\bar{a}) = 1 \text{ in } H\}$ is arithmetical. The statement "For all $i < \omega$, $t_i(\bar{x}) = 1$ iff $i \in R$", which says that $\bar{x}$ generates a group isomorphic to H (by an isomorphism taking each x_k to a_k), can be expressed in G as follows:

(12) $$\forall y(N(y,g) \wedge \theta_R(y,g) \rightarrow \bigwedge_{i<\omega}(y = g^i \rightarrow t_i(\bar{x}) = 1)) \wedge$$
$$\forall y(N(y,g) \wedge \neg\theta_R(y,g) \rightarrow \bigwedge_{i<\omega}(y = g^i \wedge t_i(\bar{x}) = 1 \rightarrow g = 1)).$$

The infinitary conjunctions are of recursive sets of strict quantifier-free Horn formulas, so that by Theorem 3.3.7 they can be replaced by $\exists_1$ formulas. This makes (12) into a first-order formula $\chi'(\bar{x},g)$ with parameter g. We have:

(13) H is embeddable in G iff $G \models \exists z\bar{x}(z \text{ has order } \infty \wedge \chi'(\bar{x},z))$.

The statement "z has infinite order" can be written

(14) $\exists w(w \neq 1 \wedge \bigwedge_{i>0}(z^i = 1 \rightarrow w = 1))$

and a final application of Theorem 3.3.7 turns this into an $\exists_1$ formula I(z) of L. Thus we can take χ_H to be the sentence $\exists z\bar{x}(I(z) \wedge \chi'(\bar{x},z))$. □

Close inspection shows that if H in the corollary has r.e. word problem, then χ_H can be chosen to be an $\exists_3$ sentence. (Cf. section 1.2.)

Some of the results of this section go through for other classes besides the class of groups. In the early 1970s, after Paul Cohn had discovered a good analogue of HNN extensions for skew fields (cf. section 5.5 of Cohn (1977)), there were high hopes that e.c. groups and e.c. skew fields could be studied in close parallel. But for skew fields we still lack an analogue of Fact 3.3.1, in spite of the efforts of Macintyre (1979). The upshot is that no general result like Theorem 3.3.7 is known for skew fields, and so advances have to be piecemeal, chasing the tail of their group-theoretic analogues.

Exercises for 3.3.

1. Show that there is an e.c. group of cardinality ω_1 which is not the union of any countable chain of proper subgroups. [Using Exercise 4 of section 3.2 and Fact 3.3.3, build a chain $(A_i : i < \omega_1)$ of countable e.c. groups so that each A_i lies in a 2-generator subgroup of A_{i+1}.]

2. (a) Show that in groups, the resultant of $\exists z([z,x]=1 \wedge [z,y]\neq 1)$ is equivalent to $\{y\neq x^n : n \in Z\}$. [Use free products with amalgamation.]
(b) Show that in any e.c. group G, if $\bar{a}$ is a tuple of elements then $\langle\bar{a}\rangle_G = C_G C_G(\bar{a})$. ($C_G(X)$ is the set of all elements that commute with every

$x \in X$.) (c) Show that the theory of groups is not companionable. [Use Exercise 15 of section 3.2 with (b) above.]

3. Let G be an e.c. group. (a) Show that if $A_1, \ldots, A_n$ are finitely generated subgroups of G then $C_G(A_1 \cap \ldots \cap A_n) = \langle C_G(A_1) \cup \ldots \cup C_G(A_n) \rangle_G$. (b) Suppose A is the intersection of finitely many finitely generated subgroups of G, and A satisfies the minimum condition on subgroups. Show that for every group H with $C_G(A) \subseteq H \subseteq G$, there is a group $K \subseteq A$ such that $C_G(K) \subseteq H \subseteq N_G(K)$. (c) Show that if A is a finite and characteristically simple subgroup of G then $N_G(A)$ is a maximal proper subgroup of G. (d) Show that if $A \subseteq B$ are finite subgroups of G, A is a maximal subgroup of B and the identity is the only automorphism of B which induces the identity on A, then $C_G(B)$ is maximal in $C_G(A)$. (e) Suppose H is a maximal subgroup of G and A is a finitely generated subgroup of H. Show that $C_G(A) \not\subseteq H$ iff there are $h_1, \ldots, h_n \in H$ such that $A^{h_1} \cap \ldots \cap A^{h_n} = \{1\}$.

4. Let G be an e.c. group. (a) Show that if H, K are finitely generated groups, $H \subseteq G$, $H \subseteq K$ and K is embeddable in G, then there is an embedding of K into G which is the identity on H. [Use Fact 3.3.2.] (b) Show that G is not finitely generated. (c) Show that if K is the union of a chain $(H_i)_{i<\omega}$ of finitely generated groups with solvable word problems, then K is embeddable in G. (d) Show that every countable abelian group is embeddable in G.

5. Show that in an e.c. group, every element is (a) a commutator and (b) the product of two elements of order 2. [Adapting the proof of Lemma 3.3.5(a), show that if b has order 2 then every element is of form $b^g.b^h$ for suitable g, h. Note that the dihedral group D_n is generated by elements c, d of order 2 such that cd has order n.] Show also (c) that in (1), the formula $\exists uvw\ y = x^u.x^v.x^w$ can be replaced by $\exists uv\ y = x^u.x^v$. [Cf. Magnus (1974) section II.5 and p. 105 for the relevant groups.]

6. Show that Theorem 3.3.7 can be effectivised as follows. There is a recursive function f such that for every number e, if e is the r.e. Gödel number of a set $\Phi(\bar{x})$ of strict quantifier-free Horn formulas, then f(e) is the Gödel number of a positive primitive formula $\phi(\bar{x})$ whose resultant is equivalent to Φ. (Of course this involves chasing up the proof of Higman's theorem.)

7. Show that the following are expressible in an e.c. group G by first-order formulas of the stated quantifier complexity, allowing a parameter

$g \neq 1$: (a) "x is in the normal closure of $\bar{y}$ in $\langle\bar{z}\rangle_G$" ($\forall_1$). (b) "$\langle\bar{x}\rangle_G$ is finite" ($\forall_1$). (c) "$\langle\bar{x}\rangle_G$ is in V" where V is any recursively axiomatised group variety ($\exists_1$). (d) "$\langle\bar{x}\rangle_G$ is soluble (or nilpotent)" ($\exists_2$). (e) "$\langle\bar{x}\rangle_G$ is recursively presented" ($\exists_3$). (f) "$\langle\bar{x}\rangle_G$ has solvable word problem" ($\exists_3$).

8. Show that for every sentence ϕ of first-order arithmetic there is a first-order sentence ϕ^* of the language of groups such that for every e.c. group G, $G \models \phi^*$ iff ϕ is true in the natural numbers. *(As Martin Ziegler pointed out to me, this implies that every e.c. group knows the truth or otherwise of the Riemann hypothesis. But they won't tell us.)*

9. With reference to the proof of Corollary 3.3.9: (a) Show that if R is a Σ^0_n relation for some $n > 1$, then θ_R can be chosen to be an $\exists_{n-1}$ formula. (b) Show that if H is a group with r.e. word problem then χ_H can be chosen to be $\exists_3$. (c) Show that if H is a group with Σ^0_n word problem ($n > 1$), then χ_H can be chosen to be $\exists_{n+1}$.

10. In any variety V an algebra A is called simple iff for all algebras B in V, every homomorphism α: $A \to B$ is either an embedding or a constant map. (This generalises the notion of simple groups.) (a) Prove that if every algebra in V can be embedded in a simple algebra, then every a.c. algebra in V with at least two elements is e.c. (b) Show that an a.c. group is e.c. iff it has at least two elements.

11. A class K of algebras is said to have internal maps iff for each $n \geqslant 1$ there is a term $t_n(x_1,\ldots,x_n,\bar{y})$ such that if A is an algebra in K and f: $A^n \to A$ is any map (not necessarily a homomorphism), then there is an algebra $B \supseteq A$ in K with elements $\bar{b}$ such that for every n-tuple $\bar{a}$ from A, $f(\bar{a}) = t^B_n(\bar{a},\bar{b})$. (a) Show that if K is a variety of algebras which has internal maps, and A is an algebra in K which is a.c. in K and has at least two elements, then A is simple and e.c. in K. (b) Show that the class of groups has internal maps. (c) Show that the weak second-order theory of an e.c. group G is reducible to the first-order theory of G, uniformly in G.

The remaining four exercises indicate four classes whose e.c. structures behave in some ways like e.c. groups. In all cases, more is known than I mention here. Some parts of the exercises need specialist knowledge.

12. Let k be a field. Let L be the first-order language with symbols +, ·, − and constants for all elements of k. A skew field over k is a

skew field containing k as a subfield of its centre. Each skew field over k can be regarded as an L-structure in an obvious way, and there is a $\forall_2$ theory T in L which axiomatises the class of skew fields over k, up to isomorphism. Let K be an e.c. skew field over k. Show: (a) If $\bar{a}$, b are respectively a tuple of elements and an element of K, then b is in the skew subfield over k generated by $\bar{a}$ iff every element of K which commutes with all of $\bar{a}$ commutes with b. (b) k is the centre of K. (c) The theory of skew fields over k is not companionable. (d) If $A \subseteq B$ are skew fields over k, both finitely generated over k, $A \subseteq K$ and B is embeddable in K over k, then there is an embedding of B in K over k which is the identity on A. (e) No e.c. skew field over k is finitely generated over k. (f) The property "x is transcendental over k" is definable in K by an $\exists_1$ formula of L. (g) If b is any element of K which is transcendental over k, then first-order arithmetic can be interpreted in K using b^n to represent n.

13. Let L be the first-order language with just the symbol $\cdot$. Let T be the theory in L consisting of just the axiom of associativity. Models of T are known as <u>semigroups</u>. (a) Show that the class of semigroups has internal maps. (b) Show that if G is an e.c. semigroup and a, b are elements of G, then a is in the subsemigroup generated by b iff $G \models \forall x(xbx=bx \rightarrow xax=ax)$. (c) Show that Theorem 3.3.7(b)⇒(a) holds with semigroups in place of groups. (d) Show that every countable e.c. group G is embeddable as a maximal subgroup in some e.c. semigroup A, and that A is unique up to isomorphism over G.

14. A <u>lattice-ordered group</u> is a group which is also a lattice, such that for all elements a, b, c, if $a \leqslant b$ then $ac \leqslant bc$ and $ca \leqslant cb$. Let G be an e.c. lattice-ordered group. Show: (a) G is simple and divisible. (b) Every element of G is conjugate to its square, and hence is a commutator. (c) If $1 < a$ in G then there are elements b, c in G such that $1 < b < a$ and $1 < c < a$ and $b \wedge c = 1$. (d) If a, b are both > 1 in G, then b is in the subgroup generated by a iff every element which commutes with a also commutes with b. (e) The theory of lattice-ordered groups is not companionable.

15. A <u>locally finite</u> group is a group whose finitely generated subgroups are all finite. Show: (a) The class of locally finite groups is inductive but is not the class of models of any first-order theory.

(b) A locally finite group G is an e.c. locally finite group iff (i) every finite group is embeddable in G and (ii) any two isomorphic finite subgroups of G are conjugate in G. [Show first that any isomorphism between finite subgroups of G extends to an inner automorphism of G, assuming (i) and (ii).] (c) Up to isomorphism there is just one countable e.c. locally finite group. (d) If G is an e.c. locally finite group and b an element $\neq 1$ in G, then every element of G is a product of two conjugates of b; hence G is simple. (e) The class of finite groups has internal maps. (f) If G is an e.c. locally finite group, then first-order arithmetic is interpretable in the first-order theory of G, using the conjugacy class of any element of order n+1 to represent the number n.

3.4 *ROBINSON FORCING*

In Chapter 2 we began to study how a model can be built by assembling an increasing chain of "conditions". Now we shall revert to that question, and to the setting of section 2.3. L is a fixed countable first-order language and W is a countable set of new constant symbols (the witnesses). T is a fixed theory in L, and to save embarrassment we shall assume that T has models.

We define a notion of forcing N for L(W) as follows: $p \in N$ iff p is a finite set of basic sentences of L(W) and $T \cup p$ has a model. Elements of N will be known simply as conditions. This notion of forcing was studied first by Abraham Robinson, in 1969, and so the forcing that goes with N is called Robinson forcing.

LEMMA 3.4.1. *(a)* N *is a notion of forcing for* L(W), *and* N *has the properties (6)-(8) of section 2.3.*

(b) The property "The compiled structure A^+ *is a model of* $T_\forall$*" is enforceable.*

(c) Let $\psi(x_1,\ldots,x_n)$ *be a formula of* $L(W)_{\omega_1\omega}$ *in which at most finitely many witnesses occur, and let* q *be a condition. Then* q *forces* $\forall x_1\ldots x_n\psi$ *iff for some tuple* $(c_1,\ldots,c_n)$ *of distinct witnesses not occurring in* q *or in* ψ, q *forces* $\psi(c_1,\ldots,c_n)$.

(d) Let p *be a condition and* ϕ *an* $\exists_1$ *sentence of* L(W), *and suppose that* $T \cup p \cup \{\phi\}$ *has a model. Then there is a condition* $q \supseteq p$ *such that* $q \vdash \phi$.

Proof. (a) is left to the reader. Then (c) follows from

Exercise 2 of section 2.3.

To prove (b), let $\forall\bar{x}\psi$ be a sentence in $T_\forall$, with $\psi(\bar{x})$ quantifier-free. We show that for every tuple $\bar{c}$ of witnesses, $\psi(\bar{c})$ is enforceable. In a play of $G(\psi(\bar{c});\text{odds})$, let player $\forall$ offer p_0. Since $T \cup p_0$ has models and $T \vdash \psi(\bar{c})$, $T \cup p_0 \cup \{\psi(\bar{c})\}$ has a model B. Express $\psi(\bar{c})$ in disjunctive normal form as $\bigvee_i \bigwedge_j \theta_{ij}$. Then for some i, $B \models \bigwedge_{j<n} \theta_{ij}$. Fixing this i, put $q = p_0 \cup \{\theta_{ij} : j < n\}$. Then $q \vdash \psi(\bar{c})$, so that player $\exists$ can ensure that $A^+ \models \psi(\bar{c})$ by putting $p_1 = q$ and then playing so that $A^+ \models \bigcup \bar{p}$ (which she can do by Lemma 2.3.1).

This argument adapts to prove (d) too. □

Theorem 2.3.4 characterised the relation "q forces ϕ". In the case where ϕ is atomic, the next result gives a better characterisation than Theorem 2.3.4′ (but only for Robinson forcing).

THEOREM 3.4.2. *Let* p *be a condition and* $\psi(\bar{x})$ *a quantifier-free formula of* L(W). *Then* p *forces* $\forall\bar{x}\psi$ *if and only if* $T \cup p \vdash \forall\bar{x}\psi$.

Proof. By Lemma 3.4.1(c) and the lemma on constants (Lemma 1.2.2), we can replace $\forall\bar{x}\psi$ by $\psi(\bar{c})$ where $\bar{c}$ is a tuple of distinct witnesses not occurring in ψ or p.

Suppose first that $T \cup p \nvdash \psi(\bar{c})$. Then $T \cup p \cup \{\neg\psi(\bar{c})\}$ has a model, and so by Lemma 3.4.1(d) there is a condition $q \supseteq p$ such that $q \vdash \neg\psi(\bar{c})$. Then player $\forall$ wins $G(\psi(\bar{c});\text{odds})$ by putting $p_0 = q$ and playing to make $A^+ \models q$. Hence p doesn't force $\psi(\bar{c})$. This proves left to right.

Secondly suppose that $T \cup p \vdash \psi(\bar{c})$. Then $T \vdash \bigwedge p \rightarrow \psi(\bar{c})$, and by the lemma on constants it follows that $T_\forall \vdash \bigwedge p \rightarrow \psi(\bar{c})$. If player $\forall$ begins a play of $G(\psi(\bar{c});\text{odds})$ by playing $p_0 \supseteq p$, then let player $\exists$ play so that A^+ is a model of $T \cup p_0$. She can do this by Lemma 2.3.1 and Lemma 3.4.1(b) (and of course the conjunction lemma, Lemma 2.3.3(e)). Clearly she wins the game. This proves right to left. □

COROLLARY 3.4.3. *Let* T *be a* $\forall_2$ *theory in* L. *Then the property "The compiled* L*-structure* A *is an e.c. model of* T*" is enforceable.*

Proof. By the equivalence of (a) and (c) in Theorem 3.2.3, the property in question can be rewritten as "A is an e.c. model of $T_\forall$". Now by Lemma 3.4.1 it is enforceable that A is a model of $T_\forall$. Also by

Lemma 2.3.1 it is enforceable that every element of A is named by infinitely many witnesses, and hence it is enforceable that every tuple of elements of A is named by a tuple of distinct witnesses. So by Corollary 3.2.4 it remains only to show that the following is enforceable, for any tuple $\bar{c}$ of distinct witnesses and any $\forall_1$ formula $\phi(\bar{x})$ of L:

(1) If $A^+ \models \phi(\bar{c})$ then there is some $\exists_1$ formula $\psi(\bar{x})$ of L such that $A^+ \models \psi(\bar{c})$ and $T_\forall \vdash \forall\bar{x}(\psi \rightarrow \phi)$.

Let P be the property (1), and suppose that player $\forall$ has chosen p_0 in a play of G(P;odds). Write $\bigwedge p_0$ as $\psi(\bar{c},\bar{d})$ where $\bar{d}$ lists the distinct witnesses which occur in p_0 but not in $\bar{c}$. Now there are two cases.

Case 1: $T \cup p_0 \vdash \phi(\bar{c})$. Then $T_\forall \vdash \forall\bar{x}(\exists\bar{y}\psi(\bar{x},\bar{y}) \rightarrow \phi(\bar{x}))$ by the lemma on constants, and player $\exists$ can ensure that $A^+ \models \exists\bar{y}\psi(\bar{c},\bar{y})$ by playing to make $A^+ \models p_0$. Thus player $\exists$ wins G(P;odds).

Case 2: $T \cup p_0 \nvdash \phi(\bar{c})$. Then by Lemma 3.4.1(d), player $\exists$ can choose $p_1 \supseteq p_0$ so that $p_1 \vdash \neg\phi(\bar{c})$; playing so that $A^+ \models p_1$, she wins G(P;odds). □

In short, every model that we get by Robinson forcing is an e.c. model. This was the justification for sections 3.1-3.3 above. From now onwards, the question is whether we can twist the screw tighter and force the model to have some further properties. For example one might ask what infinitary sentences are enforceable. That turns out to be a good question:

THEOREM 3.4.4. *Let* $\phi_i(\bar{x})$ $(i < \omega)$ *be* $\exists_1$ *formulas of* L *and let* p *be a condition. Then the following are equivalent:*

(a) p *doesn't force* $\forall\bar{x}\bigvee_{i<\omega}\phi_i$.

(b) There exist distinct witnesses $\bar{c}$ *not in* p, *and a condition* $q \supseteq p$, *such that for all* $i < \omega$, q *forces* $\neg\phi_i(\bar{c})$.

(c) There is an $\exists_1$ *formula* $\psi(\bar{x})$ *of* L(W) *such that* $T \cup p \cup \{\exists\bar{x}\psi\}$ *has a model and* $T \vdash \forall\bar{x}(\psi \rightarrow \bigwedge_{i<\omega}\neg\phi_i)$.

Proof. (a) ⇒ (b): Suppose (b) fails. Let $\bar{c}$ be distinct witnesses not in p; by Lemma 3.4.1(c) it suffices to prove that p forces $\bigvee_{i<\omega}\phi_i(\bar{c})$. If player $\forall$ begins with $p_0 \supseteq p$, then by assumption there is

$i < \omega$ such that p_0 doesn't force $\neg\phi_i(\bar{c})$. So by Theorem 2.3.4(f), player $\exists$ can extend p_0 to a condition p_1 which forces $\phi_i(\bar{c})$, and then she can make $\phi_i(\bar{c})$ true.

(b) $\Rightarrow$ (c): Take $\psi(\bar{c})$ to be $\bigwedge q$, and use Theorem 3.4.2.

(c) $\Rightarrow$ (a): Assume (c). Let $\bar{c}$ be distinct witnesses not in p or ψ. Then $T \cup p \cup \{\psi(\bar{c})\}$ has a model, so by Lemma 3.4.1(d) there is a condition $q \supseteq p$ such that $q \vdash \psi(\bar{c})$ and hence $T_\forall \cup q \vdash \bigwedge_{i<\omega}\neg\phi_i(\bar{c})$. Player $\forall$ can then falsify $\forall\bar{x}\bigvee_{i<\omega}\phi_i$ by playing to make $A^+(\bar{p})$ a model of $T_\forall \cup q$.

□

A $\forall$-set in the language L is a set $\Phi(\bar{x})$ of $\forall_1$ formulas of L. We say that the tuple $\bar{a}$ in an L-structure A realises a set $\Phi(\bar{x})$ of formulas iff $A \models \bigwedge\Phi(\bar{a})$. We say that Φ is realised in A iff some tuple $\bar{a}$ in A realises Φ; if no tuple in A realises Φ, we say that A omits Φ.

It is common to talk of "omitting the $\forall$-type Φ". However, careful writers reserve the word "type" for a set of formulas which can be realised by elements in a suitable extension of A, and we make no such assumption on Φ here.

Let T be a theory in L and $\Phi(\bar{x})$ a set of formulas of L. By a support of Φ we mean an $\exists_1$ formula $\psi(\bar{x})$ of L such that $T \cup \{\exists\bar{x}\psi\}$ has a model and $T \vdash \forall\bar{x}(\psi \rightarrow \bigwedge\Phi)$. (In Chapter 5 we shall use the word "support" in a broader sense, to match a broader kind of forcing.)

COROLLARY 3.4.5 (Omitting $\forall$-types theorem). *Let* T *be a theory in* L, *and for each* $i < \omega$ *let* Φ_i *be a* $\forall$*-set with no support. Then* T *has an e.c. model which omits all the sets* Φ_i $(i < \omega)$.

Proof. Combine Corollary 3.4.3 and Theorem 3.4.4. Theorem 3.4.4 with $p = \emptyset$ says that the property "A^+ omits Φ" is enforceable so long as there is no $\exists_1$ formula $\psi(\bar{x})$ of L(W) such that $T \cup \{\exists\bar{x}\psi\}$ has a model and $T \vdash \forall\bar{x}(\psi \rightarrow \bigwedge\Phi)$; ψ may contains witnesses, but these can be eliminated with existential quantifiers. □

Macintyre found a striking application for Corollary 3.4.5. Recall that by Corollary 3.3.8(a), if H is a finitely generated group with solvable word problem, then H is embeddable in every e.c. group. B. H. Neumann announced this result in about 1970. Macintyre promptly replied with the converse:

THEOREM 3.4.6. *Let* H *be a finitely generated group whose word problem is not solvable. Then there is an e.c. group* G *in which* H *is not embeddable.*

Proof. T shall be the theory of groups. Let $\bar{h}$ be a tuple which generates H, and let $\Phi(\bar{x})$ be the set of all equations and inequations $\phi(\bar{x})$ such that $\phi(\bar{h})$ is true in H. It suffices to construct an e.c. group which omits Φ. By Corollary 3.4.5 we can do this if we show that Φ has no support.

Suppose then that $\psi(\bar{x})$ is a support of Φ. We infer that for every equation $\phi(\bar{x})$ in L,

(2) $\qquad \phi \in \Phi \quad \text{iff} \quad T \vdash \forall\bar{x}(\psi \rightarrow \phi).$

(Right to left is because $T \cup \{\exists\bar{x}\psi\}$ has a model and Φ contains at least one of ϕ, $\neg\phi$.) This shows that the set of equations $\phi(\bar{h})$ true in H is recursively enumerable. Similarly

(3) $\qquad \phi \notin \Phi \quad \text{iff} \quad T \vdash \forall\bar{x}(\psi \rightarrow \neg\phi).$

This shows that the set of equations $\phi(\bar{h})$ false in H is also recursively enumerable, and hence that H has solvable word problem. Contradiction. So Φ has no support, as required. □

At heart this theorem is nothing to do with groups. It applies equally well to any variety with a recursively enumerable set of axioms.

One corollary of Theorem 3.4.6 is that not all e.c. groups are elementarily equivalent. For by a celebrated result of P. Novikov and W. Boone (cf. Lyndon & Schupp (1977) section IV.7) there is a finitely generated group H whose word problem is r.e. but not solvable. By Theorem 3.2.1 there is an e.c. group G_1 containing H as a subgroup. By Theorem 3.4.6 there is an e.c. group G_2 in which H is not embeddable. Hence by Corollary 3.3.9, there is a sentence χ_H of L which is true in G_1 but false in G_2. By Exercise 9 of section 3.3, χ_H is an $\exists_3$ sentence. In one way this result is best possible, because by Corollary 3.2.6, all e.c. groups agree on $\forall_2$ sentences. (But in another way Theorem 4.1.9 will prove something stronger.)

THEOREM 3.4.7. *Suppose* T *has the joint embedding property. Then for every sentence* ϕ *of* $L_{\omega_1\omega}$ *(without witnesses), either* ϕ *or* $\neg\phi$ *is enforceable.*

Proof. Let ϕ be a counterexample. Then (cf. Theorem 2.3.4(f) and Exercise 5 of section 2.3) there are conditions p, q such that p forces ϕ and q forces $\neg\phi$. Since no witnesses occur in ϕ, we can change the witnesses in q and thus assume without loss that no witness occurs both in p and in q (cf. Exercise 1 of section 2.3). Since p and q are conditions, they are satisfiable in models A, B of T. Let C be a model of T such that both A and B are embeddable in C. Then since p and q have no witnesses in common, $p \cup q$ is satisfiable in C and hence is a condition. But $p \cup q$ forces $\phi \wedge \neg\phi$, which is impossible. □

The set of all sentences of L (not $L_{\omega_1\omega}$) which are enforceable, for a theory T in L, is called the finite forcing companion of T and written T^f. By Theorem 3.4.7, if T has the joint embedding property then T^f is a complete first-order theory. It follows, for example, that by forcing with the theory of groups we shall never be able to construct two groups which are not elementarily equivalent. This is no contradiction to the result mentioned after Theorem 3.4.6, because one of the groups used there was not reached by forcing.

Exercises for 3.4. Forcing is assumed to be Robinson forcing.

1. (a) Let $\phi(x_1,\ldots,x_n)$ be a formula of $L(W)_{\omega_1\omega}$ and p a condition. Prove that p forces $\exists\bar{x}\phi$ iff for every condition $q \supseteq p$ there are a condition $r \supseteq q$ and witnesses $\bar{c}$ such that r forces $\phi(\bar{c})$. (b) Show that if p is a condition and ϕ is a sentence of $L(W)_{\omega_1\omega}$, then p forces ϕ iff $\bigwedge p \to \phi$ is enforceable.

2. Assume L and L(W) are recursive languages and T is an r.e. theory in L. All sentences ϕ below are assumed to be first-order sentences in L(W). Show: (a) The relation $\{\langle p,\phi\rangle : \phi \text{ a } \forall_1 \text{ sentence and } p \Vdash \phi\}$ is Σ^0_1. (b) For every positive n, the relation $\{\langle p,\phi\rangle : \phi \text{ a } \forall_{2n+1} \text{ sentence and } p \Vdash \phi\}$ is Π^0_{2n+1}. [Consider $p \Vdash \forall\bar{x}\exists\bar{y}\psi(\bar{x},\bar{y}) \Leftrightarrow \forall\bar{c}\forall q(q \text{ a condition} \supseteq p \to \exists\bar{d}\exists r(r \text{ a condition} \supseteq q \wedge r \Vdash \psi(\bar{c},\bar{d})))$.] (c) The set T^f is 1-1 reducible to the first-order theory of the natural numbers.

A structure A *is said to be* residually finite *iff for every primitive formula* $\phi(\bar{x})$ *and tuple* $\bar{a}$ *from* A *such that* $A \models \phi(\bar{a})$, *there is a surjective homomorphism* f: A → B *such that* B *is finite and* B ⊨ $\phi(f\bar{a})$. *The structure* A *is said to be* *locally finite* *iff every finitely generated substructure of* A *is finite. A variety* V *is said to be* *universally finite* *iff every finitely presented algebra in* V *is residually finite.*

3. Show: (a) If T is a set of axioms for a variety V with only a finite number of function symbols and constants, then for forcing with respect to T, the following are equivalent: (i) V is universally finite; (ii) The property "The compiled structure A is locally finite" is enforceable. (b) The variety of abelian groups is universally finite; more generally, so is any variety of nilpotent groups, and so is the variety of soluble groups of length ≤ 2. (c) There is a finitely presented soluble group of length 3 which is not residually finite. (d) The variety of commutative rings with 1 is universally finite. (e) If T is the theory of fields, in the language with symbols +, ·, −, 0, 1, then the property "The compiled structure A is the algebraic closure of a finite field" is enforceable. (f) The variety of all lattices is universally finite, but there exist varieties of lattices which are not.

4. Let T be an r.e. $\forall_2$ theory in a recursive language L. Let $\bar{x}$ be a tuple of variables, S an r.e. set of $\forall_1$ formulas $\phi(\bar{x})$ of L, and X a non-r.e. subset of S. Let Φ be the set $X \cup \{\neg\phi : \phi \in S \setminus X\}$. Show that it is enforceable that the compiled structure A omits Φ.

The next exercise shows a way of using forcing to get e.c. models of T *that may not be reachable by forcing directly with* T.

5. Let T be a $\forall_2$ theory in a first-order language L and D a countable model of T. Let T' be T ∪ diag(D). (a) Show that if A is an e.c. model of T' then A|L is an e.c. model of T containing a copy of D. (b) Show that if D is an e.c. model of T, then in forcing with T' it is enforceable that A|L ≅ D.

In the next four exercises, L *is the language of groups and* T *is the theory theory of groups.*

6. Let Ψ be the set of primitive sentences ψ of L with at most one inequation, such that ψ is true in some group. (a) Show that there is a $\forall$-set $\Phi(x,y,z)$ of formulas of L such that for every e.c. group G and all elements a, b, g of G with $g \neq 1$, $G \models \bigwedge\Phi(a,b,g)$ iff $\langle a,b\rangle_G \models \psi$ for all $\psi \in \Psi$. (b) Show that the $\forall$-set Φ has no support (with respect to the theory T). [By Theorem 3.3.7, for each ψ of the stated form there is a $\forall_1$

formula $\psi^*(x,y,z)$ such that $G \models \psi^*(a,b,g)$ iff $\langle a,b\rangle_G \models \psi$. By Exercise 6 of section 3.3, the map $\psi \mapsto \psi^*$ can be chosen recursive. If Φ had a support then the $\forall_1$ sentence problem for groups would be solvable, contradicting the Novikov-Boone theorem.]

7. Let P be the property "G has a 2-generator subgroup H such that every 2-generator subgroup of G is embeddable in H". Show: (a) $\neg P$ is enforceable. (b) There is an e.c. group with property P. [For (a) use the previous exercise. For (b), take from Higman (1961) and Fact 3.3.3 a finitely presented 2-generator group D such that every finitely presented group is embeddable in D, and use Exercise 5.]

8. Show that it is enforceable that every finitely generated simple subgroup of the compiled group G has solvable word problem.

9. For a group G with a tuple $\bar{g}$ of generators, write W(G) for the word problem of G, i.e. $\{i < \omega : t_i(\bar{g})=1 \text{ in } G\}$. As noted in section 3.3, W(G) is determined by G up to recursive isomorphism. If H is a group with the tuple $\bar{h}$ of generators, we say that W(G) is *-<u>reducible</u> to W(H) iff for some $\exists_1$ first-order sentence ϕ using parameters $\bar{g}$ and $\bar{h}$, such that $T \cup \text{diag}(H) \cup \{\phi\}$ has a model, we have for each $i < \omega$:

$$i \in W(G) \Leftrightarrow T \cup \text{diag}^+H \cup \{\phi\} \vdash t_i(\bar{g})=1,$$

$$i \notin W(G) \Leftrightarrow T \cup \text{diag}^+H \cup \{\phi,\chi\} \vdash t_i(\bar{g})\neq 1, \text{ for some inequation } \chi \in \text{diag}(H).$$

(a) Show that the relation "W(G) is *-reducible to W(H)" is independent of the choice of generators $\bar{g}$, $\bar{h}$. (b) Show that the following are equivalent: (i) For every e.c. group A, if H is embeddable in A then so is G; (ii) W(G) is *-reducible to W(H). (c) Show that if W(G) is *-reducible to W(H) then W(G) is enumeration reducible to W(H), but the converse fails in general. (Cf. Rogers (1967) p. 146 for enumeration reducibility.) [For (i) ⇒ (ii) in (b), assuming (ii) fails, apply Corollary 3.4.5 to $T \cup \text{diag}(H)$ to construct an e.c. group containing H but omitting G; use Exercise 4 of section 3.1 and Exercise 5 above.]

10. Let T be a theory in L. Show that $T_\forall = (T^f)_\forall \subseteq T^f = (T_\forall)^f = (T^f)^f$.

REFERENCES FOR CHAPTER 3

3.1: In the wake of B. H. Neumann (1943), B. H. and Hanna Neumann, Graham Higman and Philip Hall published a clutch of papers about extending groups to add solutions of various equations. Higman et al. (1949) is probably the best known of these papers. Theorems 3.1.3 and 3.1.4 are implicit in Neumann (1943), though it was A. I. Mal'tsev (1956; 1958) who shifted arguments of this type from varieties to quasivarieties (a quasivariety is the class of models of a strict universal Horn theory with no relation symbols). Abraham Robinson's group at Yale round about 1970 made the extension to arbitrary first-order theories. Corollary 3.1.2 is from Łoś (1955) and Tarski (1954).

Exercise 3(b): Craig (1953). Exercise 4: McKinsey (1943). Exercise 7 is from Robinson's group at Yale in the early 1970s; cf. Wheeler (1979). Exercise 9: Neumann (1943). Exercise 10: Eklof & Sabbagh (1970). Exercise 11: Cherlin (1973; 1976). Exercise 12: Podewski & Reineke (1979). Exercise 13: Belyaev & Taitslin (1979). Exercise 14: Eklof & Mez (198- a); the case $\mu = 1$, R a field is due to Bokut' (1963). Exercise 15: Krivine (1964).

3.2: One could reasonably trace the notion of existential closure back to Cardan (1545) and his Renaissance colleagues who introduced the complex numbers. Modern developments were (i) a notion equivalent to existential closure as we defined it, in Hilbert's Nullstellensatz (Hilbert 1893), (ii) study of algebraic or similar closures in classes other than the class of fields (Hensel 1907; Artin & Schreier 1926; W. Scott 1951), and (iii) a proof of the existence of existentially closed extensions (Steinitz 1910). E.c. models were introduced into model theory by Rabin (1962), Lindström (1964) and Robinson (1971 a,b,c); Robinson's group called them existentially complete. Simmons (1972) was an early model-theoretic survey. The notion of model-completeness was introduced by Robinson (1956) as a helpful way of describing certain phenomena in algebra and number theory; cf. Macintyre (1977) for a survey.

Theorem 3.2.1 belongs by rights to Steinitz (1910). Models of ZF containing fields with no algebraic closure were given by Pincus (1972) and Hodges (1976). Theorem 3.2.3, or strictly Corollary 3.2.4, lies at the heart of Rabin (1962). Theorem 3.2.7 was circulated in a preprint of Bacsich and Fisher in about 1971, cf. Bacsich & Rowlands Hughes (1974) and

Bacsich (1975). The proof in the text, via Lemma 3.2.8, is stolen from stability theory, cf. Lascar & Poizat (1979). Lemma 3.2.9 is from Robinson (1956) and Theorem 3.2.10 from Lindström.

Exercise 2 was suggested by remarks in Blass (1979). Exercise 5: Henkin (1956). Exercise 7: cf. Hirschfeld & Wheeler (1975) Lemma 7.2. Exercise 9: the Kaiser hull is from Kaiser (1969). In Exercise 11, (b) is Robinson's (1956) definition of model-completeness. Exercise 12: Eklof & Sabbagh (1970). Exercises 13 and 14 came out of Robinson's group at Yale; on 13(b) cf. Eklof & Sabbagh (1970), and on quantifier elimination see Feferman (1968), Shoenfield (1971) and L. Blum as reported in Sacks (1972) Chapter 17 for closely related ideas. Exercise 16: Cherlin (1973). Exercise 17: Podewski & Reineke (1979). Exercise 18: (a) Mal'tsev (1940) and Cohn (1962); (b) Eklof & Sabbagh (1970); (c) and (d) Mez (1982). Exercise 19: (a) Eklof & Sabbagh (1970); (b) Sabbagh (1971). Exercise 20: it was this example that started Abraham Robinson off. Exercise 21: Wheeler (1978) uses this to illustrate criteria for the existence of model companions.

3.3: Fact 3.3.1 is due to G. Higman (1961). Fact 3.3.2 is from Higman et al. (1949). Long before the logicians went to work on e.c. groups, Neumann (1952) showed that every such group is simple (our Theorem 3.3.6(a)). In (1973) he added our Corollary 3.3.8(a) together with Exercise 4(b). Then came Macintyre's paper (1972 b) which gave us Lemma 3.3.5(a), Corollary 3.3.8(b), Exercise 2(b) and Exercises 5 and 8 together with a considerable amount of the inspiration behind later results. Lemma 3.3.5(b) is from Belegradek (1978); Theorem 3.3.6(b) and Corollary 3.3.9 are from Belegradek (1978) and Ziegler (1980). Theorem 3.3.7 is due to Ziegler (1980).

Exercise 1: Sabbagh (1975). Exercise 2(c): Eklof & Sabbagh (1970). Exercise 3(a-c): Hickin & Macintyre (1980); 3(d,e): Hickin (198-). Exercise 7 is based on Belegradek (1974 b; 1978) and Ziegler (1980). Exercise 9: Ziegler (1980). Exercise 10: Bacsich (1972) and Sabbagh (1976). Exercise 11: Belegradek (1981) and Trofimov (1975). Exercise 12: (a-e) rest on Theorem 5.5.1 of Cohn (1977), cf. Hirschfeld & Wheeler (1975); (f,g) Boffa & van Praag (1972 a) and Hirschfeld & Wheeler (1975). Exercise 13: (a) Bokut' (1963); (b) Belyaev (1977); (c) Belyaev (1978); (d) Belyaev (1982). Exercise 14: Glass & Pierce (1980 b), cf. Glass (1981). Exercise 15: (b-d) P. Hall (1959 b); (e,f) Hodges (198- a).

3.4: In 1969 Abraham Robinson introduced Robinson forcing as a way of constructing models "inspired entirely by the corresponding procedure in Set Theory" (Robinson (1971 a), cf. Barwise & Robinson (1970)). Robinson's definitions were simple but ad hoc, and many people were puzzled by the lack of motivation. However, Macintyre (1972 a,b) was able to use Robinson's machinery to prove Theorem 3.4.6 and other interesting results not mentioning forcing. Corollary 3.4.5 was implicit in Macintyre (1972 a), and Keisler (1973) and Simmons (1973) made it explicit. In (1980) Ziegler gave a rationale in terms of games. Henrard (1973) has a different approach to T^f.

Exercise 3: Tulipani (198-) studies universally finite varieties. Exercise 3(b): P. Hall (1959 a); the abelian case is in Robinson (1951). Exercise 3(c) is the most ridiculously unfair exercise in the book; this was a well-known open problem for twenty-two years before Kharlampovich (1981) solved it. Exercise 3(d): G. Baumslag (1971) Theorem 5.3. Exercise 3(e): Robinson (1951). Exercise 3(f): Evans (1969) for the positive part and Baker (1969) for the negative. Exercises 6-8: Macintyre (1972 b). Exercise 9: Ziegler (1980) and C. F. Miller III (unpublished). Exercise 10: Barwise & Robinson (1970).

4 CHAOS OR REGIMENTATION

The aggregates of sound have no necessary direction.

John Cage (1958)

Il y a suffisamment d'inconnu.

Pierre Boulez (1952)

This chapter continues our study of Robinson forcing. But many of the results adapt to other kinds of forcing too. Our main question is how much control player ∃ has in the construction.

A few decades ago, avant garde musicians found themselves divided into two camps. Leading one camp, Pierre Boulez was insisting that a composer must have absolute control over how his music should sound. In the opposite corner of the ring John Cage was preaching the virtues of randomness in composition and performance.

It turns out that there are Boulez theories and Cage theories. Forcing with a Boulez theory T, player ∃ can determine exactly what the compiled model $A(\bar{p})$ is like, up to isomorphism. If T is a Cage theory, then for every enforceable property P there are 2^{ω} pairwise non-isomorphic models of T with property P; player ∃ can never pin down $A(\bar{p})$ precisely, however hard she tries. There are some theories between these extremes. But we shall see that every theory which is not a Cage theory is in a sense the common part of a family of at most countably many Boulez theories.

Even when we know that a theory falls on the Boulez side, there is still work for us to do. First, can we give a good structural description of the unique enforceable model? Second, how hard does player ∃ have to work in order to be sure of reaching this model? There are ways of measuring this.

Section 4.1 imagines player ∃ trying to construct two or more non-isomorphic models of T simultaneously. She may not succeed. But if she does fail, then we shall see in section 4.2 that she can do the opposite: as soon as she has seen player ∀'s first move, she can determine exactly what the compiled structure will be. Section 4.4 illustrates this second situation with some examples of nilpotent groups of class 2. Section 4.3 is an interlude on finite-generic models.

4.1 MASS PRODUCTION

As in the previous chapter, L is a countable first-order language, W is a countable set of witnesses and T is a theory in L. We assume T is a $\forall_2$ theory with models, and we build models of T by Robinson forcing. But now we propose to construct two models at the same time, comparing them as we go along.

EXAMPLE. Suppose the signature of L consists of just countably many 1-ary relation symbols R_i $(i < \omega)$, and let T be the empty theory. Let A be any L-structure. Then A is a model of T. If b is an element of A, define the colour of b to be the set $\{i < \omega : A \models R_i(b)\}$. Imagine now that players ∀ and ∃ are working together to construct L-structures A and B simultaneously by Robinson forcing. At any stage during the construction, just finitely many elements of each structure A and B have been mentioned, and only a finite amount of the colour of each element has been specified. So player ∃ can use her moves to make sure that no element of A has the same colour as any element of B. Note that although player ∃ can enforce that the structures A and B are completely different from each other, there is no way that she can make any particular colour appear in either structure.

To make matters precise, we introduce a new kind of game. Really it is the old game, but played on two boards at once. Let R be a property of pairs of L(W)-structures (B_1,B_2), and X a subset of ω. In the game G(R;X), the players between them construct two chains $(p_i^1)_{i<\omega}$ and $(p_i^2)_{i<\omega}$ of conditions, and at the end of the game, $A_1^+(\bar{p}^1)$ and $A_2^+(\bar{p}^2)$ are respectively the canonical models of the least =-closed sets containing all the atomic sentences in $\bigcup_{i<\omega} p_i^1$ and $\bigcup_{i<\omega} p_i^2$. The chains are constructed simultaneously: a player chooses p_i^1 at the same time as p_i^2. Player ∃ chooses p_i^1 and p_i^2 if $i \in X$; otherwise player ∀ chooses them.

Finally player ∃ wins the play $(\bar{p}^1, \bar{p}^2)$ iff the pair $(A_1^+(\bar{p}^1), A_2^+(\bar{p}^2))$ of compiled structures, or (A_1^+, A_2^+) for short, has the property R.

As before, we say that the property R is <u>enforceable</u> iff player ∃ has a winning strategy in G(R;odds), or equivalently in any standard game G(R;X). Also as before, the same definitions make sense if R is a property of pairs $(\bigcup \bar{p}^1, \bigcup \bar{p}^2)$ rather than of pairs of L(W)-structures. It is clear what it would mean to say that a pair (q_1, q_2) of conditions forces R, though we shall not need the notion.

LEMMA 4.1.1. *(a) If* P *and* Q *are enforceable properties of single* L(W)-*structures, then the property* "A_1^+ *has property* P *and* A_2^+ *has property* Q" *is enforceable.*

(b) (Conjunction lemma) *If* P *is the conjunction of countably many enforceable properties of pairs of structures, then* P *is enforceable.*

<u>Proof</u>. (a) is obvious, and (b) is proved as Lemma 2.3.3(e). □

Thinking back to the example above, we want a suitable generalisation of the "colour" of an element. Experience shows that the following notion will serve the purpose. The definition is relative to a fixed theory T in a first-order language L.

Let $\bar{a} = (a_1, \ldots, a_n)$ be an n-tuple of elements of an L-structure A. The <u>∃-type</u> of $\bar{a}$, $\exists\text{-tp}_A(\bar{a})$, is the set of all $\exists_1$ formulas $\phi(\bar{x})$ of L such that $A \models \phi(\bar{a})$. For a fixed $n \geqslant 0$, $S_n^\exists$ is the set of all sets of form $\exists\text{-tp}_A(\bar{a})$ as A ranges over models of T and $\bar{a}$ ranges over n-tuples of elements of A. (We say: $S_n^\exists$ is the set of all <u>existential types</u> of n-tuples of elements in models of T.)

A <u>maximal ∃-type</u> is a maximal element of some $S_n^\exists$. This notion depends only on $T_\forall$, according to the next lemma:

LEMMA 4.1.2. *The maximal ∃-types are exactly the sets of form:* $\exists\text{-tp}_A(\bar{a})$ *for some e.c. model* A *of* $T_\forall$ *and some tuple* $\bar{a}$ *of elements of* A.

<u>Proof</u>. Suppose first that Φ is a maximal ∃-type. Then there exist a model A of T and an n-tuple $\bar{a}$ of elements such that $\Phi = \exists\text{-tp}_A(\bar{a})$. By Theorem 3.2.1 there is an e.c. model B of $T_\forall$ with $A \subseteq B$. Let Ψ be $\exists\text{-tp}_B(\bar{a})$. Then $\Phi \subseteq \Psi$ by the lemma on preservation (Lemma 1.2.4(b)). By

Corollary 3.1.2 there is a model C of T with $B \subseteq C$. Let Δ be $\exists\text{-tp}_C(\bar{a})$; then similarly $\Psi \subseteq \Delta$, and so $\Phi \subseteq \Delta$. Since Φ is maximal in $S_n^{\exists}$, it follows that $\Phi = \Psi = \exists\text{-tp}_B(\bar{a})$ as required.

Conversely let A be an e.c. model of $T_\forall$, $\bar{a}$ an n-tuple of elements of A and $\Phi = \exists\text{-tp}_A(\bar{a})$. By Corollary 3.1.2 there is a model B of T with $A \subseteq B$; since A is an e.c. model of $T_\forall$ and B is a model of $T_\forall$, this implies $\exists\text{-tp}_A(\bar{a}) = \exists\text{-tp}_B(\bar{a})$ and hence $\Phi \in S_n^{\exists}$. If Φ is not maximal, then there exist a model C of T with elements $\bar{c}$, and an $\exists_1$ formula $\phi(\bar{x})$ of L, such that $C \models (\phi \wedge \bigwedge\Phi)(\bar{c})$ but $A \models \neg\phi(\bar{a})$. By Corollary 3.2.4, since $A \models \neg\phi(\bar{a})$, there is some $\exists_1$ formula $\psi(\bar{x})$ of L such that $A \models \psi(\bar{a})$ and $T \vdash \forall\bar{x}(\psi \rightarrow \neg\phi)$. But then $\psi \in \Phi$, and so $C \models \psi(\bar{c})$ and hence $C \models \neg\phi(\bar{c})$; contradiction. Hence Φ is maximal. □

As in section 3.4, we say that a set Φ of formulas is realised by $\bar{a}$ in the L-structure A iff $A \models \bigwedge\Phi(\bar{a})$. By Theorem 3.2.3, one corollary of the lemma above is that in an e.c. model of a $\forall_2$ first-order theory, every tuple of elements realises a maximal $\exists$-type. In fact this is a characterisation of e.c. models; cf. Exercise 1 below.

In our example above, the colour of an element was not determined by any finite set of formulas. That leads to the following notion. We say that a formula $\psi(\bar{x})$ of L isolates the $\exists$-type Φ iff ψ is an $\exists_1$ formula and Φ is the unique maximal $\exists$-type containing ψ. One can show that if Φ is a maximal $\exists$-type and ψ is a support of Φ then ψ isolates Φ; the converse fails in general. (Cf. Exercise 2.) We say Φ is isolated iff some formula isolates Φ.

THEOREM 4.1.3. *(We assume Robinson forcing relative to a fixed $\forall_2$ theory T.) The property "Every maximal $\exists$-type which is realised both in A_1 and in A_2 is isolated" is enforceable.*

Proof. By Lemmas 2.3.1 and 4.1.1(a) it is enforceable that every element of A_1^+ or A_2^+ is named by infinitely many witnesses. So by the conjunction lemma (Lemma 4.1.1(b)) we can take a fixed n and disjoint n-tuples $\bar{c}$, $\bar{d}$ of distinct witnesses; the theorem follows if we show that the following property of (A_1^+, A_2^+) is enforceable:

(1) If the elements named by $\bar{c}$ in A_1^+ realise the same maximal $\exists$-type Φ in A_1^+ as do the elements named by $\bar{d}$ in A_2^+, then Φ is isolated.

Let R be this property, and let us play G(R;odds). Suppose player ∀ has chosen p_0^1 and p_0^2. Write $\bigwedge p_0^1$ and $\bigwedge p_0^2$ respectively as $\psi_1(\bar{c},\bar{b}_1)$ and $\psi_2(\bar{d},\bar{b}_2)$ where $\bar{b}_1$ (resp. $\bar{b}_2$) lists the witnesses which occur in p_0^1 (resp. in p_0^2) but not in $\bar{c}$ (resp. $\bar{d}$). There are two cases to consider.

<u>Case 1</u>: $\exists\bar{y}\psi_1(\bar{x},\bar{y})$ isolates some maximal ∃-type $\Psi(\bar{x})$. Let player ∃ play to make A_1 an e.c. model of T (by Corollary 3.4.3) and A_1^+ a model of $\bigcup \bar{p}^1$ (by Lemma 2.3.1). Then according to Lemma 4.1.2, the elements named by $\bar{c}$ in A_1^+ realise a maximal ∃-type, and this type contains $\exists\bar{y}\psi_1(\bar{x},\bar{y})$, so it must be Ψ.

<u>Case 2</u>: $\exists\bar{y}\psi_1(\bar{x},\bar{y})$ doesn't isolate any maximal ∃-type. Now certainly $\exists\bar{y}\psi_1(\bar{x},\bar{y})$ lies in at least one maximal ∃-type (e.g. by the argument of Case 1). So it must lie in at least two distinct maximal ∃-types Ψ and Ψ'. Likewise $\exists\bar{z}\psi_2(\bar{x},\bar{z})$ lies in at least one maximal ∃-type Θ. Since Ψ and Ψ' are distinct, at least one of them is different from Θ; let it be Ψ. Then we claim that

(2) There are $\chi(\bar{x}) \in \Psi$ and $\theta(\bar{x}) \in \Theta$ such that $T \vdash \forall\bar{x}\neg(\chi \wedge \theta)$.

For otherwise, using the compactness theorem, there would be a model B of T with elements $\bar{a}$ such that $B \models \bigwedge(\Psi \cup \Theta)(\bar{a})$. Since Ψ and Θ are maximal, this would imply Ψ = Θ.

Now let player ∃ take χ and θ as in (2), let her use Lemma 3.4.1(d) to find conditions $p_1^1 \supseteq p_0^1$ and $p_1^2 \supseteq p_0^2$ such that $p_1^1 \vdash \chi(\bar{c})$ and $p_1^2 \vdash \theta(\bar{d})$; let her play (p_1^1, p_1^2) and then use the rest of the game to make sure that $A_1^+ \models (T \cup p_1^1)$ and $A_2^+ \models (T \cup p_1^2)$ (by Lemma 2.3.1 and Corollary 3.4.3). Necessarily then the elements named by $\bar{c}$ and $\bar{d}$ respectively in A_1^+ and A_2^+ realise distinct ∃-types in A_1 and A_2.

In either case, player ∃ ensures that (1) holds. □

There is a useful variant of this theorem. We define $\text{qf-tp}_A(\bar{a})$ to be the set of quantifier-free formulas of L satisfied by $\bar{a}$ in A. Then we have S_n^{qf} by analogy to $S_n^{\exists}$. Using compactness it is easy to see that the elements of S_n^{qf} are the sets $\Psi(x_0,\ldots,x_{n-1})$ of quantifier-free formulas of L which are maximal with the property that $T \cup \{\exists\bar{x}\bigwedge\Psi\}$ has a model; we call such sets Ψ <u>maximal qf-types</u>. (Note that every set in S_n^{qf} is maximal in S_n^{qf}.) We say that an $\exists_1$ (NB: not necessarily quantifier-free) formula $\psi(\bar{x})$ of L <u>isolates</u> the maximal qf-type Φ iff Φ is the unique maximal qf-type $\Psi(\bar{x})$ such that $T \cup \{\exists\bar{x}(\psi \wedge \bigwedge\Psi)\}$ has a model. One can show that ψ isolates the maximal qf-type Φ if and only if ψ is a support of Φ (cf. Exercise 3).

THEOREM 4.1.4. *(Assuming Robinson forcing,) the property "Every maximal qf-type which is realised both in* A_1 *and in* A_2 *is isolated" is enforceable.*

Proof. Much as Theorem 4.1.3, but a little easier. □

Before I go to applications, let me make the theorems above much stronger. This is one of those many places in model theory where we can get continuum many goodies for the same price as two.

We recast our games G(P;X) once more. Now we shall allow each player, if he wishes, to split the game into a finite number of games on separate boards, by making several alternative choices at the same move. Either player can split the game at any of his moves, so that the final outcome is a branching tree of positions, for example:

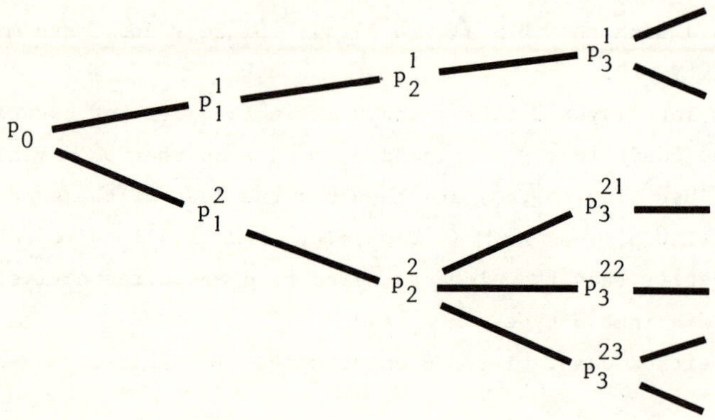

The superscripts show where we are in the branching; every time a player extends p_i^r in n alternative ways, he calls them $p_{i+1}^{r,1}, \ldots, p_{i+1}^{r,n}$. So in general the superscripts on conditions will be sequences of natural numbers. If r, s are two such sequences, then we write $r \subseteq s$ to mean that r is an initial segment of s.

At the end of a play, conditions will have been chosen with

various different superscripts. Let S be the set of infinite sequences s of natural numbers such that each finite initial segment of s occurs as a superscript on a condition in the play. Then each $s \in S$ gives rise to a chain $\bar{p}^s = (p_i^r : i < \omega \text{ and } r \subseteq s)$. Let A_s^+ be the canonical model of the least =-closed set containing all atomic sentences in $\bigcup \bar{p}^s$. We define the <u>outcome</u> of this play to be the indexed family $(A_s^+ : s \in S)$ of L(W)-structures.

Let P now be a property of indexed families $(B_j : j \in J)$ of L(W)-structures. (For example P might be the property that each B_j is existentially closed, or that $B_i \not\cong B_j$ whenever $i \neq j$.) The game G(P;X) is played as above: player $\exists$ has the i-th move iff $i \in X$, and either player may split the game at any of his moves. Suppose the outcome of a play is the indexed family $(A_s^+ : s \in S)$. Then player $\exists$ wins this play iff this indexed family has property P. We say P is <u>enforceable</u> iff player $\exists$ has a winning strategy in G(P;odds), or equivalently in G(P;X) whenever X is an infinite coinfinite subset of $\omega \setminus \{0\}$.

THEOREM 4.1.5. *Let* R *be an enforceable property of ordered pairs of* L(W)-*structures. Let* P *be the property which an indexed family* $(B_j : j \in J)$ *of* L(W)-*structures has iff for all* $j \neq k$ *in* J, (B_j, B_k) *has property* R. *Then* P *is enforceable.*

<u>Proof</u>. Let σ be a winning strategy for player $\exists$ in the game G(R;odds) on pairs of structures. We shall show how player $\exists$ can win G(P;odds). At each odd-numbered stage in the game she will in fact make a finite number of successive choices, but to preserve the form of the game she will only mention the final choice to player $\forall$, and the rest she will keep secret. At the end of the game, if s and t are distinct elements of the index set S, then some subsequence of player $\exists$'s choices (including the secret ones) will form a play of G(R;odds) in which the pair (A_s^+, A_t^+) is compiled and player $\exists$ uses σ. Player $\exists$ will never split the game.

Suppose player $\forall$ has just chosen the conditions p_i^r for some even i. We write S_i for the set of all the superscripts r which occur for this i. Player $\exists$ will now privately list the set of ordered pairs of distinct elements of S_i, say as $(s_1, t_1), \ldots, (s_k, t_k)$. (Of course the list depends on i, but I avoid writing horrors like $s_{k_i}^i$.) Secretly, she will choose conditions

(3) $$p^r_i = q^r_{i,0} \subseteq q^r_{i,1} \subseteq \cdots \subseteq q^r_{i,k}$$

for each $r \in S_i$. Her public choice of p^r_{i+1} will be $q^r_{i,k}$.

If (s,t) is (s_m,t_m) for some m ($1 \leqslant m \leqslant k$), we shall write $q^r_{i,(s,t)}$ for $q^r_{i,m}$; $Q^r_{i,(s,t)}$ will mean $q^r_{i,m-1}$. For each even $j \leqslant i$, there is a unique initial segment r' of r such that player $\forall$ chose a condition $p^{r'}_j$ at his j-th move; we write r/j for this segment r'. For a fixed pair (s,t), let h be the least even integer such that $s/h \neq t/h$; taking (s,t) as fixed, we write q^r_j for $q^{r/j}_{j,(s/j,t/j)}$, for each even j ($h \leqslant j \leqslant i$).

In this notation we can describe player $\exists$'s plan of action. She will play so that for each pair (s,t):

(4) (Q^s_h, Q^t_h), (q^s_h, q^t_h), (Q^s_{h+2}, Q^t_{h+2}), (q^s_{h+2}, q^t_{h+2}), ..., (Q^s_i, Q^t_i), (q^s_i, q^t_i) is a position in a play of G(R;odds) in which player $\exists$ uses her winning strategy σ.

If we assume inductively that player $\exists$ followed this plan of action at all earlier moves, it is not hard to see how she can keep to it when she chooses (3). For all pairs (s,t), when $Q^s_{i,(s,t)}$ and $Q^t_{i,(s,t)}$ have been chosen, (4) says at once what $q^s_{i,(s,t)}$ and $q^t_{i,(s,t)}$ should be. When $r \notin \{s,t\}$, she can choose $q^r_{i,(s,t)}$ to be $Q^r_{i,(s,t)}$.

Now at the end of a play, suppose s and t are distinct elements of the index set S. We can define s/j, t/j for even j just as above. If h is the first even integer such that $s/h \neq t/h$, then by (4),

(5) (Q^s_h, Q^t_h), (q^s_h, q^t_h), (Q^s_{h+2}, Q^t_{h+2}), (q^s_{h+2}, q^t_{h+2}), ... is a play of G(R;odds) in which player $\exists$ uses her winning strategy σ.

Now clearly $\bigcup \bar{p}^s = \bigcup_{2i \geqslant h} q^s_{2i}$ and $\bigcup \bar{p}^t = \bigcup_{2i \geqslant h} q^t_{2i}$. So player $\exists$ has won a game of G(R;odds) in which (A^+_s, A^+_t) was constructed, and it follows that (A^+_s, A^+_t) has property R as required. □

An application:

THEOREM 4.1.6. *There is a family* $(G_\alpha : \alpha < 2^\omega)$ *of countable groups such that:*

(a) if H *is a finitely generated group and there are* $\alpha \neq \beta$

such that H *is embeddable in both* G_α *and* G_β*, then* H *has solvable word problem;*

(b) *each* G_α *is an e.c. group, and hence contains a finitely generated subgroup with unsolvable word problem;*

(c) *all the groups* G_α *are elementarily equivalent.*

Moreover if P *is any enforceable property of groups, then we can require that all the groups* G_α *have* P.

Proof. In the splitting game player ∀ can keep splitting each condition into two, so that the outcome is a family $(A_\alpha : \alpha < 2^\omega)$. So by Lemma 4.1.1 and Theorem 4.1.5, the theorem follows if we show that each of (a), (b) and (c) is an enforceable property of pairs of groups. By Lemma 4.1.1(a), Corollary 3.3.8(b) and Corollary 3.4.3, (b) is enforceable for pairs. By Lemma 4.1.1 and Theorem 3.4.7, (c) is enforceable for pairs. Finally (a) is enforceable by Theorem 4.1.4 and the proof of Theorem 3.4.6 (cf. Exercise 3 below). □

In a nutshell, Theorem 4.1.6 says that however hard player ∃ tries to pin down the compiled e.c. group, player ∀ can make it vary between continuum many possibilities.

It is interesting to note that player ∃ can pin down the theory Th(G) of the compiled group G. One naturally asks how many pairwise non-elementarily-equivalent e.c. groups there can be. After Theorem 3.4.6 we noted that there are at least two, using the Novikov-Boone theorem. In fact our methods can be coaxed into yielding continuum many. But this needs more than a combinatorial trick for turning 2 into 2^ω. We call on two further pieces of information, one from group theory:

FACT 4.1.7. *For every Turing degree* d *there exists a finitely presented group whose word problem has degree* d.

Proof. Clapham (1964). □

And one from recursion theory:

FACT 4.1.8. *For every countable partially ordered set* Π *there is a family of r.e. degrees which is isomorphic to* Π *under the relation of Turing reducibility.*

Proof. Sacks (1963) §4 Corollary 3. □

THEOREM 4.1.9. *There is a family* $(G_\alpha : \alpha < 2^\omega)$ *of countable e.c. groups such that if* $\alpha \neq \beta$ *then there is an* $\exists_3$ *sentence of the first-order language of groups which is true in* G_α *but not in* G_β.

<u>Proof</u>. Let Π be a free lattice on countably many generators. By Fact 4.1.8 there is a family of r.e. degrees which is isomorphic to Π as partially ordered set, under the partial ordering of Turing reducibility. Let $(d_i : i < \omega)$ be the degrees which correspond to the generators of Π, and for each $i < \omega$ let H_i be a finitely presented group whose word problem has degree d_i (by Fact 4.1.7). Let $(X_\alpha : \alpha < 2^\omega)$ be a family of subsets of ω such that for all $\alpha \neq \beta$, $X_\alpha \setminus X_\beta$ is not empty. For each $\alpha < 2^\omega$, let K_α be the direct product of the groups H_i with $i \in X_\alpha$.

Let T be the theory of groups. For each $\alpha < 2^\omega$, we force with $T \cup \text{diag}(K_\alpha)$ to construct an e.c. group G_α with K_α as a subgroup (cf. Exercise 5 of section 3.4). We claim that it is enforceable that for all $j \notin X_\alpha$, H_j is not embeddable in G_α. Given the claim, the theorem follows from Corollary 3.3.9 as augmented by Exercise 9(b) of section 3.3 (and of course the conjunction lemma, as always).

We prove the claim. Suppose $j \notin X_\alpha$, let $\bar{h}$ be a tuple of generators of H_j, and let $\Phi(\bar{x})$ be the set of all equations and inequations $\phi(\bar{x})$ such that $\phi(\bar{h})$ is true in H_j. We prove the claim by showing that it is enforceable that Φ is omitted. By Theorem 3.4.4, it suffices to show that Φ has no support.

For contradiction, suppose that Φ has a support $\psi(\bar{x})$. Thus $T \cup \text{diag}(K_\alpha) \cup \{\exists\bar{x}\psi\}$ has a model and

(6) $$T \cup \text{diag}(K_\alpha) \vdash \forall\bar{x}(\psi \to \bigwedge\Phi).$$

Now ψ is in the language of $T \cup \text{diag}(K_\alpha)$, so that it may mention finitely many elements of K_α. Write K_α as $K'_\alpha \times K''_\alpha$, where K'_α is the product of a finite number of groups H_i $(i \in X_\alpha)$ and every element named in ψ lies in K'_α. I assert that

(7) $$T \cup \text{diag}(K'_\alpha) \vdash \forall\bar{x}(\psi \to \bigwedge\Phi).$$

For otherwise there exists a model A of $T \cup \text{diag}(K'_\alpha)$ containing elements $\bar{a}$ such that $A \models \psi(\bar{a})$ but $A \models \neg\phi(\bar{a})$ for some $\phi \in \Phi$. Form the group $A \times K''_\alpha$. Now $K_\alpha \subseteq A \times K''_\alpha$, so $A \times K''_\alpha \models T \cup \text{diag}(K_\alpha)$. Also ψ and $\neg\phi$ are $\exists_1$ formulas and

A is a subgroup of $A\times K''_\alpha$, so $A\times K''_\alpha \models (\psi \wedge \neg\phi)(\bar{a})$. This contradicts (6). So (7) is proved.

Suppose K'_α is $H_{i_1}\times\ldots\times H_{i_k}$ with $i_1, \ldots, i_k \in X_\alpha$. Since an equation $\phi(\bar{x})$ is in Φ iff $\neg\phi \notin \Phi$, (7) implies that Φ is Turing reducible to $\mathrm{diag}(K'_\alpha)$ and hence to the degree $d_{i_1}\vee\ldots\vee d_{i_k}$ (cf. Exercise 4). Since $j \notin X_\alpha$, d_j is not Turing reducible to the sup of $d_{i_1}, \ldots, d_{i_k}$ in the image of Π, and a fortiori it is not Turing reducible to $d_{i_1}\vee\ldots\vee d_{i_k}$. Here we reach a contradiction. □

Also one might naturally ask how many pairwise non-isomorphic e.c. groups there can be in an uncountable cardinality λ. On general principles the maximum possible number is 2^λ. In fact one can show that this maximum number is always achieved. The proof uses very little information about e.c. groups. I quote the relevant lemma of Shelah without any hint of the proof (it lies in infinitary combinatorics):

LEMMA 4.1.10. *Let* L *be a countable language,* $\phi(\bar{x},\bar{y})$ *a quantifier-free formula of* L, *and* ψ *a sentence of* $L_{\omega_1\omega}$. *Suppose that for every ordinal* α, ψ *has a model* A_α *containing tuples* $\bar{a}_i$ $(i < \alpha)$ *such that for all* $i, j < \alpha$, $A_\alpha \models \phi(\bar{a}_i,\bar{a}_j)$ *iff* $i < j$. *Then for every uncountable cardinal* λ *there is a family* $(B_\beta : \beta < 2^\lambda)$ *of models of* ψ, *all of cardinality* λ, *such that for all distinct* $\beta, \gamma < 2^\lambda$:

(a) B_β *is not isomorphic to* B_γ;

(b) if λ *is regular, then* B_β *is not embeddable in* B_γ;

(c) B_β *is* $L_{\infty\omega}$*-equivalent to* B_γ *(cf. section 4.2 below).*

Proof. Morley (1965) combined with section VIII.2 of Shelah (1978 a). Cf. Hodges (198- d) section 7.10 for details. □

Exercise 9 below describes a quantifier-free formula which well-orders arbitrarily long sequences of tuples in groups. From this, Theorem 3.2.1 and Exercise 6 of section 3.2 it follows that Lemma 4.1.10 applies to e.c. groups: for every uncountable λ there is a family of 2^λ e.c. groups of cardinality λ satisfying (a)-(c) of the lemma.

We paid in very little algebra to get this result, and unfortunately it gives us very little algebra back in return. For example, define the skeleton of a group to be the set of all isomorphism types of finitely generated groups which are embeddable in it. By analysing the proof of Lemma 4.1.10 one can show that all the models B_β have the same skeleton, independent of the cardinality λ, and this skeleton is countable; but just from this proof we have no idea what the skeleton is.

By an argument that seems (at the moment) to work only for groups, Shelah and Ziegler were able to show:

THEOREM 4.1.11. *Let G be any countable e.c. group. Then e.c. groups can be chosen as in the conclusion of Lemma 4.1.10, for all uncountable cardinalities λ, so that each of the e.c. groups has the same skeleton as G.*

<u>Proof</u>. Shelah & Ziegler (1979). □

Exercises for 4.1:

1. Let T be a $\forall_2$ first-order theory and A a model of T. Show that A is existentially closed iff for every tuple $\bar{a}$ of elements of A, $\exists\text{-tp}_A(\bar{a})$ is maximal. *(In this exercise and Lemma 4.1.2, the language need not be countable.)*

2. (a) Show that if Φ is a maximal $\exists$-type and $\psi(\bar{x})$ is a support of Φ, then ψ isolates Φ. (b) Give an example of a theory T, a maximal $\exists$-type Φ and a formula $\psi(\bar{x})$ such that ψ isolates Φ but is not a support of Φ. [For (b) consider the theory in the hint to (c) of Exercise 13, section 3.2.]

3. Let Φ be a maximal qf-type and $\psi(\bar{x})$ an $\exists_1$ formula. Show that ψ is a support of Φ iff ψ isolates Φ.

4. Suppose $H_1, \ldots, H_k$ are finitely generated groups whose word problems have degrees of unsolvability $d_1, \ldots, d_k$ respectively. Show that the word problem of $H_1 \times \ldots \times H_k$ is Turing reducible to $d_1 \vee \ldots \vee d_k$.

By a theorem of J. Harrison (cf. Barwise (1975) Theorem IV.4.1), every Σ^1_1 set of reals which contains a non-hyperarithmetical real has cardinality 2^ω. This allows a swifter approach to Theorem 4.1.9 above, though it gives a weaker result:

5. Let T be an r.e. (or even a hyperarithmetical) $\forall_2$ theory in a recursive first-order language L. Let X be the set of all theories of form Th(A) with A an e.c. model of T. Show: (a) X is a Σ^1_1 set. [Use Exercise 7 of section 3.2.] (b) If there is a theory $U \in X$ such that

complete second-order arithmetic is 1-1 reducible to U, then $|X| = 2^\omega$.
(c) If T is either the theory of groups or the theory of skew fields, then the hypothesis of (b) holds. [For groups, cf. Exercise 14 of section 5.3 below.]

6. Show that if T in Exercise 5 is r.e., then X is a Π^0_3 set. [$U \in X$ iff U is complete, $T_\forall \subseteq U_\forall$ and a certain countable family of $\forall$-sets have no supports in the sense of section 5.1 below.]

7. Let T be a $\forall_2$ theory in a countable first-order language L. Show that if T has uncountably many maximal $\exists$-types, then T has 2^ω maximal $\exists$-types. (Show in fact that for each $n \geqslant 0$, if $|S^\exists_n| > \omega$ then $|S^\exists_n| = 2^\omega$.)

8. Show that if L is a countable language and λ is an uncountable cardinal, then there are at most 2^λ pairwise non-isomorphic L-structures of cardinality λ.

9. (a) Let G be a non-abelian group. Show that if λ is a cardinal then the cartesian power G^λ contains elements a_i, b_i $(i < \lambda)$ such that for all $i,j < \lambda$, $a_i b_j = b_j a_i$ iff $i < j$. (b) Deduce that every non-abelian variety of groups contains 2^λ pairwise non-isomorphic groups in every uncountable cardinality λ.

10. Show that if B is an infinite boolean algebra and λ is an uncountable cardinal, then there is a family of 2^λ pairwise non-isomorphic boolean algebras of cardinality λ, all elementarily equivalent to B.

4.2 ATOMIC MODELS

Like fashions in society, we swing from the highly permissive theories of section 4.1 to some thoroughly authoritarian ones. Our aim is to prove an anti-moderate theorem: there is no halfway house between the extremes.

Our first question must be to ask how one shows that two structures are isomorphic. Of course there may be many ways of doing it. But model theorists have a machine which they like to wheel out on these occasions, called <u>back-and-forth</u>.

Let L be a language and A an L-structure. Recall from section 1.2 that if $\bar{a}$ is a sequence $(a_i : i < \gamma)$ of elements of A, then $(A,\bar{a})$ is the structure A expanded by adding new constants c_i to name the elements a_i. If A and B are L-structures, $\bar{a}$ and $\bar{b}$ are sequences of elements from A and B respectively, and $\bar{a}$ and $\bar{b}$ have the same length, then we can compare

$(A,\bar{a})$ and $(B,\bar{b})$. For example, assuming that L is a first-order language,

(1) $(A,\bar{a}) \equiv (B,\bar{b})$ iff for every formula $\phi(\bar{x})$ of L, $A \models \phi(\bar{a}) \Leftrightarrow B \models \phi(\bar{b})$.

(If $\bar{a}$ and $\bar{b}$ have length γ, then $\bar{x}$ here is assumed to be a tuple of variables from $\{x_i : i < \gamma\}$. Thus if $\phi(\bar{x})$ is the formula $x_{13}=x_\omega$, then $A \models \phi(\bar{a})$ says that $a_{13} = a_\omega$; and likewise with B and $\bar{b}$.)

Refining (1), we introduce three new relations $\equiv_0$, $\Rightarrow_1$ and $\equiv_1$ between structures of form $(A,\bar{a})$:

(2) $(A,\bar{a}) \equiv_0 (B,\bar{b})$ iff for every quantifier-free formula $\phi(\bar{x})$ of L, $A \models \phi(\bar{a}) \Leftrightarrow B \models \phi(\bar{b})$.

(3) $(A,\bar{a}) \Rightarrow_1 (B,\bar{b})$ iff for every $\exists_1$ formula $\phi(\bar{x})$ of L, $A \models \phi(\bar{a}) \Rightarrow B \models \phi(\bar{b})$.

(4) $(A,\bar{a}) \equiv_1 (B,\bar{b})$ iff for every $\exists_1$ formula $\phi(\bar{x})$ of L, $A \models \phi(\bar{a}) \Leftrightarrow B \models \phi(\bar{b})$.

Note that $\equiv$ implies $\equiv_1$, $\equiv_1$ implies $\Rightarrow_1$ and $\Rightarrow_1$ implies $\equiv_0$. We write $A \equiv_1 B$ to mean $(A,\emptyset) \equiv_1 (B,\emptyset)$, and similarly with the other relations.

LEMMA 4.2.1. *If* $\bar{a}$ *and* $\bar{b}$ *respectively list all the elements of* A *and* B, *and* $(A,\bar{a}) \equiv_0 (B,\bar{b})$, *then there is an isomorphism from* A *to* B *which takes* $\bar{a}$ *to* $\bar{b}$.

Proof. Immediate from the definition of an isomorphism. □

Let A and B be L-structures. By a back-and-forth system from A to B we mean a family I of pairs $(\bar{a},\bar{b})$ consisting of a sequence $\bar{a}$ from A and a sequence $\bar{b}$ from B, such that

(5) If $(\bar{a},\bar{b}) \in I$ then $(A,\bar{a}) \equiv_0 (B,\bar{b})$.

(6) $(\emptyset,\emptyset) \in I$.

(7) If $(\bar{a},\bar{b}) \in I$ and c is any element of A, then there is an element d of B such that $(\bar{a}c,\bar{b}d) \in I$.

(8) If $(\bar{a},\bar{b}) \in I$ and d is any element of B, then there is an element c of A such that $(\bar{a}c,\bar{b}d) \in I$.

Clauses (7) and (8) explain the name of these systems. We say that A is back-and-forth equivalent to B iff there exists a back-and-forth system from A to B. We need only remember two facts about this notion:

LEMMA 4.2.2. *If* A *and* B *are both at most countable, and* A *is back-and-forth equivalent to* B*, then* $A \cong B$.

Proof. Let I be a back-and-forth system from A to B. Starting from $(\emptyset,\emptyset)$, build up sequences $\bar{a}$ and $\bar{b}$ of length ω, so that (i) for every $n < \omega$, $(\bar{a}|n, \bar{b}|n) \in I$, and (ii) every element of A occurs in $\bar{a}$, and every element of B occurs in $\bar{b}$. This is possible by (6)-(8) and the bound on the size of A and B. Since atomic formulas only have finitely many variables each, it follows by (5) that $(A,\bar{a}) \equiv_0 (B,\bar{b})$ and so $A \cong B$ by Lemma 4.2.1. □

LEMMA 4.2.3. (Karp's Theorem) *Suppose* A *and* B *are* L*-structures. Then* A *is back-and-forth equivalent to* B *if and only if for every sentence* ϕ *of* $L_{\infty\omega}$, $A \models \phi \Leftrightarrow B \models \phi$.

Proof. Cf. Theorem VII.5.3 in Barwise (1975). □

When the conclusion of Lemma 4.2.3 holds, we say that A is $L_{\infty\omega}$-equivalent to B. The lemma shows, what was easy enough to verify directly, that back-and-forth equivalence is an equivalence relation on structures.

EXAMPLE. Recall from section 4.1 that the skeleton of a group is the set of all isomorphism types of finitely generated subgroups of the group. We assert that two e.c. groups G and H are $L_{\infty\omega}$-equivalent if and only if they have the same skeleton. The easy direction is from left to right: if K is a finitely generated group, then one can write down a sentence θ_K of $L_{\infty\omega}$ which is true in exactly those groups which have a subgroup isomorphic to K. In the other direction, let I be the set of all pairs of tuples, $(\bar{a},\bar{b})$, such that $(G,\bar{a}) \equiv_0 (H,\bar{b})$. Then (5) holds automatically. Considering what quantifier-free sentences there can be in the language of groups, (6) is clear. To show (7), suppose $(G,\bar{a}) \equiv_0 (H,\bar{b})$ and c is an element of G. Assuming that G and H have the same skeleton, there is an embedding $f: \langle\bar{a},c\rangle_G \to H$. In some group $H' \supseteq H$ there is an element h such that for each a_i in $\bar{a}$, $(b_i)^h = (fa_i)^h$; since H is an e.c.

group, there is such an element h already in H. Choose d in H so that $d^h = fc$. Then $(G,\bar{a},c) \equiv_0 (H,\bar{b},d)$ as required. The proof of (8) is similar.

By Lemma 4.2.2 it follows that each countable e.c. group is determined up to isomorphism by its skeleton.

Let us connect these notions with those of section 4.1 above. As before, L is a countable first-order language and T is a $\forall_2$ theory in L. A model A of T is said to be <u>∃-atomic</u> iff every tuple of elements of A realises an isolated maximal ∃-type. Note that by Exercise 1 of section 4.1, every ∃-atomic model of T is existentially closed.

LEMMA 4.2.4. *Let* A *and* B *be finite or countable* ∃*-atomic models of the theory* T. *Then* $A \cong B$ *if and only if* $A \equiv_1 B$.

<u>Proof</u>. Left to right is obvious. For right to left, assume $A \equiv_1 B$, and let I be the set of all pairs of tuples $(\bar{a},\bar{b})$ such that $(A,\bar{a}) \equiv_1 (B,\bar{b})$. We claim that I is a back-and-forth system from A to B.

Clauses (5) and (6) are clear. To prove (7), assume $\bar{a}$, $\bar{b}$ are tuples, $(A,\bar{a}) \equiv_1 (B,\bar{b})$ and c is an element of A. By assumption $\exists\text{-tp}_A(\bar{a},c)$ is an isolated maximal ∃-type. Let $\psi(\bar{x},y)$ be an $\exists_1$-formula of L which isolates it. Then $A \models \exists y\psi(\bar{a},y)$, and so $B \models \exists y\psi(\bar{b},y)$ since $\exists y\psi$ is an $\exists_1$ formula. Hence there is d in B such that $B \models \psi(\bar{b},d)$, and thus $\bar{b}d$ realises $\exists\text{-tp}_A(\bar{a},c)$ in B. Since this type is a maximal ∃-type, it follows that $(A,\bar{a},c) \equiv_1 (B,\bar{b},d)$ as required. The same argument gives (8).

Hence I is a back-and-forth system from A to B, as claimed. It follows by Lemma 4.2.2 that A is isomorphic to B. □

We shall say that <u>isolated maximal ∃-types are dense</u> (for the theory T) iff for each $n \geq 0$ and each $\exists_1$ formula $\phi(x_0,\ldots,x_{n-1})$ of L, if $T \cup \{\exists\bar{x}\phi\}$ has a model, then there is an isolated maximal ∃-type $\Phi(\bar{x})$ of T which contains ϕ.

LEMMA 4.2.5. *Suppose that isolated maximal* ∃*-types are dense for the* $\forall_2$ *theory* T. *Then the property "*∃*-atomic model of* T*" is enforceable.*

<u>Proof</u>. By the conjunction lemma it suffices to show that for each tuple $\bar{c}$ of witnesses, it is enforceable that the tuple of elements

of the compiled structure A^+ named by $\bar{c}$ realises an isolated maximal $\exists$-type. Let player $\forall$ open with a choice p_0. Write $\bigwedge p_0$ as $\phi(\bar{c},\bar{d})$ where $\phi(\bar{x},\bar{y})$ is a quantifier-free formula of L and $\bar{d}$ are the witnesses which occur in p_0 but not in $\bar{c}$. Then $T \cup \{\exists\bar{x}(\exists\bar{y}\phi(\bar{x},\bar{y}))\}$ has a model, so by assumption there is an isolated maximal $\exists$-type $\Phi(\bar{x})$ which contains $\exists\bar{y}\phi(\bar{x},\bar{y})$. Let $\psi(\bar{x})$ isolate Φ. By Lemma 3.4.1(d), player $\exists$ can choose $p_1 \supseteq p_0$ so that $p_1 \vdash \psi(\bar{c})$. By Lemma 2.3.1 and Corollary 3.4.3, she can play so that $A^+ \models p_1$ and A is an e.c. model of T. Then by Lemma 4.1.2, the tuple of elements named by $\bar{c}$ in A^+ will realise a maximal $\exists$-type containing ψ, and this $\exists$-type must be Φ. □

Now we can separate the sheep from the goats.

THEOREM 4.2.6 (Dichotomy theorem). *Let* L *be a countable first-order language and* T *a* $\forall_2$ *theory in* L. *Then one of the following two situations holds for Robinson forcing with* T*:*

(a) For every enforceable property P *of* L*-structures, there are continuum many pairwise non-isomorphic finite or countable models of* T *which have property* P.

(b) There is a set K *of at most countably many* $\exists$*-atomic models of* T *such that the property "*A *is isomorphic to a model in* K*" is enforceable. The cardinality of* K *is the number of isolated maximal elements of* $S_0^\exists$*. Player* $\forall$ *can choose by his first move which of the models in* K *will be isomorphic to* A.

Proof. The deciding question is whether isolated maximal $\exists$-types are dense for T.

Suppose first that they are not. Then there is an $\exists_1$ formula $\phi(\bar{x})$ of L such that $T \cup \{\exists\bar{x}\phi\}$ has a model but there is no isolated maximal $\exists$-type containing ϕ. Now we set players $\forall$ and $\exists$ to play the splitting game of section 4.1. Player $\forall$ will choose p_0 so that $p_0 \vdash \phi(\bar{c})$ for some tuple $\bar{c}$ of witnesses. At all his later moves he will split each condition into two, so that the outcome of the play is a family $(A_\alpha^+ : \alpha < 2^\omega)$ of L(W)-structures. Write $\bar{a}_\alpha$ for the tuple of elements named by $\bar{c}$ in A_α^+. By Theorem 4.1.5, player $\exists$ can ensure that each A_α^+ is a model of $\phi(\bar{c})$, and that A_α is an e.c. model of T with property P, where P is any enforceable property of L-structures. Then by choice of ϕ, each $\bar{a}_\alpha$ will realise in A_α

a maximal ∃-type Φ_α which is not isolated. Hence by Theorem 4.1.4, player ∃ can arrange that whenever $\alpha \neq \beta$, A_β omits Φ_α and so $A_\alpha \not\cong A_\beta$.

On the other hand suppose that isolated maximal ∃-types are dense for T. Then by Lemma 4.2.5 it is enforceable that A is an ∃-atomic model of T. Let K be a set consisting of representatives of the isomorphism types of finite or countable ∃-atomic models of T, one representative for each isomorphism type. By Lemma 4.2.4 and the downward Löwenheim-Skolem theorem, the number of elements of K is the number of $\equiv_1$-types of ∃-atomic models of T. For an ∃-atomic model B, the set of all $\exists_1$ sentences of L which are true in B is an isolated element of $S_0^\exists$. So player ∀ can determine it by taking an isolating sentence ψ and then choosing p_0 so that $p_0 \vdash \psi$. □

EXAMPLE of (b) in the theorem. Let T be the theory of fields. It is enforceable that the compiled model A is an e.c. field, that is to say, an algebraically closed field. By Exercise 3(e) of section 3.4, it is enforceable that A is the algebraic closure of a finite field. Player ∀ can use his first move to choose any prime characteristic.

I should add two remarks which refine Theorem 4.2.6. First, the continuum many models in (a) are non-isomorphic in a very strong sense: if $\alpha \neq \beta$ then A_α is not even embeddable in A_β. This is because any ∃-type realised in A_α is also realised in every structure in which A_α is embedded, by the lemma on preservation (Lemma 1.2.4(b)).

Second, $S_0^\exists$ has a unique maximal element if and only if T has the joint embedding property (cf. Exercise 4 below). Hence when T has the joint embedding property, we can sharpen (b) as follows: there is an e.c. model B of T such that it is enforceable that the compiled structure A is isomorphic to B. We call such a model B an <u>enforceable model</u> of T.

As a general principle, whenever we can show that a certain class of structures has few members, we should hope to be able to find a good structure theory for the structures in the class. This is vague, but it has to be.

In particular suppose (b) holds in Theorem 4.2.6. What can we say about the structures in the class K? First, each structure in K answers to some maximal ∃-type in $S_0^\exists$ which is isolated by a sentence ψ.

If we replace T by $T \cup \{\psi\}$, this replaces K by a 1-element class. So the isolating sentences ψ serve as invariants of a sort, to label the elements of K.

We are reduced to the case where T has an enforceable model. In this case, let us define a relation < between isolated maximal $\exists$-types, as follows:

(9) $\Phi(\bar{x}) < \Psi(\bar{x},y)$ iff for some $n < \omega$, $\Phi(\bar{x})$ is an isolated maximal $\exists$-type $\in S^{\exists}_{n}$ and $\Psi(\bar{x},y)$ is an isolated maximal $\exists$-type $\in S^{\exists}_{n+1}$, and for all $\psi(\bar{x},y) \in \Psi$, $\exists y\psi(\bar{x},y) \in \Phi$.

For each isolated maximal $\exists$-type $\Phi(\bar{x})$, choose an isolating formula $\psi_{\Phi}(\bar{x})$. For each such Φ, let θ_{Φ} be the following sentence of $L_{\omega_1\omega}$:

(10) $$\forall\bar{x}(\psi_{\Phi}(\bar{x}) \rightarrow \forall y \bigvee_{\Phi<\Psi}\psi_{\Psi}(\bar{x},y) \wedge \bigwedge_{\Phi<\Psi}\exists y\psi_{\Psi}(\bar{x},y)).$$

Let T^* be the theory which consists of (i) all the sentences θ_{Φ} as Φ ranges over the isolated maximal $\exists$-types of T, (ii) all sentences of the form

(11) $\forall\bar{x}(\psi_{\Phi}(\bar{x}) \rightarrow \chi(\bar{x}))$ where χ is a quantifier-free formula $\in \Phi$,

and (iii) the single sentence which isolates the unique maximal element of $S^{\exists}_{0}$.

THEOREM 4.2.7. *Suppose the countable* $\forall_2$ *first-order theory* T *has an enforceable model* B. *Then:*

(a) B *is a model of the theory* T^* *described above;*

(b) any two models of T^* *are back-and-forth equivalent.*

Proof. (a) is clear from the definition of T^*. For (b), suppose that C and D are any two models of T^*. Define I to be the set of pairs of tuples $(\bar{a},\bar{b})$ such that for some isolated maximal $\exists$-type Φ of T, $C \models \psi_{\Phi}(\bar{a})$ and $D \models \psi_{\Phi}(\bar{b})$. We claim that I is a back-and-forth system from C to D. Clause (5) holds because of the sentences (11), (6) is because both C and D satisfy the sentence isolating the maximal element of $S^{\exists}_{0}$, and finally (7) and (8) are by the sentences (10). □

Thus by Lemma 4.2.2, $\bigwedge T^*$ is a sentence of $L_{\omega_1\omega}$ which characterises B in the theorem up to isomorphism among finite or countable L-structures. This shows at least that we have a canonical way of describing enforceable models. (Sentences of $L_{\omega_1\omega}$ which have a unique countable model up to isomorphism are called Scott sentences. Cf. Barwise (1975) section VII.6.)

There is a case that deserves special mention. Suppose T is a countable $\forall_2$ first-order theory which has, for each $n \geqslant 1$, just finitely many maximal types in $S_n^\exists$, and only one maximal type in $S_0^\exists$. Then every maximal $\exists$-type of T is isolated; so by the proof of Theorem 4.2.6, T has an enforceable model. The disjunction and conjunction in (10) are finite, so that T^* reduces to a $\forall_2$ first-order theory. By Lemma 4.2.2 and Theorem 4.2.7(b), T^* has up to isomorphism just one finite or countable model. By Theorem 3.2.10 (if T^* has an infinite model, and a trivial argument if not), it follows that T^* is model-complete. This is perhaps the extreme case of regimentation.

Exercises for 4.2:

1. Let L be a signature. Verify directly from the definition that back-and-forth equivalence is an equivalence relation on the class of L-structures.

2. Let T be a countable $\forall_2$ first-order theory and A an at most countable e.c. model of T. Show: (a) If A is $\exists$-atomic then A is embeddable in every e.c. model B of T such that $A \Rrightarrow_1 B$. (b) If A is embeddable in every e.c. model of T then A is $\exists$-atomic. [Use Theorem 4.1.3.]

3. Let T be a countable $\forall_2$ first-order theory with models. (a) Show that if T has the joint embedding property and an $\exists$-atomic model then isolated maximal $\exists$-types are dense for T. (b) Show that if for all $n < \omega$, $S_n^\exists$ has at most countably many maximal elements, then T has an $\exists$-atomic model.

4. (a) Show that if A and B are L-structures such that $A \Rrightarrow_1 B$ then A is embeddable in some elementary extension of B. *(This generalises (a)⇒(b) of Exercise 5 in section 3.2; see Exercises 4, 5 of section 5.3 for further developments.)* (b) Let T be a first-order theory which has models. Show that T has the joint embedding property if and only if $S_0^\exists$ has just one maximal element. [Left to right is by the lemma on preservation; right to left is by part (a).]

In a class K *of structures (for example the class of models of some theory), a structure* B *of cardinality* λ *is said to be* _universal_ *iff every structure of cardinality* $\leqslant \lambda$ *in* K *is embeddable in* B.

5. Let T be a countable $\forall_2$ first-order theory with infinite models. Show that T has a countable universal model if and only if T has the joint embedding property and for every $n \geqslant 1$, $S_n^{\exists}$ has at most countably many maximal elements.

6. (a) Show that up to isomorphism there are just countably many finitely generated commutative rings with 1. [Use Hilbert's basis theorem.] (b) Show that there is no countable universal commutative ring with 1. [Let Π be any set of primes, and consider an element x such that $\pi \in \Pi \Rightarrow \pi$ divides x, and $\pi \notin \Pi \Rightarrow$ for some z, $\pi z = 0$ and $zx \neq 0$. Meditate on $\exists$-tp(x).]

7. (a) Give an example of a countable $\forall_2$ first-order theory T such that $S_1^{\exists}$ has uncountably many maximal elements but isolated maximal $\exists$-types are dense for T. (b) Give an example of a countable $\forall_2$ first-order theory T such that $S_1^{\exists}$ has uncountably many elements but just one maximal element. (c) Show that there is an infinite group G such that T = Th(G) is not ω-categorical but $S_n^{\exists}$ is finite for each $n < \omega$.

8. Let T be the theory of abelian groups. (a) Show that T has an enforceable model, viz. $\bigoplus_{p \text{ prime}} Z(p^{\infty})^{(\omega)}$, where $G^{(\lambda)}$ is the direct sum of λ copies of G. (b) Writing B for the enforceable model of T, show that Th(B) is not ω-categorical. [By compactness it has models which are not torsion.] (c) Show that T has just countably many countable e.c. models up to isomorphism, viz. $B \oplus Q^{(\alpha)}$ $(\alpha \leqslant \omega)$.

9. Show that there is a $\forall_2$ first-order theory with the joint embedding property and exactly two countable e.c. models up to isomorphism. [Arrange that P_n^A $(n < \omega)$ are infinite and disjoint, while dom $A \setminus \bigcup_{n<\omega} P_n^A$ is either empty or infinite.]

10. Suppose $2 \leqslant n < \omega$. Let L be the first-order language with relation symbols < (2-ary) and $P_1, \ldots, P_n$ (1-ary). Consider the L-structure A where: dom A is the subset $[0,1] \cup [2,3] \cup \ldots$ of the rationals, ordered as usual by $<^A$, and $P_1^A, \ldots, P_n^A$ partition dom A into n dense subsets so that $\omega \subseteq P_1^A$. Let T be the set of all $\forall_1$ sentences of L which are true in A. Show that T has up to isomorphism exactly n+1 elementary equivalence classes of e.c. models, and that T has the joint embedding property.

The next four exercises study what I called "the extreme case of regimentation". Exercise 13 finds continuum many examples of this extreme case.

11. Let L be a first-order language with finitely many relation and constant symbols and no function symbols. (a) Show that for each $n < \omega$ there are only finitely many pairwise logically non-equivalent quantifier-free formulas $\phi(x_0,\ldots,x_{n-1})$ of L. (b) Let S be a set of finite L-structures. Show that there is a $\forall_1$ theory T in L such that if A is any L-structure, then $A \models T$ iff every finite substructure of A is isomorphic to some structure in A.

12. Let L, S and T be as in Exercise 11. Suppose (i) S is closed under substructures, (ii) S has the joint embedding property (i.e. if B, C $\in$ S then there is D in S such that B, C are both embeddable in D), (iii) S has the amalgamation property (i.e. if $e_1: B \to C_1$ and $e_2: B \to C_2$ are embeddings between structures in S, then there are a structure D in S and embeddings $f_i: C_i \to D$ such that $f_1e_1 = f_2e_2$), and (iv) S contains arbitrarily large finite structures. (a) Show that T has up to isomorphism just one countable e.c. model A. (b) Show that S is up to isomorphism the set of finite substructures of A. (c) Show that every isomorphism between finite substructures of A extends to an automorphism of A. (d) Show that every countable model of T is embeddable in A. (e) Show that Th(A) is ω-categorical, and that for every $n \geq 1$ and formula $\phi(x_0,\ldots,x_{n-1})$ of L there is a quantifier-free formula $\psi(x_0,\ldots,x_{n-1})$ of L which is equivalent to ϕ modulo Th(A).

13. Let L be the first-order language with just the 2-ary relation symbol R. For each $n < \omega$, let A_n be the L-structure with domain $\{0,\ldots,n+2\}$ and R^A defined by: $R^A(a,b)$ iff $b \neq a+1 \pmod{n+3}$. For each set $X \subseteq \omega$, let S_X be the set of all finite L-structures B such that no structure A_n $(n \in X)$ is embeddable in B. Show that S_X meets conditions (i)-(iv) of Exercise 12.

14. Show that the conclusions of Exercise 12 remain true if we make the following modification: allow L to have finitely many function symbols besides the finitely many relation and constant symbols, but add (v) there is a function $f: \omega \to \omega$ such that for every structure B in S, if B is generated by n elements then B has cardinality $\leqslant f(n)$.

Very roughly, Robinson forcing generalises the setup of Exercise 12 by replacing quantifier-free formulas by $\exists_1$ *formulas. We get a different generalisation if we keep the atomic formulas but drop the first-order theory* T*:*

15. Let L be a countable language and S a countable set of finitely generated L-structures. Suppose that S satisfies (i)-(iii) of Exercise 12 and contains countable or arbitrarily large finite structures. Show that there exists an L-structure A such that (a) S is up to isomorphism the set of finitely generated substructures of A, and (b) every isomorphism between finitely generated substructures of A extends to an automorphism of A. Show that (a) and (b) characterise A up to isomorphism. Show that if S is a recursive set and the structures in S are all finite, then A can be chosen to be a recursive structure.

16. Let S be the set of all finitely generated recursively presented groups. Show that S meets the conditions at the beginning of Exercise 15. Show that the L-structure A of Exercise 15, for this S, is an e.c. group.

4.3 FINITE-GENERIC MODELS

This section is a brief obeisance to history. When Abraham Robinson first introduced his forcing, he used it to build what he called generic structures. (One usually adds finite, to distinguish them from the results of Robinson's "infinite forcing", cf. section 5.3 below.) Roughly speaking, a model is finite-generic iff its elements have all properties which are both enforceable and first-order definable. I doubt it would have entered anybody's head to define such a class of structures if Paul Cohen's generic models of set theory had not been sitting on the mantelpiece. But perhaps this is the main justification for studying finite-generic structures: try a curious definition and you may learn something new.

To be more precise, L is again a countable first-order language, W is a countable set of witnesses and T is a $\forall_2$ theory in L. We assume T has a model. We shall say that an L(W)-structure B^+ is finite-generic$^+$ iff:

(1) every element of B^+ is named by infinitely many witnesses, and

(2) for every sentence ϕ of L(W), if $B^+ \models \phi$ then there is a condition $p \subseteq \mathrm{diag}(B^+)$ which forces ϕ.

In (2), diag(B^+) is the set of all basic sentences of L(W) which are true in B^+.

THEOREM 4.3.1. *It is enforceable that the compiled structure* A^+ *is finite-generic*$^+$.

Proof. (1) is by Lemma 2.3.1. For each sentence ϕ of L(W), player $\exists$ can use Theorem 2.3.4(f) to throw in a condition which forces either ϕ or $\neg\phi$, and then she can play so as to make ϕ or $\neg\phi$ true accordingly. By Lemma 2.3.1 again, she can also play so that $\bigcup\bar{p} \subseteq$ diag(A^+). Hence she can make (2) hold. □

In section 3.4 we met the finite forcing companion T^f of T; T^f is the set of all sentences of L which are enforceable. T^f is intimately connected with finite-generic$^+$ structures:

THEOREM 4.3.2. *For every* L-*structure* B, *if* B *is at most countable then the following are equivalent:*

(a) B *is* $B^+|L$ *for some finite-generic*$^+$ B^+.

(b) $B \models T^f$; *and for every* L-*structure* C, *if* $B \subseteq C$ *and* $C \models T^f$ *then* $B \preccurlyeq C$.

(c) $B \models T_\forall$; *and for every* L-*structure* C, *if* $B \subseteq C$ *and* $C \models T^f$ *then* $B \preccurlyeq C$.

Clauses (b) and (c) are equivalent for L-*structures* B *of any cardinality.*

Proof. (a) ⇒ (b): Assume (a). If $\phi \in T^f$ then no condition forces $\neg\phi$; so by (2), $B^+ \models \phi$ and hence $B \models \phi$. Suppose next that $B \subseteq C$ and $C \models T^f$. Expand C to an L(W)-structure C^+ so that $B^+ \subseteq C^+$. Because of (1), to show that $B \preccurlyeq C$ it will suffice to take any formula $\phi(\bar{x})$ of L and tuple $\bar{c}$ of distinct witnesses, and show that if $B^+ \models \phi(\bar{c})$ then $C^+ \models \phi(\bar{c})$. Now if $B^+ \models \phi(\bar{c})$, then by (2) there is a condition $p \subseteq$ diag(B^+) which forces $\phi(\bar{c})$. We can write $\bigwedge p$ as $\theta(\bar{c},\bar{d})$ where $\theta(\bar{x},\bar{y})$ is a formula of L and $\bar{d}$ lists without repetition those witnesses which are in p but not in $\bar{c}$. Since p forces $\phi(\bar{c})$, $\theta(\bar{c},\bar{d}) \to \phi(\bar{c})$ is enforceable (Exercise 1(b) of section 3.4); hence (Lemma 3.4.1(c)) $\forall\bar{x}\bar{y}(\theta \to \phi)$ is enforceable and lies in T^f. We infer that $C^+ \models \theta(\bar{c},\bar{d}) \to \phi(\bar{c})$, since $C^+ \models T^f$. But $B^+ \models \theta(\bar{c},\bar{d})$ and $B^+ \subseteq C^+$, so that $C^+ \models \theta(\bar{c},\bar{d})$ since θ is quantifier-free. Therefore $C^+ \models \phi(\bar{c})$ as required.

(c) ⇒ (a) for B at most countable: Assume (c). Since B is at most countable, we can expand it to an L(W)-structure B^+ so that (1) holds. By the diagram lemma (Lemma 1.2.3), the second half of (c) implies that for every sentence ϕ of L(W), if $B^+ \models \phi$ then

(3) $$\mathrm{diag}(B^+) \cup T^f \vdash \phi.$$

So by the compactness theorem there are a finite $p \subseteq \mathrm{diag}(B^+)$ and an enforceable sentence χ such that $p \cup \{\chi\} \vdash \phi$. Then p is a condition; for otherwise if $\bigwedge p$ is $\theta(\bar{c})$ for some formula $\theta(\bar{x})$ of L and distinct witnesses $\bar{c}$, we would have $T \vdash \forall\bar{x}\neg\theta$, whence $T_\forall \vdash \forall\bar{x}\neg\theta$, contradicting that $B^+ \models T_\forall$. Since $p \cup \{\chi\} \vdash \phi$ and χ is enforceable, p must force ϕ. This proves (2).

(b) ⇒ (c) holds because $T_\forall \subseteq T^f$ (Exercise 10 of section 3.4). Conversely suppose B satisfies (c). Since $T_\forall = (T^f)_\forall$ (Exercise 10 of section 3.4 again), Corollary 3.1.2 says that there is a model C of T^f such that $B \subseteq C$. But then $B \preccurlyeq C$ and so $B \models T^f$. □

We shall call an L-structure B finite-generic (for T) iff B satisfies the equivalent conditions (b), (c) of Theorem 4.3.2, whether or not B is at most countable. By that theorem and Theorem 4.3.1, finite-generic structures always exist. But it is not at all obvious that uncountable ones exist. In fact sometimes they don't; cf. Exercise 11 below.

COROLLARY 4.3.3. *Let* B *be finite-generic for* T. *Then* B *is an e.c. model of* T.

Proof. By Theorem 3.2.3 it suffices to show that B is an e.c. model of $T_\forall$. Since B is finite-generic, $B \models T_\forall$. Suppose now that $B \subseteq C$, $C \models T_\forall$, $\phi(\bar{x})$ is an $\exists_1$ formula of L, $\bar{b}$ is a tuple of elements of B and $C \models \phi(\bar{b})$. We have to show that $B \models \phi(\bar{b})$. Since $T_\forall = (T^f)_\forall$ by Exercise 10 of section 3.4, Corollary 3.1.2 says that there is some $D \supseteq C$ such that $D \models T^f$. Since ϕ is an $\exists_1$ formula, $D \models \phi(\bar{b})$. Since B is finite-generic, $B \preccurlyeq D$. Hence $B \models \phi(\bar{b})$ as required. □

By the corollary, the finite-generic models of T are a subclass of the e.c. models of T. How useful is this subclass for classifying the e.c. models of T? There is very little that one can say in general about this. Sometimes all the e.c. models of T are finite-generic. (This happens if, but not only if, T is companionable; cf. Exercises 7 and 8 below.) In the next section we shall examine a case where T has continuum many countable e.c. models but only one of them is finite-generic. Of course if T satisfies (a) in Theorem 4.2.6, then there are continuum many pairwise non-isomorphic finite-generic models of T, since finite-genericity is an enforceable property by Theorem 4.3.1.

Suppose T has the joint embedding property. Then by Theorem 3.4.7, T^f is a complete theory, and so all finite-generic models of T are elementarily equivalent. On the other hand, by Corollary 3.2.6, any two e.c. models of T satisfy the same $\forall_2$ sentences. The next theorem clarifies the difference:

THEOREM 4.3.4. *Suppose* T *has the joint embedding property, and let* ϕ *be an* $\exists_3$ *sentence of* L. *If* ϕ *is true in some finite-generic model of* T, *then* ϕ *is true in every e.c. model of* T.

Proof. Write ϕ as $\exists\bar{x}\forall\bar{y}\exists\bar{z}\psi(\bar{x},\bar{y},\bar{z})$ with ψ quantifier-free. If ϕ is true in some finite-generic model of T, then $\phi \in T^f$ and so ϕ is true in some finite-generic$^+$ structure B^+. For some tuple $\bar{b}$ of witnesses, $B^+ \models \forall\bar{y}\exists\bar{z}\psi(\bar{b},\bar{y},\bar{z})$. Since B^+ is finite-generic$^+$, there is some condition $p \subseteq \mathrm{diag}(B^+)$ which forces $\forall\bar{y}\exists\bar{z}\psi(\bar{b},\bar{y},\bar{z})$. So by Lemma 3.4.1(c) there is a tuple $\bar{c}$ of distinct witnesses not occurring in p or $\bar{b}$, such that p forces $\exists\bar{z}\psi(\bar{b},\bar{c},\bar{z})$.

Now let A be any e.c. model of T. Since T has the joint embedding property and A is e.c., we can expand A to a structure A^+ by interpreting the witnesses which occur in p and $\bar{b}$, so that $A^+ \models \bigwedge p$. We claim that $A^+ \models \forall\bar{y}\exists\bar{z}\psi(\bar{b},\bar{y},\bar{z})$. The claim implies the theorem.

Expand A^+ to A^* by interpreting the witnesses $\bar{c}$ in an arbitrary way. It suffices to show that $A^* \models \exists\bar{z}\psi(\bar{b},\bar{c},\bar{z})$. We assert that there is a model D^* of T, $D^* \supseteq A^*$, such that $D^* \models \exists\bar{z}\psi(\bar{b},\bar{c},\bar{z})$. For if not, then by the diagram lemma (Lemma 1.2.3),

$$(4) \qquad T \cup \mathrm{diag}(A^*) \vdash \forall\bar{z}\neg\psi(\bar{b},\bar{c},\bar{z}).$$

By the compactness theorem there is a finite set $q \subseteq \mathrm{diag}(A^*)$ such that $T \cup q \vdash \forall\bar{z}\neg\psi(\bar{b},\bar{c},\bar{z})$. Without loss we can assume that $p \subseteq q$, and we can replace the new constants occurring in q by witnesses not occurring in p, $\bar{b}$ or $\bar{c}$. Then by Theorem 3.4.2, q is a condition which forces $\forall\bar{z}\neg\psi(\bar{b},\bar{c},\bar{z})$.

But this is impossible since $p \subseteq q$ and p forces $\exists\bar{z}\psi(\bar{b},\bar{c},\bar{z})$. It follows that D^* exists as asserted. But then $A^* \models \exists\bar{z}\psi(\bar{b},\bar{c},\bar{z})$ since A is an e.c. model of T. This proves the claim. □

Theorem 4.3.4 goes some way towards explaining the feeling one has when working with finite-generic models, that they belong at the "thin" end of the e.c. models of T.

Exercises for 4.3:

Throughout these exercises, L *is a countable first-order language and* T *is a* $\forall_2$ *theory in* L *which has a model.*

1. Show that if B^+ is finite-generic$^+$, p is a condition $\subseteq \text{diag}(B^+)$ and ϕ is a sentence of $L(W)$ such that p forces ϕ, then $B^+ \models \phi$.

2. Show that T^f is the set of those sentences of L which are true in every finite-generic model of T. [Use Theorem 4.3.1.]

3. Show that there is a countable set of $\forall$-sets, $\{\Phi_i : i < \omega\}$, such that the finite-generic models of T are precisely those models of $T_\forall$ which omit every Φ_i $(i < \omega)$. [For each formula $\phi(\bar{x})$ of L, choose a tuple $\bar{c}$ of distinct witnesses and consider any condition p which forces $\phi(\bar{c})$. Write $\bigwedge p$ as $\theta_p(\bar{c},\bar{d})$ where $\theta_p(\bar{x},\bar{y})$ is a formula of L and $\bar{d}$ are the distinct witnesses which are in p but not in $\bar{c}$. Write $F(\phi)$ for the set of all formulas $\exists\bar{y}\theta_p(\bar{x},\bar{y})$ where p ranges over the conditions forcing $\phi(\bar{c})$. Show that an L-structure B is a finite-generic model of T iff (i) $B \models T_\forall$, (ii) for every formula $\phi(\bar{x})$ of L, $B \models \forall\bar{x} \bigvee(F(\phi) \cup F(\neg\phi))$, and (iii) for every formula $\chi(\bar{x},y)$ of L and formula $\psi(\bar{x}) \in F(\exists y\chi)$, $B \models \forall\bar{x}(\psi \to \bigvee\{\exists y\theta : \theta \in F(\chi)\})$. For the 'if', show that (i)-(iii) imply that for every formula $\phi(\bar{x})$ of L and all tuples $\bar{b}$ in B, $B \models \phi(\bar{b})$ iff $B \models \psi(\bar{b})$ for some $\psi \in F(\phi)$, by induction on the complexity of ϕ.]

4. From Exercise 3 or otherwise, show: (a) There is a sentence of $L_{\omega_1\omega}$ which axiomatises the class of finite-generic models of T. (b) The union of a chain of finite-generic models of T is finite-generic. (c) If $A \subseteq B$, B is a finite-generic model of T and for every $\exists_1$ formula $\phi(\bar{x})$ of L and tuple $\bar{a}$ from A, $B \models \phi(\bar{a}) \Rightarrow A \models \phi(\bar{a})$, then A is finite-generic. (d) If A is an e.c. model of T and A can be extended to a finite-generic model of T, then A is finite-generic. (e) If A, B are finite-generic models of T containing tuples $\bar{a}$, $\bar{b}$ such that $(A,\bar{a}) \equiv_1 (B,\bar{b})$, then $(A,\bar{a}) \equiv (B,\bar{b})$.

5. Show that if T has the joint embedding property and there are up to isomorphism fewer than 2^{ω} countable finite-generic models of T, then T has an ∃-atomic finite-generic model. [Use Theorem 4.2.6.]

6. Let λ be the number of elementary equivalence classes of finite-generic models of T. Show that if $\lambda < 2^{\omega}$ then $\lambda \leqslant \omega$.

7. Show that the following are equivalent: (a) T is companionable. (b) T^f is a model-companion of T. (c) Every model of T^f is finite-generic.

8. Let R be a countable ring with 1 and let T be the theory of left R-modules. Show that if A is an e.c. model of T then (a) $A \models T^f$, and in fact (b) A is finite-generic. [For (a), use Corollary 3.2.6 to show that all e.c. models of T are elementarily equivalent. Then for (b), use Theorem 3.4.7 and the fact that an embedding of an e.c. model of T into an elementarily equivalent model is elementary. Note that by the Baur-Monk quantifier elimination theorem, every formula of L is equivalent modulo T to a boolean combination of $\forall_2$ sentences and $\exists_1$ formulas.]

9. Let T be the theory of abelian lattice-ordered groups. (Cf. Exercise 14 of section 3.3 for the definition of lattice-ordered groups.) A model G of T is called <u>archimedean</u> iff a, b ∈ G and a > 0 imply that $na \not\leqslant b$ for some positive integer n; G is <u>hyperarchimedean</u> iff every homomorphic image of G is archimedean. Write $\bar{C}(X,A)$ for the model of T consisting of all continuous maps from X to A with compact support, where X is a disjoint union of countably many Cantor sets, and A is a divisible totally ordered subgroup of the additive group of reals with the discrete topology. (a) Show that T has an enforceable model, namely $\bar{C}(X,Q)$. (b) Show that the finite-generic models of T are exactly the hyperarchimedean e.c. models of T. (c) Show that every group $\bar{C}(X,A)$ as defined above is a finite-generic model of T, and hence that T has continuum many non-isomorphic countable finite-generic models. (d) Show that the theory of totally ordered abelian groups is companionable.

10. Let T be the theory of groups. Show: (a) There is a family of continuum many countable finite-generic groups, none of which is embeddable in any other group in the family. (b) If G is a finite-generic group then no finitely generated subgroup of G has arithmetical but unsolvable word problem. [For (b), combine Corollary 3.3.9 with the proof of Theorem 3.4.6.]

11. Let T be a $\forall_2$ theory which says: E is an equivalence relation, each equivalence class carries the structure of an abelian group in which no element has order a prime > 2, and the elements $\{b_i : i < \omega\}$ (named by constants) are such that the equivalence class containing b_i has exactly 2^i elements. Show: (a) The property "Each equivalence class contains one of the b_i" is enforceable. (b) T^f contains a sentence which expresses "Each abelian group contains an element not divisible by 2". (c) If A is a model of T^f and a an element of A which is not divisible by 2 and not in the equivalence class of any b_i, then A can be extended to a model of T^f in which a is divisible by 2. (d) In every finite-generic model of T, each equivalence class contains an element b_i; hence all finite-generic models of T are countable. (e) There are continuum many non-isomorphic finite-generic models of T. [On (e): the player who gets to the class of b_3 first can decide whether this class contains an element of order 8.]

12. Show that T can be found so that T has arbitrarily large finite finite-generic models but no infinite finite-generic model.

13. Show that T can be found (with countable language still) so that T has finite-generic models of all infinite cardinalities $\leqslant 2^\omega$ but no higher cardinalities.

14. (a) Show that T can be found so that for a certain set Σ of $\exists_1$ sentences, some e.c. model of T is a model of Σ, but no finite-generic model of T is a model of Σ. [Fields!] (b) Show that if T is as in (a), then $S_0^\exists$ must contain infinitely many maximal elements.

15. Show that if every at most countable model of T can be extended to a finite-generic model of T, then every model of T can be extended to a finite-generic model of T.

4.4 E.C. NILPOTENT GROUPS OF CLASS 2

The message of Exercise 8 in section 4.2 was that e.c. abelian groups are rather a dull lot. By contrast we have been able to grow rich pastures of e.c. groups. Somewhere between these extremes one might hope to find a theory of groups with an enforceable model that is complicated enough to be interesting. This is just what we find when we look at nilpotent groups of class 2. The enforceable model is in fact the unique countable finite-generic nilpotent group of class 2. Part of the interest of this example (at least for model theorists) is that although the group in question can be characterised in purely algebraic terms, the uniqueness proof has a distinctly model-theoretic flavour, and one feels only the most puritanical anti-logician would set about removing this.

At the same time we shall find that the class of all countable e.c. nilpotent groups of class 2 has the same kind of luxuriance as the class of all countable e.c. groups.

A nilpotent group of class 2 is a group G which satisfies the law

(1) $[[x,y],z] = 1$ for all $x, y, z \in G$.

Further information, including several equivalent definitions and an explanation of what "class 2" means, can be found in Schenkman (1965) and Kargapolov & Merzljakov (1979). I abbreviate "nilpotent of class 2" to N_2.

We write G' for the commutator subgroup of G and Z(G) for the centre of G. By (1), $G' \subseteq Z(G)$. Also by standard calculations with commutators one can infer the following from (1):

(2) $[x,y].[x,z] = [x,yz]$ and $[x,z].[y,z] = [xy,z]$.

(3) $[x^n,y] = [x,y^n] = [x,y]^n$ for every integer n.

(4) $[w(x_1,\ldots,x_n),y] = w([x_1,y],\ldots,[x_n,y])$ for every term w in the language of groups.

Note in particular the multiplicative bilinear laws (2) and (3); we shall use them constantly.

In the first-order language L of groups, the class of N_2 groups is axiomatised by a theory T which consists of the axioms for groups and the identity (1). Thus this class of groups is a variety, and in particular it has free products. We write G*H for the free product in this variety (it is sometimes known as the second Golovin product). Fact 4.4.1 below will describe it.

We write G^{ab} for the abelianisation of G, i.e. G/G'; G^{ab} will usually be taken as an additive group. For each element g of G, we write $g^@$ for the image gG' of g in G^{ab}. Given abelian groups A and B, we write $A\otimes B$ for their tensor product over the integers (cf. Fuchs (1970) Ch. X.)

FACT 4.4.1. *Let* G *and* H *be* N_2 *groups. Then* G *and* H *can be identified with subgroups of* G*H *in such a way that:*

(a) the set C *of products of commutators* [h,g] *with* $h \in H$ *and* $g \in G$ *is a subgroup of the centre of* G*H;

(b) every element of G*H *is uniquely of form* g.h.c *with* $g \in G$, $h \in H$ *and* $c \in C$;

(c) C *is isomorphic to* $H^{ab} \otimes G^{ab}$ *by an isomorphism taking each commutator* [h,g] *to* $h^{@} \otimes g^{@}$.

<u>Proof</u>. (a) and (b) are exercises in commutators. For (c), see MacHenry (1960) or Wiegold (1959) Lemma 3.9. □

Fact 4.4.1 gives a complete description of G*H. If $g_1, g_2 \in G$, $h_1, h_2 \in H$ and $c_1, c_2 \in C$, then $(g_1h_1c_1).(g_2h_2c_2)$ is equal to $g_1g_2.h_1h_2.[h_1,g_2]c_1c_2$ by (a) and the definition of commutators. The subgroup C is called the <u>cartesian subgroup</u> of G*H (since it is what you factor out to get the cartesian product G×H).

Let T be the $\forall_1$ theory which axiomatises the class of N_2 groups. We shall use Robinson forcing to construct models of T. But first it would be sensible to describe the e.c. models of T, and readers of Chapter 3 will not be surprised if I start by calculating some resultants.

LEMMA 4.4.2. *Let* G *be an* N_2 *group,* $X = \{g_i : i \in I\}$ *a family of elements of* G *and* $\{n_i : i \in I\}$ *a family of positive integers. Then the following are equivalent:*

(a) There is an N_2 *group* H *such that* $H \supseteq G$ *and for each* $i \in I$ *there are elements* $h_i, k_i \in H$ *such that* $[h_i,k_i]^{n_i} = g_i$.

(b) $X \subseteq Z(G)$.

<u>Proof</u>. (a) ⇒ (b): If (a) holds, then for each $i \in I$, $g_i \in H' \subseteq Z(H)$, so g_i is central in G.

(b) ⇒ (a): For each $i \in I$, let F_i be the free N_2 group generated by elements x_i and y_i, and form the direct product $G \times \bigoplus_{i \in I} F_i$. Let N be the normal subgroup of $G \times \bigoplus F_i$ generated by the elements $g_i.[x_i,y_i]^{-n_i}$. Assuming (b), we claim that $G \cap N = \{1\}$. The claim implies that G is embedded in $(G \times \bigoplus F_i)/N$ by the map $a \mapsto aN$, and since $g_iN = [x_iN,y_iN]^{n_i}$ in $(G \times \bigoplus F_i)/N$, this gives (a).

To prove the claim, note that since each g_i is central in G and each $[x_i,y_i]$ is central in F_i, any element d of N is a product of elements

(5) $(g_i)^{m_i}.[x_i,y_i]^{-n_i m_i}$ for some integers m_i.

Suppose that $d \in G$. Then for each $i \in I$, $[x_i,y_i]^{-n_i m_i} = 1$. Since n_i is positive and x_i, y_i generate F_i freely, this implies that $m_i = 0$ and so $d = 1$ as required. □

Lemma 4.4.2 was untypically easy, because the added elements h_i, k_i could be assumed to commute with everything in G. The next example reverts to the recipe of the proof of Theorem 3.1.3 above: we add free solutions of a system of equations. Fact 4.4.1 comes into play.

LEMMA 4.4.3. *Let* G *be an* N_2 *group,* g *an element of* G *and* n *a positive integer. Then the following are equivalent:*

(a) There is an N_2 *group* $H \supseteq G$ *containing an element* h *such that* $h^n = 1$ *and* $[h,g] \neq 1$.

(b) g *is not an* n*-th power modulo* G'.

Proof. (a) ⇒ (b): Assume (a), and suppose for contradiction that there are $b \in G$ and $c \in G'$ such that $g = b^n c$. Then using (1)-(3) and computing in H, we have

(6) $$1 \neq [h,g] = [h,b^n].[h,c] = [h^n,b].1 = [1,b] = 1,$$

contradiction.

(b) ⇒ (a): Assume (b). Let Z_n be the cyclic group of order n, with generator h. Form the N_2 free product $G*Z_n$. By Fact 4.4.1 we can regard G as a subgroup of $G*Z_n$. By Fact 4.4.1(c) (and (H) on p. 255 of Fuchs (1970)), $[h,g] = 1$ iff n divides $g^@$ in G^{ab}. By (b) this latter is false, and so $[h,g] \neq 1$. □

We can begin to describe e.c. N_2 groups:

THEOREM 4.4.4. *Let* G *be an e.c.* N_2 *group. Then:*

(a) Every element of Z(G) *is a commutator, and so* Z(G) = G'.

(b) G' *is a divisible abelian group containing infinitely*

many elements of each prime order.

(c) G^{ab} *contains infinitely many elements of each finite order.*

(d) For every element g *of* G *and every positive integer* n, *exactly one of the following holds: (i)* g *is an* n*-th power; (ii) there is an element* h *such that* $h^n = 1$ *and* $[h,g] \neq 1$.

(e) If G *is periodic and countable then* G^{ab} *is a direct sum of cyclic groups.*

Proof. By Lemma 4.4.2, if g is a central element of G then for every positive integer n, g is the n-th power of a commutator in some group $H \supseteq G$. Since G is e.c., the commutator can be found already in G. This proves (a) and the divisibility of G'. For the rest of (b), let p be a prime and K the direct product of k cyclic groups of order p. Then K lies in the centre of G×K, and hence by Lemma 4.4.2 there is $H \supseteq G\times K$ in which all the elements of K are commutators. It follows that the sentence "There are at least k commutators which have order p" is true in H, and hence also in G since G is e.c. Thus we have (b).

For (c) let n, k be positive integers and K a direct product of k cyclic groups of order n. Let C be cyclic of infinite order. Let S be the sentence "There are at least k elements x such that: $x^n = 1$ but for each m ($1 \leqslant m < n$) there is y such that $[x^m,y] \neq 1$". Written out formally, S is an $\exists_1$ sentence. Now S is true in K*C (by Fact 4.4.1(c)), hence also in G×(K*C), and hence in G too since G is e.c. This proves (c).

(d) follows from (b), Lemma 4.4.3 and the fact that G is e.c.

For (e) we quote the well-known fact that a countable abelian p-group without elements of infinite height is a direct sum of cyclic groups (cf. Fuchs (1970) Theorem 17.3). Assuming that G is periodic, we show that if p is any prime, then no element $g^@ \neq 0$ of p-power order in G^{ab} has infinite p-height. If $g^@ \neq 0$, then by (a), g is not central in G, so there is an element h of G such that $[h,g] \neq 1$. Now by assumption h has finite order n; say $n = p^k m$ with m prime to p. Then by Lemma 4.4.3, $g^@$ is not divisible by n in G^{ab}. Since all elements of p-power order are divisible by m, $g^@$ has p-height < k. □

In order to say any more about e.c. N_2 groups, we have to separate the periodic from the non-periodic. Between these two classes there is a great gulf fixed. We shall use non-periodic groups to prove:

THEOREM 4.4.5. *There is a family* $\{G_\alpha : \alpha < 2^\omega\}$ *of countable e.c.* N_2 *groups, such that if* $\alpha \neq \beta$ *then* G_α *is not embeddable in* G_β.

On the other hand we shall show:

THEOREM 4.4.6. *There is up to isomorphism just one periodic countable e.c.* N_2 *group.*

The proofs of these two results will take us up to the end of the chapter. Before embarking on either, I should record where N_2 groups stand in relation to the dichotomy of Theorem 4.2.6:

THEOREM 4.4.7. *Forcing with the theory* T *of* N_2 *groups, the property "The compiled structure* A *is a periodic e.c.* N_2 *group" is enforceable.*

<u>Proof</u>. By Corollary 3.4.3 it is enforceable that A is an e.c. N_2 group. It is enforceable that A is periodic because all finitely generated nilpotent groups are residually finite (cf. Exercise 3 of section 3.4.) □

It follows that the unique group of Theorem 4.4.6 is in fact the enforceable model of T. By general principles (cf. Exercise 2(a) of section 4.2) we infer that the countable periodic e.c. N_2 group is embeddable in every e.c. N_2 group. We also infer that the natural candidate for a proof of Theorem 4.4.5, namely case (a) in Theorem 4.2.6, won't work.

Let us prove Theorem 4.4.5 by a different route. Clause (d) of Theorem 4.4.4 encapsulates the main idea. Let g be an element of an N_2 group. Writing Π for the set of all prime numbers, let us say that g <u>represents</u> the set $X \subseteq \Pi$ iff for each $p \in \Pi$,

(7) $p \in X$ iff g is a p-th power modulo G'.

If G happens to be an e.c. N_2 group, then Theorem 4.4.4(d) tells us that for each prime p, either (i) g is a p-th power modulo G', or (ii) there is an element of G of order p which fails to commute with g. These two options are both expressed by $\exists_1$ formulas, and by Lemma 4.4.3 they are mutually exclusive in all N_2 groups. It follows that if G is an e.c. N_2 group and g represents X in G, then g still represents X in any N_2 group $H \supseteq G$. (Incidentally we are capitalising here on the fact that the amalgamation property fails for the theory T of N_2 groups; cf. Exercise 7.)

Note that if g represents X and infinitely many primes are not in X, then g must have infinite order modulo G'. The reason is that in an abelian group an element of finite order k is q-divisible for all primes q not dividing k.

Let us say that X is <u>represented</u> in G iff some element of G represents X. To prove the theorem, it suffices to construct the groups G_α so that whenever $\alpha \neq \beta$, some set $X \subseteq \Pi$ is represented in G_α but not in G_β. For this, let $\{X_n : n < \omega\}$ be a partition of Π into countably many infinite sets. For each $\alpha < 2^\omega$, choose a subset E_α of $\{X_n : n < \omega\}$ so that if $\alpha \neq \beta$ then $E_\alpha \setminus E_\beta$ is not empty. We shall construct G_α so that for each $n < \omega$, X_n is represented in G_α iff $X_n \in E_\alpha$.

If X is any set of prime numbers, we define Q(X) to be the additive group of all rational numbers whose denominators are not divisible by any prime not in X. We write B(X) for the group $\bigoplus\{Z_p : p \text{ a prime } \notin X\}$. We put $G(X) = Q(X)*B(X)$, and we write g(X) for the element 1 in Q(X). By construction and Fact 4.4.1, if $p \in X$ then g(X) is a p-th power in G(X), whereas if $p \notin X$ then some element of order p in G(X) fails to commute with g(X). So by Lemma 4.4.3, if $p \notin X$ then g(X) is not a p-th power in any N_2 group $H \supseteq G(X)$.

If E is a set of subsets X of Π, we write G(E) for the direct product $\bigoplus\{G(X) : X \in E\}$.

LEMMA 4.4.8. *Suppose* E *is a set of at most countably many subsets of* Π, *and* $Y \subseteq \Pi$ *is such that (i)* Y *and* $\Pi \setminus Y$ *are both infinite, and (ii) there is no set* $X \in E$ *such that* $Y \setminus X$ *is finite. Then there is a countable e.c.* N_2 *group* $A \supseteq G(E)$ *such that no element of* A *represents* Y.

<u>Proof</u>. We use forcing in the style of Exercise 5 of section

3.4 above. In other words, putting $\Delta = \mathrm{diag}(G(E))$, we force with $T \cup \Delta$. By the exercise just mentioned, the compiled structure A is a countable e.c. N_2 group $\supseteq G(E)$. Let c be a witness, and let P be the property "The element named by c in A^+ doesn't represent Y". By general principles of forcing (Lemma 2.3.1 and Lemma 2.3.3(e)), it suffices to show that P is enforceable.

Let players $\forall$ and $\exists$ play G(P;odds). Because the letters p, q have been pre-empted by the prime numbers, we now write π_0 for player $\forall$'s opening move. Let K be the N_2 group presented by the set of equations in $\Delta \cup \pi_0$. Since π_0 is a condition, there is a model B of $T \cup \Delta \cup \pi_0$. By the choice of K there is a homomorphism from K to B, and so K is in fact a model of the whole of $T \cup \Delta \cup \pi_0$. Thus without loss we can suppose that $G(E) \subseteq K$. Also without loss we can suppose that the witness c occurs in π_0, so that c is an element of K. Now there are three cases to consider.

<u>Case 1</u>: c has finite order modulo K'. In this case π_0 forces that c has finite order modulo A', so player $\exists$ can make this true in A. Then c will be a q-th power modulo A' for all but finitely many primes q. So by assumption (i) of the lemma, player $\exists$ wins.

<u>Case 2</u>: c has infinite order modulo K', but there is a prime $q \in Y$ such that c is not a q-th power modulo K'. Then by Lemma 4.4.3 there is an N_2 group $H \supseteq K$ containing an element h of order q such that $[h,c] \neq 1$. Let player $\exists$ play the condition $\pi_1 = \pi_0 \cup \{d^q=1, [d,c]\neq 1\}$ for some witness d not in π_0; then she prevents c from being a q-th power modulo A', and so again she wins.

<u>Case 3</u>: c has infinite order modulo K', and for every prime $q \in Y$, c is a q-th power modulo K'. We shall finish the proof of the lemma by showing that this case never occurs.

By the case assumption, $c^@$ has infinite order in K^{ab}, and for every prime $q \in Y$, $c^@$ is q-divisible in K^{ab}. Now K^{ab} is finitely generated over $G(E)^{ab}$, so that $K^{ab}/G(E)^{ab}$ is a direct sum of cyclic groups. Since $c^@+G(E)^{ab}$ is divisible by the infinitely many primes in Y (here we use the lemma assumption (i) again), it must be a torsion element in $K^{ab}/G(E)^{ab}$. Hence for some positive integer m, $mc^@ \in G(E)^{ab}$.

Now each group $G(X)^{ab}$ is of form $Q(X)\times\bigoplus\{Z_p : p \notin X\}$, and $G(E)^{ab}$ is $\bigoplus_{X\in E} G(X)^{ab}$. Hence there are $X_1, \ldots, X_n \in E$ and integers $k_1, \ldots, k_n$ such that

(8) $$mc^@ = k_1 g(X_1)^@ + \dots + k_n g(X_n)^@ + t$$

where t is a torsion element of $G(E)^{ab}$. Since $c^@$ has infinite order, we can assume that $k_1 \neq 0$. By assumption (ii) of the lemma, $Y \smallsetminus X_1$ is infinite and so there is a prime $q \in Y \smallsetminus X_1$ such that t is q-divisible in $G(E)^{ab}$. Since $q \in Y$, c is a q-th power modulo K' in K. Then by (8), the element $g = g(X_1)^{k_1} \dots g(X_n)^{k_n}$ is a q-th power modulo K' in K. But by the construction of G(E), since $q \notin X_1$ there is an element of order q in K which commutes with $g(X_2), \dots, g(X_n)$ but not with $g(X_1)^{k_1}$, and hence not with g either. By our stalwart Lemma 4.4.3 it follows that g is not a q-th power modulo K' in K; contradiction. □ Lemma

For each $\alpha < 2^\omega$, let G_α be the e.c. N_2 group A constructed from E_α by the lemma. Since $G(E_\alpha) \subseteq G_\alpha$, every set $X \in E_\alpha$ is represented in G_α. If $\alpha \neq \beta$ then there is some set $X_n \subseteq \Pi$ which is in E_α but not in E_β. By the lemma, X_n is represented in G_α but not in G_β. We saw how this implies that G_α is not embeddable in G_β. □ Theorem 4.4.5

With a little care we can choose the groups G_α so that whenever $\alpha \neq \beta$ there is an $\exists_3$ sentence which is true in G_α but not in G_β; cf. Theorem 4.1.9 above. Also "nilpotent of class 2" can be replaced by "nilpotent of class k" for any $k \geqslant 2$. These improvements are sketched in the exercises below.

We turn to periodic groups. The proof of Theorem 4.4.6 will fall into two parts. In the first part we describe some algebraic axioms, and we show that any two countable groups satisfying these axioms must be isomorphic. The second part will show that periodic e.c. N_2 groups obey the axioms.

Let A be an abelian group and $\bar{a} = (a_0, \dots, a_{n-1})$ a tuple of elements of A. We say that $\bar{a}$ is <u>independent</u> in A iff whenever an equation $\Sigma_{i<n} k_i a_i = 0$ holds, then $k_0 a_0 = \dots = k_{n-1} a_{n-1} = 0$. This is equivalent to saying that $\langle\bar{a}\rangle_A$ is a direct sum of cyclic groups and the a_i generate the cyclic summands.

Let G be an N_2 group and $\bar{g} = (g_0, \dots, g_{n-1})$ a tuple of elements of G. We say that $\bar{g}$ is <u>basic</u> in G iff (i) $\bar{g}^@$ is independent in G^{ab}, (ii) $\langle\bar{g}^@\rangle_{G^{ab}}$ is a pure subgroup of G^{ab}, and (iii) $\langle\bar{g}^@\rangle_{G^{ab}}$ is a torsion group. A theorem of Kulikov (cf. Theorem 27.5 of Fuchs (1970)) says that in the presence of (iii), (ii) means that $\langle \bar{g}^{@} \rangle_{G^{ab}}$ is a direct summand of G^{ab}.

Consider the following four axioms on a group G:

Axiom I. G is periodic.

Axiom II. Every element of Z(G) is a commutator, and every commutator is in Z(G).

Axiom III. Z(G) is a divisible group containing infinitely many copies of $Z(p^{\infty})$ for each prime p.

Axiom IV. Suppose $\bar{g}$ is basic in G, m is a positive integer and for each g_i in $\bar{g}$, b_i is an element of Z(G) such that $b_i^m = b_i^{k_i} = 1$ where k_i is the order of $g_i^@$; then there is an element h such that $h^@$ has order m, each $[g_i,h]$ is b_i, and the tuple $\bar{g}h$ is basic.

The second part of Axiom II says that G is an N_2 group.

LEMMA 4.4.9. *Let* G *and* H *be countable groups which obey Axioms I-IV. Then* $G \cong H$.

Proof. By Axioms I-III there is an isomorphism $f: G' \to H'$. To climb up from f to an isomorphism between G and H, we shall use the back-and-forth idea of section 4.2 above, slightly modified. Let I be the set of all pairs $(\bar{g},\bar{h})$ of tuples $\bar{g}$, $\bar{h}$ from G, H respectively, such that for some $n < \omega$, $\bar{g} = (g_0,\ldots,g_{n-1})$, $\bar{h} = (h_0,\ldots,h_{n-1})$ and

(9) $\bar{g}$, $\bar{h}$ are basic in G, H respectively;

(10) for each $i < n$, $g_i^@$ and $h_i^@$ have the same order; and if $g_i^k \in G'$ then $f(g_i^k) = h_i^k$;

(11) for all $i, j < n$, $f([g_i,g_j]) = [h_i,h_j]$.

It is clear that $(\emptyset,\emptyset) \in I$.

We claim that if $(\bar{g},\bar{h}) \in I$, then there is an embedding $e: \langle \bar{g} \rangle_G \to H$ such that $e\bar{g} = \bar{h}$. For this it suffices to show that if $t(\bar{x})$ is any term in the language of groups, then $t^G(\bar{g}) = 1$ iff $t^H(\bar{h}) = 1$. By symmetry we need only prove the implication from left to right.

Let $t(\bar{x})$ be a term in the language of groups. Since we are in N_2 groups by Axiom II, $t(\bar{x})$ can be brought to the form

(12) $$x_0^{\alpha_0} \ldots x_{n-1}^{\alpha_{n-1}} . \Pi_{i<j} [x_i, x_j]^{\beta_{ij}} \qquad (\alpha_i, \beta_{ij} \text{ integers}).$$

Suppose that $t^G(\bar{g}) = 1$. Then $t^G(\bar{g})^@ = 0$ in G^{ab}, and since $\bar{g}$ is basic it follows that

(13) $$g_0^{\alpha_0}, \ldots, g_{n-1}^{\alpha_{n-1}} \in G'.$$

But then by (10) and (11) we infer that $f(t^G(\bar{g})) = t^H(\bar{h})$, and so $t^H(\bar{h}) = 1$. The claim is proved.

The claim above corresponds to (5) of section 4.2. Our next claim corresponds to (7) and (8) of that section. We claim that:

(14) If $(\bar{g}, \bar{h}) \in I$ and $\bar{g}g_n$ is a basic tuple in G, then there is h_n in H such that $(\bar{g}g_n, \bar{h}h_n) \in I$; and likewise with G and H transposed.

By symmetry we need only show the first part. By Axiom I, $g_n^@$ has finite order m. For each $i < n$, let a_i be $[g_i, g_n]$, and let b_i be $f(a_i)$. Writing k_i for the order of $g_i^@$, we certainly have $a_i^m = a_i^{k_i} = 1$, and since f is an isomorphism from G' to H', the same equations hold with b_i in place of a_i. By (10), k_i is the order of $h_i^@$ too. Thus the hypothesis of Axiom IV holds in H with $\bar{h}$ in place of $\bar{g}$. So the conclusion holds: there is an element h in H such that $h^@$ has order m, each $[h_i, h]$ is $b_i = f([g_i, g_n])$, and the tuple $\bar{h}h$ is basic.

Thus we have all of (9)-(11) for $(\bar{g}g_n, \bar{h}h)$, except that $f(g_n^m)$ need not be h^m. Suppose in fact that $g_n^m = a$ and $h^m = b$. Then $b^{-1}.f(a)$ is an element of H', and by Axiom III it is an m-th power in H'. Choose c in H' so that $c^m = b^{-1}.f(a)$, and put $h_n = h.c$. Then $h_n^m = f(a)$ as required, and since c is in H', nothing else in (9)-(11) is affected. So $(\bar{g}g_n, \bar{h}h_n) \in I$. This proves (14).

Now by the argument of Theorem 4.4.4(e) and Axioms I and II, each of G^{ab}, H^{ab} is a direct sum of countable p-groups without elements of infinite height. It follows that if $\bar{g}$ is basic in G and a is an element of G, then there is a tuple $\bar{d}$ in G such that $\bar{g}\bar{d}$ is basic and $a \in \langle \bar{g}\bar{d} \rangle_G$;

and likewise in H. From this and the claims above, the reader can easily cobble together a back-and-forth argument which finds an isomorphism from G to H. (Cf. Exercise 12 below.) □

We turn to the second and harder half of the proof of Theorem 4.4.6, which is to show (Lemma 4.4.13 below) that Axioms II-IV are true in every e.c. N_2 group.

The following axiom - a sort of poor man's Axiom IV - will form a useful stepping stone:

> Axiom V. Suppose g is an element of G and m is a positive integer such that g is not an m-th power modulo Z(G); then there is an element h of order m such that $[g,h] \neq 1$.

Axiom V follows quite easily from Axioms II-IV, but we never need to know this.

LEMMA 4.4.10. *If* G *is an e.c.* N_2 *group then* G *obeys Axioms II, III and V.*

Proof. Theorem 4.4.4(a,b,d). □

Axiom IV is the troublesome one, and we need two blows to kill it. The first blow is to show that whenever the hypothesis of the axiom holds in an N_2 group G, an element h as in the conclusion can be found in some N_2 group $H \supseteq G$. The second is to pull h down from H into G when G is existentially closed. We find H and h by the following rather formidable lemma. (At base it is the calculation of yet another resultant.)

LEMMA 4.4.11. *Let* G *be an* N_2 *group,* $\bar{g}$ *a basic* n*-tuple in* G, $\bar{b}$ *an* n*-tuple from* G' *and* m *a positive integer. Then the following are equivalent:*

(a) There is an N_2 *group* $H \supseteq G$, *containing an element* h *such that* h *has order* m, $h^j \notin H'$ *for all* j $(1 \leq j < m)$, $b_i = [g_i, h]$ *for all* $i < n$, *and* $\bar{g}h$ *is basic in* H.

(b) For all $i < n$ *and all integers* k, *if* $g_i^k \in G^m G'$ *then* $b_i^k = 1$.

Proof. (a) ⇒ (b): Assume that (a) holds. Suppose g is in $\bar{g}$, and $g^k = a^m c$ for some integer k, some $a \in G$ and some $c \in G'$. Then by (1)-(3) we have for the corresponding b in $\bar{b}$:

$$(15) \qquad b^k = [g,h]^k = [g^k,h] = [a^m c,h] = [a,h^m][c,h] = 1.$$

This proves (b).

(b) ⇒ (a): Assume (b). Let F be the infinite cyclic group generated by an element x, and in the N_2 free product G*F let N be the normal subgroup generated by the elements x^m and $b_i^{-1}[g_i,x]$ ($i < n$). The idea is to make $(G*F)/N = H$. For this to work, we need to prove (i) that $G \cap N = \{1\}$, (ii) that if j is an integer such that x^j is a product of commutators in G*F modulo N, then $m|j$, (iii) that $(\bar{g}h)^@$ is independent in $(G*F/N)^{ab}$, and (iv) that $(\bar{g}h)^@$ generates a pure subgroup of $(G*F/N)^{ab}$.

The first part of all four proofs is the same, viz. to characterise N. Bearing in mind that the elements $b_i^{-1}[g_i,x]$ are products of commutators, so that they commute with everything, we can write an arbitrary element d of N as a product of p+q+r+s elements of the following forms, in some order:

(16) $b_i'^{-1}[g_i',x]$ ($1 \leq i \leq p$, where for each i there is $i' < n$ such that $b_i' = b_{i'}$, $g_i' = g_{i'}$),

(17) $[g_j',x]^{-1}b_j'$ ($1 \leq j \leq q$, where for each j there is $j' < n$ such that $b_j' = b_{j'}$, $g_j' = g_{j'}$),

(18) $a_k^{-1}.x^m.a_k$ ($1 \leq k \leq r$, $a_k \in G$),

(19) $c_\ell^{-1}.x^{-m}.c_\ell$ ($1 \leq \ell \leq s$, $c_\ell \in G$).

(Note that by the N_2 laws, the elements (18) and (19) are unchanged under conjugation by x.) After rearrangement and some commutator calculations using (1)-(4) (I recommend a scratch pad), this yields

$$(20) \qquad d = w(\bar{b}).x^{m(r-s)}.[x,w(\bar{g})c^m]$$

for some term w in the language of groups and some element c of G. In particular, d is a product of a power of x and some commutators.

(i) Suppose $d \in N$. Then by Fact 4.4.1(c), $w(\bar{g})c^m \in G'$, and so $w(\bar{g})^@$ is divisible by m in G^{ab}. Now $w(\bar{g})^@$ can be written as $\Sigma k_i g_i^@$ for some integers k_i. Since $\bar{g}$ is basic in G, we infer that each $k_i g_i^@$ is divisible by m in G^{ab}. By assumption (b) it follows that for each $i < n$, $b_i^{k_i} = 1$, which implies that $w(\bar{b}) = 1$. Now by Fact 4.4.1(b) and (20), $d = w(\bar{b}) = 1$. Hence $G \cap N = \{1\}$ as required.

Henceforth we write H for $(G*F)/N$, and h for xN.

(ii) We have $H' = (G*F)'N/N$. Assume that $x^j \in (G*F)'N$. We must show that $m|j$. (In other words, if $h^j \in H'$ then $h^j = 1$.) By Fact 4.4.1 and the fact that F is abelian, a typical element of $(G*F)'$ is of form gc where $g \in G'$ and c is in the cartesian subgroup. By (20) a typical element of N is of form $g'.x^{mr}.c'$ where $g' \in G'$, c' is in the cartesian subgroup and r is an integer. So our assumption takes the form

(21) $$x^j = gc.g'x^{mr}c' = gg'.x^{mr}.c'' \quad \text{for some } c'' \in \text{cartesian subgp.}$$

By Fact 4.4.1(b), $x^j = x^{mr}$. Since x freely generates F, $j = mr$ as required.

(iii) Regarding G as a subgroup of H, $G \cap H' = G'$. (For $\subseteq$, note that if the typical element $gg'.x^{mr}.c''$ of $(G*F)'N$ in (ii) is in G then it is gg', which is in G'.) Hence G^{ab} is naturally embedded in H^{ab}. In particular if g is in G, then we can form $g^@$ in G^{ab} or in H^{ab} indifferently.

We must check that $(\bar{g}h)^@$ is independent in H^{ab}. By assumption $g^@$ is independent in G^{ab} and hence also in H^{ab}. So it suffices to show that there are no non-trivial relationships between $h^@$ and any element $g^@$ of G^{ab}. Suppose then that $ig^@ + jh^@ = 0$. We infer that

(22) $$g^i x^j \in (G*F)'N.$$

Then just as in (ii), we use (20) and Fact 4.4.1(b) to deduce that $m|j$ and hence $jh^@ = 0$.

(iv) We check that in H^{ab} the elements $(\bar{g}h)^@$ generate a pure subgroup. Let g be any element in the subgroup of G generated by $\bar{g}$, and suppose that for some integers j and k, $g^@+jh^@$ is divisible by k in H^{ab}.

This means that for some $a \in G$, some integer i and some element c of the cartesian subgroup of $G*F$,

$$(23) \qquad (gx^j)(ax^i c)^{-k} \in (G*F)'N.$$

Then by (20) and Fact 4.4.1 as before,

$$(24) \qquad ga^{-k} \in G' \quad \text{and} \quad m|(j-ik).$$

Since $\langle\bar{g}^@\rangle_{G^{ab}}$ was a pure subgroup of G^{ab} by assumption, it follows that $g^@$ is already divisible by k within $\langle\bar{g}^@\rangle_{G^{ab}}$. Also by the second part of (24), $jh^@ = k(ih^@)$. This completes the argument. □ Lemma

Let G be an e.c. N_2 group. We want to show that G obeys Axiom IV. Suppose $\bar{g} = (g_0,\ldots,g_{n-1})$ is basic in G, $\bar{b} = (b_0,\ldots,b_{n-1})$ is a tuple from $Z(G) = G'$, each g_i has order k_i, m is a positive integer, and for each $i < n$, $b_i^m = b_i^{k_i} = 1$. We verify (b) of Lemma 4.4.11. Let $i < n$ and let k be an integer such that $g_i^k \in G^m G'$. Then m divides $kg_i^@$ in G^{ab}. Since $\bar{g}$ is basic in G, it follows that $k = \alpha m + \beta k_i$ for some integers α, β. Hence $b_i^k = 1$ as required.

Since (b) of the lemma holds, we infer (a) of the lemma. Thus there exists an N_2 group $H \supseteq G$ which satisfies "There is an element h such that $h^@$ has order m, $b_i = [g_i,h]$ for all $i < n$, and $\bar{g}h$ is basic". Our object is to show that the sentence in quotation marks is true already in G. Since G is existentially closed, it will be enough if we express the sentence as an $\exists_1$ formula.

It is not immediately obvious that we can do this. In general the statement "$h^@$ has order m" is not even first-order. The next lemma says that we can always find an $\exists_1$ formula which works, but only after extending H to make certain instances of Axioms II and V true.

LEMMA 4.4.12. *Let* H *be an* N_2 *group,* b *an element of finite order* m *modulo* H' *and* $\bar{h}$ *a basic tuple in* H. *Then there is an* $\exists_1$ *first-order formula* $\phi(x,\bar{y})$ *such that*

(a) $F \models \phi(b,\bar{h})$ *for some* N_2 *group* $F \supseteq H$, *and*

(b) *for any group* G *satisfying Axioms II and V, if* a, $\bar{g}$ *are an element and a tuple such that* $G \models \phi(a,\bar{g})$, *then* a *has order* m *modulo* G' *and* $\bar{g}$ *is basic in* G.

Proof. We begin with the special case where H already satisfies Axioms II and V. In this case, F in (a) can be taken to be H.

First ϕ should state the order of $b^@$ and of each element $h_i^@$. This can be done with existential formulas, since by Axiom II, $h \in H'$ iff $\exists xy\ h = [x,y]$, and $h \notin H'$ iff $\exists x\ [x,h] \neq 1$.

Next, ϕ should state that every equation $\Sigma k_i h_i^@ = 0$ implies $k_0 h_0^@ = k_1 h_1^@ = \ldots = 0$. Because the $h_i^@$ have finite order, there are only finitely many equations to consider, and the resulting formulas can be made existential by Axiom II again.

Third and finally, ϕ should say that the elements $h^@$ generate a pure subgroup of H^{ab}. For example if $\Sigma k_i h_i^@$ is not divisible by j in $\langle \bar{h}^@ \rangle_{H^{ab}}$, then it is not divisible by j in H^{ab} either. Lemma 4.4.3 and Axiom V allow us to say this latter with an existential formula.

Since G in (b) is assumed to satisfy Axioms II and V, the formula ϕ has the intended meaning in G too.

We turn to the general case. Here we use the same formula ϕ as before, so that part (b) of the lemma is secure. Possibly $\phi(b,\bar{h})$ is not true in H. We make it true by adding elements to H, as follows. First suppose that $\phi(b,\bar{h})$ expresses that some element h is in H', by saying that h is a commutator. Then $h \in H' \subseteq Z(H)$, so that Lemma 4.4.2 shows how to extend H to an N_2 group H_1 in which h is a commutator. The relevant clause of ϕ now holds in H_1. Since $h \in H'$, the construction of H_1 implies that H^{ab} is a direct summand of H_1^{ab}, so that the assumption of the lemma remains true in H_1. Thus we can replace H by H_1 and move on to the other clauses of ϕ.

Next suppose that $\phi(b,\bar{h})$ expresses that some element h is not in H', by saying that h is not central. If we take an infinite cyclic group Z and put $H_2 = H*Z$, then just as in Lemma 4.4.3(b)⇒(a), h is not central in H_2, and again H^{ab} is a direct summand of H_2^{ab}. Finally when ϕ expresses that some element is not an n-th power modulo H', we use the construction of Lemma 4.4.3(b)⇒(a) and apply the same reasoning as above. After a finite number of these extensions of H, we reach the required group F.

□ Lemma

Pulling things together, we infer from Lemma 4.4.10, Lemma

4.4.12 and the discussion after Lemma 4.4.11 that:

LEMMA 4.4.13. *If* G *is an e.c.* N_2 *group then* G *obeys Axioms II, III and IV.* □

Theorem 4.4.6 follows at once from Lemma 4.4.9 and Lemma 4.4.13. □ Theorem 4.4.6

Exercises for 4.4:

1. Let G be an N_2 group and a, b elements of G. Show that the following are equivalent: (a) There is an N_2 group $H \supseteq G$ with an element h such that $[a,h] = 1 \neq [b,h]$. (b) For all integers n, $b \notin a^n G'$.

2. Let G be an N_2 group and a, b, c elements of G. Show that the following are equivalent: (a) There is an N_2 group $H \supseteq G$ with an element h such that $[a,h] = [b,c]$. (b) For all integers n, if $a^n \in G'$ then $[b,c]^n = 1$.

3. Let G be an N_2 group and a, b, c, d elements of G such that a has infinite order modulo G'. Show that the following are equivalent: (a) There is an N_2 group $H \supseteq G$ with an element h such that $[a,h] = [b,c]$ and $[d,h] \neq 1$. (b) For all integers n, if $a^n \in dG'$ then $[b,c]^n \neq 1$.

4. Show that for each of the following there is a first-order formula which expresses it in any e.c. N_2 group G; find the formula in the form stated. (a) x is a power of y modulo G' ($\forall_1$). (b) x has infinite order modulo G' ($\forall_2$). (c) [x,y] has infinite order ($\forall_2$). [For (c), say that x has infinite order modulo G', and if z is a power of x modulo G', such that $[z,y] = 1$, then $z \in G'$.]

5. Let G be an e.c. N_2 group with elements g, h such that [g,h] has infinite order. Show that the first-order theory of the ring of integers can be interpreted in Th(G), in such a way that each integer n is interpreted as g^n (modulo G'). Show in particular: (a) The formulas representing equality and addition can be chosen to be $\forall_1$. (b) The formula representing multiplication can be chosen to be $\exists_2$. (c) For any r.e. set Y of integers, the statement "$n \in Y$" can be represented by an $\exists_2$ formula. [For multiplication use the fact that $[g^i,h^j] = [g,h]^{ij}$. For (c) use Matiyasevich's theorem.]

Here and below, "nilpotent of class k*" includes nilpotent of class* $\leqslant$ k.

6. (a) Show that there is an $\exists_3$ first-order sentence which is true in an e.c. N_2 group G iff G contains an element of infinite order modulo G'. (b) Show that for each recursive set X of primes there is an $\exists_3$ first-order sentence which is true in an e.c. N_2 group G iff G contains elements g, h such that [g,h] has infinite order and g represents X. (c) Show that there is a family $\{G_\alpha : \alpha < 2^\omega\}$ of countable e.c. N_2 groups, such that if $\alpha \neq \beta$ then there is an $\exists_3$ first-order sentence which is true in G_α but not in G_β. (d) Show that (c) holds also if N_2 is replaced by "nilpotent of class k" for any $k \geqslant 2$. [Find sentences which say "There are elements $g_1, \ldots, g_k$ such that the commutator $[g_1,\ldots,g_k]$ of weight k has infinite order and g_1 represents X".] (e) Show that when $k \geqslant 2$, the theory of nilpotent groups of class k is not companionable.

Several arguments above are intimately connected with the fact that for any $k \geqslant 2$ *the class of nilpotent groups of class* k *does not have the amalgamation property. A structure* A *in a class* K *is called an* <u>*amalgamation base*</u> *in* K *iff (4) of section 3.2 holds, but with the final equation* (dom B) $\cap$ g(dom C) = dom A *deleted.*

7. Let G be an N_2 group. Show that the following are equivalent: (a) G is an amalgamation base in the class of N_2 groups. (b) G is a strong amalgamation base in the class of N_2 groups. (c) G' = Z(G), and if n is any positive integer and g any element of G which is not an n-th power modulo G', then for some integer k, some element h of G satisfies $h^n g^{-k} \in G'$ and $[h,g] \neq 1$.

8. Show that for every $k \geqslant 2$, if G is an e.c. nilpotent group of class k then the upper and lower central series of G coincide.

9. Show that for every $k \geqslant 2$, if G is a finite-generic nilpotent group of class k then G is periodic. [Cf. Theorem 4.4.7 and Exercise 4(b).]

10. Show that Axioms II-IV can be written as a $\forall_2$ first-order theory. [First deduce Axiom V from the other axioms, and write Axioms II, III and V in $\forall_2$ form. Then use Lemma 4.4.12 to take care of Axiom IV.]

11. (a) Show that a group G is a finite-generic N_2 group iff it satisfies Axioms I-IV. [For 'if', use the fact that every countable elementary substructure of G is a finite-generic N_2 group.] (b) Show that every periodic N_2 group can be extended to a group satisfying Axioms I-IV.

(c) Show that in every uncountable cardinality λ there are 2^λ finite-generic N_2 groups which are pairwise non-isomorphic (and pairwise non-embeddable if λ is regular). [Cf. Lemma 4.1.10; consider the N_2 group of exponent 3 presented by the equations $[a_i,a_j] = c$ $(i < j < \lambda)$.]

12. (a) If L is a language and A, B are L-structures, we define a <u>weak back-and-forth system</u> from A to B to be a family I of pairs $(\bar{a},\bar{b})$ consisting of a sequence $\bar{a}$ from A and a sequence $\bar{b}$ from B, such that (5), (6) of section 4.2 hold, together with the following weak form of (7):

> If $(\bar{a},\bar{b}) \in I$ and c is any element of A, then there are $\bar{a}'$, $\bar{b}'$ extending $\bar{a}$, $\bar{b}$ respectively, such that $(\bar{a}',\bar{b}') \in I$ and c is in $\langle\bar{a}'\rangle_A$,

and the corresponding weakening of (8). Show that if there is a weak back-and-forth system from A to B then A is back-and-forth equivalent to B. [Define J to be the set of pairs $(\bar{c},f\bar{c})$ where for some $(\bar{a},\bar{b}) \in I$, f is the embedding f: $\langle\bar{a}\rangle_A \to B$ taking $\bar{a}$ to $\bar{b}$, and $\bar{c}$ is in dom f.] (b) Show that if G and H are N_2 groups satisfying Axioms I-IV then G is back-and-forth equivalent to H. [Start by replacing the last paragraph of the proof of Lemma 4.4.9 by an appeal to Theorem 17.2 and Corollary 27.8 of Fuchs (1970).]

13. Let m be an integer > 1, and T^m the theory of N_2 groups of exponent m. (a) Show that any two countable e.c. models of T^m are isomorphic; adapt Axioms I-IV to this case. (b) Show that the theory of any countable e.c. model of T^m is ω-categorical and is a model-companion of T^m.

14. Let p be a prime $\neq 2$, let i, j, k be positive integers satisfying $i \leqslant j \leqslant k \leqslant i+j$, and let Z be a countable abelian group of exponent p^j. Call G a <u>*-group</u> iff G has the following properties: (a) G is an N_2 group, $Z(G) = Z$, G has exponent p^k and G^{ab} has exponent p^i; (b) if X is a set of elements of G, $X^@$ is a finite pure subgroup of G^{ab}, f: $X^@ \to Z(G)$ is a homomorphism, $c \in Z(G)$, m is a positive integer, exponent(fA) $\leqslant p^m \leqslant$ exponent(G^{ab}) and p^m.order(c) $\leqslant$ exponent(G), then there is $h \in G$ such that $\langle X^@,h^@\rangle_{G^{ab}}$ is pure in G^{ab}, $h^@$ has order p^m, $h^{p^m} = c$, and for each $x \in X$, $[x,h] = f(x^@)$. Show that up to isomorphism there is just one countable *-group G, and Th(G) is ω-categorical.

15. (a) Show that for each $k \geqslant 2$, the theory of torsion-free nilpotent

groups of class k is not companionable. (b) Let T be the theory of torsion-free nilpotent groups of class 2. Show that if $1 \leqslant \kappa \leqslant \omega$ then there is an e.c. model G_κ of T whose centre is the direct sum of κ copies of the rationals, and that this determines G_κ up to isomorphism. Show that $G_1 \subseteq G_2 \subseteq \ldots \subseteq G_\omega$, and that there are no other countable e.c. models of T.

16. Show that each of the following theories is not companionable. (a) For any $k \geqslant 2$, the theory of rational Lie algebras nilpotent of class $\leqslant k$. [A category equivalence reduces this to Exercise 15(a).] (b) For any infinite field K, the theory of Lie algebras over K. (c) For any $k \geqslant 2$ the theory of soluble groups of length $\leqslant k$.

REFERENCES FOR CHAPTER 4

4.1: Theorem 4.1.3 is due to Grilliot (1972) and Theorem 4.1.5 to Shelah (Theorem IV.5.16 of (1978 a)). Theorem 4.1.9 was proved by Belegradek (1978) and Ziegler (1980).

Exercise 5(a,b) is due to Steve Simpson, cf. section 7.4 of Hirschfeld & Wheeler (1975); the same section in Hirschfeld & Wheeler gives Exercise 6. Exercise 5(c): Macintyre (unpublished), cf. Hirschfeld & Wheeler (1975) chapter 16 for skew fields. Exercise 7 can be traced back to Ehrenfeucht (1958). Exercise 9: Baldwin & Saxl (1976).

4.2: The back-and-forth approach to proofs of isomorphism was used by Hausdorff (1914) to prove a theorem of Cantor about isomorphisms between dense linear orderings. It made its way into general model theory through several papers in the '50s and '60s, notably Fraissé (1954) and Vaught (1961). Lemma 4.2.2 is from Fraissé's paper and Lemma 4.2.4 is essentially from Vaught's. Lemma 4.2.3 is from Karp (1965). Scott sentences were introduced by Scott (1965). Kueker (1975) surveys back-and-forth methods with special reference to infinitary languages. Exercises 2-5 and the bulk of Theorem 4.2.7 were given by Pouzet (1972 a-c) as analogues of results in Vaught (1961); see also Simmons (1975 b, 1976) on $\exists$-atomic models. Recently Lachlan (198-) (cf. Cherlin & Lachlan (198-)) has made dramatic advances in the structural analysis of structures A as in Exercise 12 when Th(A) is stable.

Exercise 6: Saracino & Wood (1984). Exercise 7(c): Apps (198-). Exercise 9: Hirschfeld (1978), cf. Schmerl (1977). Exercise 10: Schmerl (1975). Exercise 12: Fraissé (1953). Exercise 13: Henson (1972), cf. Ash (1971), Ehrenfeucht (1972), Glassmire (1971), Schmerl (1980). Exercise 14 was used by Berline & Cherlin (1981) and Saracino & Wood (1982) to construct rings and nilpotent groups whose theories have quantifier elimination. Exercise 16: Ziegler (1980).

4.3: Finite-generic models were introduced in Robinson (1971 a) and Barwise & Robinson (1970). See Hirschfeld & Wheeler (1975) for a thorough discussion. Theorem 4.3.1 is due to Ziegler (1980). Theorem 4.3.2 and its corollary are from Barwise & Robinson (1970), as are Exercises 2, 4 and 7. Theorem 4.3.4 is related to Henrard (1973).

Exercise 5: Cusin (1974), cf. Lee & Nadel (1975). Exercise 6: cf. Fisher et al. (1975) and Simmons (1975 b). Exercise 8: Sabbagh (1971); the quantifier elimination theorem for modules is in Baur (1976). Exercise 9(a-c): Saracino & Wood (1983), developing Glass & Pierce (1980 a). Exercise 9(d) is from Robinson (1956), and Point (198-) studies its connection with the rest of Exercise 9. Exercise 10(b): Ziegler (1980). Exercise 12: Cherlin (1976) p. 105f. Exercise 13: Chikhachev (1975 a). Exercise 14(b): Fisher et al. (1975). Exercise 15: Chikhachev (1975 b), cf. Hirschfeld (1980).

4.4: The study of e.c. groups in nilpotent varieties was opened up by Saracino (1976); this paper contains Theorem 4.4.4(a), Exercises 1, 4, 6(a,e) and Exercise 9 for k = 2. The rest of Theorem 4.4.4 is from Saracino & Wood (1979), which also contains Theorem 4.4.6 and Exercises 10, 11 and 13. Theorem 4.4.5 is from Hodges (198- c). Note that while Theorem 4.4.5 is not at all sensitive to the nilpotency class (cf. Exercise 6(d)), Theorem 4.4.6 is subtler and needs the exact description of the cartesian subgroup of the 2nd Golovin product (or alternatively as in Saracino & Wood (1979), facts about split extensions of N_2 groups). Readers hoping to extend it to class 3 should read B. H. Neumann's description of the cartesian subgroup of the 3rd Golovin product, stated in Wiegold (1959) and verified by Allenby (1966).

Exercises 5 and 6(b-d): Hodges (1980, 198- c); cf. Mal'tsev (1960) for this interpretation of arithmetic. Exercise 7: Saracino (1983), cf. Maier (198-). Exercises 8, 9: Maier (1983). The group mentioned in Exercise 11(c) was studied by Felgner (1975). Exercise 14: Apps (1983). Exercise 15: Saracino (1978), cf. Baumslag & Levin (1976) and Maier (1984). Exercise 16: (a) Macintyre & Saracino (1973), (b) Macintyre (1973 b), (c) Saracino (1974 b).

5 CLASSICAL LANGUAGES

The progress of historical linguistics showed that the standard language was by no means the oldest type, but had arisen, under particular historical conditions, from local dialects. Standard English, for instance, is the modern form not of literary Old English, but of the old local dialect of London ...

Leonard Bloomfield, Language (1933)

Although I have taken Robinson's forcing as an exemplar of forcing, the root idea behind forcing and omitting types appeared much earlier in other forms. The Henkin-Orey theorem in 1956, the omitting types theorem of Engeler, Ryll-Nardzewski, Svenonius and Vaught in 1959, the representability theorem of Grzegorczyk, Mostowski and Ryll-Nardzewski in 1961 and Karp's completeness theorem for $L_{\omega_1\omega}$ in 1964 were all of them volcanoes thrown up by one and the same hot spot. (Round about 1970, recursion theorists were apt to add Gödel's compactness theorem to the list. But old hands in model theory know that <u>any</u> argument will prove Gödel's compactness theorem if one tries long enough.)

In section 5.1 we shall revisit these classical theorems and look at them from the point of view of games. Compared with Robinson forcing, the main new feature is that the forcing conditions may contain any first-order sentences, not just basic ones. The treatment will be brief because the results are well known and widely available.

In section 5.3 we shall broaden the definitions still further and allow infinite conditions. The resulting structures are saturated to some degree - again this is a classical notion. Robinson's "infinite-generic models" stand in roughly the same ratio to saturated models as finite-generic models do to atomic models. I shall say something about this, but in section 5.3 I am more concerned to lay the groundwork for Shelah's use of λ-compact models in Chapter 8.

Section 5.1 climbs up from the finite to the countable. Section 5.3 climbs up from one infinite cardinal to another, and this is altogether a different kettle of fish. It means that from section 5.3 onwards we shall have to be aware of set theory. Sometimes we shall need infinite counting principles such as Jensen's $\diamondsuit$. Section 5.2 is a sampler of the main set-theoretical ideas that will be required.

Speaking of the "old local dialect of London", volume 1 of Abraham Robinson's Selected Papers opens with a page of his address to the Mathematics Seminar at Birkbeck College, London, where he was studying for Ph.D., in February 1948. In it he advocates the use of mathematical logic "as a tool for the discovery or demonstration of actual mathematical theorems".

5.1 CLASSICAL OMITTING TYPES

L shall be a countable first-order language. W shall be a countable set of constants which are not in L; these constants are known as the witnesses. L(W) is the first-order language got from L by adding the witnesses as new constants. T shall be a theory in L, and we always assume that T has a model.

We define N to be the set of all finite sets p of sentences of L(W) such that T ∪ p has a model. The sets p in N are known as conditions.

LEMMA 5.1.1. *(a)* N *is a notion of forcing for* L(W), *and* N *has properties (6)-(8) of section 2.3.*

(b) The property "The compiled structure A^+ *is a model of* T*" is enforceable.*

(c) Let $\psi(x_1,\ldots,x_n)$ *be a formula of* $L(W)_{\omega_1\omega}$ *in which at most finitely many witnesses occur, and let* q *be a condition. Then* q *forces* $\forall x_1\ldots x_n\psi$ *iff for some tuple* $(c_1,\ldots,c_n)$ *of distinct witnesses not occurring in* q *or in* ψ, q *forces* $\psi(c_1,\ldots,c_n)$.

(d) Let q *be a condition and* ϕ *a sentence of* L(W). *Then* q *forces* ϕ *iff* $T \cup q \vdash \phi$.

Proof. (a) is left to the reader. Then (c) follows from Exercise 2 of section 2.3.

To prove (b), let ϕ be any sentence in T. In a play of $G(\phi;odds)$, let player $\forall$ offer p_0. Then $T \cup \{\phi\} \cup p_0$ has a model, and so $p_1 = p_0 \cup \{\phi\}$ is a condition. Hence player $\exists$ can ensure that $A^+ \models \phi$ by replying with p_1 and playing so that $A^+ \models \bigcup \bar{p}$ (which she can do by Lemma 2.3.1).

A similar argument shows that if $T \cup q \vdash \phi$ then q forces ϕ, proving right to left in (d). Conversely suppose $T \cup q \nvdash \phi$. Then $q \cup \{\neg\phi\}$ is a condition p_0. If player $\forall$ makes p_0 his opening move in $G(\phi;odds)$, then by Lemma 2.3.1 he can play so that $A^+ \models \neg\phi$, and hence q doesn't force ϕ. □

Part (d) of the lemma shows that when ϕ is a first-order sentence, the relation "p forces ϕ" is not interesting. However, there are plenty of infinitary sentences, such as those which say that a type is omitted.

Let Φ be a set of formulas $\phi(\bar{x})$ of L, all with the same tuple $\bar{x}$ of free variables. (We write $\Phi(\bar{x})$ for such a set.) By analogy with section 3.4, a <u>support</u> of Φ is a formula $\psi(\bar{x})$ of L such that $T \cup \{\exists\bar{x}\psi\}$ has a model and $T \vdash \forall\bar{x}(\psi \rightarrow \bigwedge\Phi)$. Before applying this notion, we should compare it with some of its cousins in section 4.1.

The <u>type</u> of a tuple $\bar{a}$ of elements of an L-structure A, $tp_A(\bar{a})$, is the set of all formulas $\phi(\bar{x})$ of L such that $A \models \phi(\bar{a})$. For a fixed $n < \omega$, we write S_n for the set of all sets of form $tp_A(\bar{a})$ as A ranges over models of T and $\bar{a}$ ranges over n-tuples of elements of A. (Thus S_n depends on T, and the following notions are all relative to T.) By the definition, the type of a tuple $\bar{a}$ is always a maximal consistent set of formulas $\phi(\bar{x})$ of L. Elements of any S_n are known as <u>complete types</u>. We say that a formula $\psi(\bar{x})$ <u>isolates</u> the complete type $\Phi(\bar{x})$ iff Φ is the unique complete type which contains ψ. A complete type is said to be <u>isolated</u>, or (more usually) <u>principal</u>, iff some formula isolates it.

It is easy to see that if Φ is a complete type, then a formula ψ of L isolates Φ if and only if ψ is a support of Φ. So a complete type is principal if and only if it has a support.

We say that a set $\Phi(\bar{x})$ of formulas is <u>realised</u> by $\bar{a}$ in A iff $A \models \bigwedge\Phi(\bar{a})$. We say that A <u>omits</u> Φ iff Φ is not realised in A by any tuple of elements.

The classical "omitting types theorem" runs as follows:

THEOREM 5.1.2. *Let* T *be a theory in a countable first-order language* L. *For each* $i < \omega$ *let* $\Phi_i(\bar{x})$ *be a set of formulas of* L *(where the tuple* $\bar{x}$ *may vary with* i*), and suppose that no set* Φ_i *has a support. Then* T *has a model which omits all the* Φ_i $(i < \omega)$.

Proof. Exactly as Corollary 3.4.5, but forcing with first-order conditions. Put briefly, suppose player $\forall$ has just played p, and player $\exists$ is trying to prevent $\bigwedge\Phi_i(\bar{c})$ from coming out true, where $\bar{c}$ is some tuple of distinct witnesses. (Every tuple of elements will be named by such a tuple $\bar{c}$; cf. Lemma 2.3.1.) All she need do is play a condition $p \cup \{\neg\phi(\bar{c})\}$ for some formula $\phi(\bar{x}) \in \Phi_i$. If no such set $p \cup \{\neg\phi(\bar{c})\}$ is a condition, then $T \cup p \vdash \phi(\bar{c})$ for each formula $\phi(\bar{x})$ in Φ_i. But then it follows by the lemma on constants (Lemma 1.2.2) that Φ_i has a support; contradiction. □

For complete types the theorem can be strengthened:

THEOREM 5.1.3. *Let* T *be a theory in a countable first-order language* L. *Then there is a family of continuum many models of* T, $(A_\alpha : \alpha < 2^\omega)$, *such that for all* $\alpha < \beta < 2^\omega$, *if a complete type* Φ *is realised both in* A_α *and in* A_β, *then* Φ *is principal.*

Proof. As Theorems 4.1.4 and 4.1.5, but using first-order conditions. □

Virtually everything in section 4.2 transfers intact to the new form of forcing, and some things become simpler. Thus for example let L be a countable first-order language and T a theory in L. In strict analogy with section 4.2, a model A of T is said to be atomic iff every tuple of elements of A realises a principal complete type. Inspection shows that this property depends entirely on A; T is irrelevant. So we can speak simply of atomic structures.

LEMMA 5.1.4. *Let* A *and* B *be finite or countable atomic structures. Then* $A \cong B$ *if and only if* $A \equiv B$.

Proof. As Lemma 4.2.4. □

We say that principal types are dense (for T) iff for each $n < \omega$ and each formula $\phi(x_0,\ldots,x_{n-1})$ of L, if $T \cup \{\exists\bar{x}\phi\}$ has a model, then there is a principal complete type $\Phi(\bar{x})$ of T which contains ϕ.

LEMMA 5.1.5. *Suppose that principal types are dense for* T. *Then the property "atomic model of* T*" is enforceable.*

Proof. Cf. Lemma 4.2.5. □

What corresponds to the joint embedding property (cf. section 3.2)? Assuming that T is a first-order theory with models, the following are equivalent: (i) T is equivalent to a theory of form Th(A), or in other words, all models of T are elementarily equivalent; (ii) if A and B are any models of T then there is a model C of T in which both A and B are elementarily embeddable. (This follows e.g. from Frayne's theorem, Corollary 4.3.13 in Chang & Keisler (1973).) A theory with property (i) is said to be complete. Completeness stands to our forcing as the joint embedding property stood to Robinson forcing. For simplicity let me state the dichotomy theorem for complete theories only:

THEOREM 5.1.6. *Let* L *be a countable first-order language and* T *a complete theory in* L. *Suppose players* ∀ *and* ∃ *are constructing a model* A *of* T *by forcing as defined above. Then one of the following two situations occurs:*

- *(a) Principal types are not dense for* T*; and for every enforceable property* P *there is a family of continuum many finite or countable models of* T*, all of which have property* P*, and none of which can be elementarily embedded in any other model in the family.*
- *(b) Principal types are dense for* T*; and* T *has an atomic model* B *such that the property "The compiled structure* A *is isomorphic to* B*" is enforceable.*

Proof. As Theorem 4.2.6. □

For some theories, (b) in the theorem holds in an extreme form: every countable model of T is atomic. A well-known theorem of Engeler, Ryll-Nardzewski and Svenonius describes this situation. Recall that T is ω-categorical (or $\aleph_0$-categorical) iff it has up to isomorphism just one countable model.

THEOREM 5.1.7. *Let* L *be a countable first-order language and* T *a theory in* L. *Assume that* T *is complete and has countable models.*

Then the following are equivalent:

(a) T *is* ω*-categorical.*

(b) *Every countable model of* T *is atomic.*

(c) *For each* $n < \omega$*, the set* S_n *of complete types is finite.*

(d) *For some countable model* B *of* T *and every* $n < \omega$*, the number of orbits of the automorphism group of* B*, acting on the set of* n*-element subsets of* dom(B)*, is finite.*

Proof. (a) implies (b) by Theorem 5.1.6. (b) implies (c); for if some set S_n was infinite, by compactness it would contain a non-principal type, and this type would be realised in some countable model of T.

Assume (c), and let A, B be countable models of T. Define I to be the set of all pairs $(\bar{a},\bar{b})$ such that $tp_A(\bar{a}) = tp_B(\bar{b})$. Using (c) it is easy to show that I is a back-and-forth system from A to B (cf. section 4.2). By Lemma 4.2.2 this implies (a). By the same lemma it also implies (d), if we put A = B and consider the two structures $(A,\bar{a})$ and $(A,\bar{b})$ with $(\bar{a},\bar{b}) \in I$.

I leave it to the reader to verify that (d) implies (c). □

Let us move on to some of the other theorems mentioned in the chapter introduction. The theorems in question are all deducible from Theorem 5.1.2 above, but I think we learn more if we make a fresh approach.

As before, L is a countable first-order language. By an infinitary rule for L we shall mean an ordered pair $\langle\Phi(\bar{x}),\theta(\bar{x})\rangle$ where $\bar{x}$ is a tuple of variables, $\Phi(\bar{x})$ is a set of formulas of L and $\theta(\bar{x})$ is a formula of L. The idea will be that Φ implies θ. We can formalise this idea either syntactically or semantically, and the content of Theorem 5.1.8 below is that these two formalisations agree.

To begin with syntax, consider any natural deduction calculus for L. An infinitary rule $\langle\Phi(\bar{x}),\theta(\bar{x})\rangle$ can be read as a natural deduction rule with perhaps infinitely many premises:

(1) $$\frac{\{\phi(\bar{t}) : \phi(\bar{x}) \in \Phi\}}{\theta(\bar{t})} \quad \text{for any tuple } \bar{t} \text{ of terms.}$$

Let Ω be an at most countable set of infinitary rules for L. We write

$T \vdash_\Omega \chi$ to mean that χ is deducible from T in the natural deduction calculus with added rules (1) corresponding to the infinitary rules in Ω. We say that a theory T is Ω-consistent iff $T \nvdash_\Omega \exists x\ x \neq x$.

Turning to semantics, let A be an L-structure and Ω an at most countable set of infinitary rules for L. We shall say that A is an Ω-structure iff for every tuple $\bar{a}$ of elements of A and every infinitary rule $\langle\Phi,\theta\rangle$ in Ω, if $A \models \phi(\bar{a})$ for every $\phi(\bar{x})$ in Φ, then $A \models \theta(\bar{a})$. An Ω-model of a theory T is a model of T which is also an Ω-structure.

EXAMPLE. Suppose the symbols of L include a 1-ary relation symbol Z and distinct constants n_i $(i < \omega)$. Take Φ to be the set of all formulas $x \neq n_i$ $(i < \omega)$, and θ to be the formula $\neg Zx$; write Ω for $\{\langle\Phi,\theta\rangle\}$. Then an Ω-structure is just an L-structure A such that every element of Z^A is of form n_i^A for some $i < \omega$.

In particular suppose T is a theory in L which contains some form of arithmetic, so that Zx is read as "x is a number", and each constant n_i stands for the number i. We assume that T implies Zn_i for each $i < \omega$. For such a theory T, Ω-models are generally known as ω-models. In fact the rule $\langle\Phi,\theta\rangle$ defined above is one form of the ω-rule of arithmetic, which was first introduced by Hilbert (1931). For this particular Ω, we shall write $\vdash_\Omega$ as $\vdash_\omega$, and ω-consistent will mean Ω-consistent. (Gödel's (1931) ω-consistency was a weaker notion.) The next theorem was originally proved with ω-models of arithmetic in mind, though people quickly saw that it had wider applications.

THEOREM 5.1.8 (Henkin-Orey theorem). *Let* L *be a countable first-order language,* T *a theory in* L *and* Ω *an at most countable set of infinitary rules for* L. *Then* T *is* Ω*-consistent if and only if* T *has an* Ω*-model.*

Proof. Right to left follows from the easy fact that if A is an Ω-structure then Th(A) is closed under $\vdash_\Omega$.

For left to right, assume that T is Ω-consistent. Take a countable set W of new constants, to be called witnesses. Just as in first-order logic, the theory T remains Ω-consistent in the language L(W). Define a condition to be a finite set p of sentences of L(W), such that $T \cup p$ is Ω-consistent. Let N be the set of all conditions. From the definitions one can verify that N is a notion of forcing for L(W). It is enforceable that the compiled structure A^+ is a model of T in which every element is named by infinitely many witnesses (cf. Lemmas 2.3.1, 5.1.1(b)).

We assert that the property "A is an Ω-structure" is enforceable. For let $\bar{c}$ be a tuple of distinct witnesses and $\langle\Phi,\theta\rangle$ an infinitary rule in Ω. Let P be the property "if $\phi(\bar{c})$ holds for all $\phi \in \Phi$, then $\theta(\bar{c})$ holds", and let players $\forall$ and $\exists$ play G(P;odds). Suppose player $\forall$ opens with p_0. As usual, there are two cases to consider.

<u>Case 1</u>: for some $\phi \in \Phi$, $p_0 \cup \{\neg\phi(\bar{c})\}$ is a condition. If player $\exists$ plays this condition, then she can guarantee to win (cf. Lemma 2.3.1 again).

<u>Case 2</u>: for all $\phi \in \Phi$, $p_0 \cup \{\neg\phi(\bar{c})\}$ is not a condition. By a familiar proof-theoretic argument it follows that $T \cup p_0 \vdash_\Omega \phi(\bar{c})$ for each $\phi \in \Phi$. Hence by (1), $T \cup p_0 \vdash_\Omega \theta(\bar{c})$. We infer that $T \cup p_0 \cup \{\theta(\bar{c})\}$ is Ω-consistent, since $T \cup p_0$ was Ω-consistent. Thus $p_0 \cup \{\theta(\bar{c})\}$ is a condition, and by playing this condition player $\exists$ puts herself in a position to win.

Either way, player $\exists$ wins G(P;odds). By the conjunction lemma (Lemma 2.3.3(e)), the conjunction of all such properties P is enforceable, and thus it is enforceable that A is an Ω-model of T. □

Don't overlook the case where Ω is empty. In this case Theorem 5.1.8 reduces to the completeness theorem for the natural deduction calculus. It is only a short step to a completeness theorem for $L_{\omega_1\omega}$ (cf. Exercise 10).

From here to the end of the section, let L and Ω be as in the Example before Theorem 5.1.8. We are about to make another comparison between syntax and semantics, but now it will be a question of definability rather than provability. We assume that T is a theory in L and that T contains all the sentences Zn_i $(i < \omega)$. Generously regarding T as a theory of arithmetic, we shall say ω-model etc. for Ω-model etc.

A subset X of ω is said to be ω-<u>representable</u> in T iff there is a formula $\theta(x)$ of L such that for all $i < \omega$,

(2) $\quad i \in X \Rightarrow T \vdash_\omega \theta(n_i)$, and $i \notin X \Rightarrow T \vdash_\omega \neg\theta(n_i)$.

Let us say that X is <u>almost ω-representable</u> in T iff there is a sentence ϕ of L such that $T \cup \{\phi\}$ is ω-consistent and X is ω-representable in $T \cup \{\phi\}$.

Let A be an L-structure. We say that the subset X of ω is <u>internal</u> for A iff there are a formula $\chi(x,\bar{y})$ of L and a tuple $\bar{b}$ from A such that for all elements a of Z^A,

$$(3) \qquad A \models \chi(a,\bar{b}) \iff \text{for some } i \in X,\ a = n_i^A.$$

This is a kind of semantic definability of X, just as (2) was a kind of syntactic definability.

THEOREM 5.1.9. *Let the language* L *and the theory* T *be as above, and let* X *be a subset of* ω. *If* X *is internal for every* ω*-model of* T, *then* X *is almost* ω*-representable in* T.

<u>Proof</u>. We force with the same notion of forcing as in the proof of the Henkin-Orey theorem above. Assuming that X is not almost ω-representable in T, we try to construct an ω-model of T for which X is not internal. By the proof of the Henkin-Orey theorem, the property "A is an ω-model of T in which every element is named by a witness" is enforceable. Hence we need only show that the following property P is enforceable, where $\psi(x,\bar{y})$ is any formula of L and $\bar{d}$ any tuple of distinct witnesses:

$$(4) \qquad \text{For some } i < \omega, \qquad A^+ \models \psi(n_i,\bar{d}) \not\Leftrightarrow i \in X.$$

Let players ∀ and ∃ play G(P;odds); let player ∀ play p_0 as his first move. Again there are two cases.

<u>Case 1</u>: For some $i < \omega$, either $i \in X$ and $p_0 \cup \{\neg\psi(n_i,\bar{d})\}$ is a condition, or $i \notin X$ and $p_0 \cup \{\psi(n_i,\bar{d})\}$ is a condition. Either way it is clear how player ∃ can win.

<u>Case 2</u>: There is no i as in Case 1. Write $\bigwedge p_0$ as $\phi(\bar{b},\bar{d})$ where $\phi(\bar{z},\bar{y})$ is a formula of L and $\bar{b}$ lists the witnesses occurring in p_0 but not in $\bar{d}$. By the case assumption, if $i \in X$ then $T \cup \{\phi(\bar{b},\bar{d})\} \vdash_\omega \psi(n_i,\bar{d})$ and if $i \notin X$ then $T \cup \{\phi(\bar{b},\bar{d})\} \vdash_\omega \neg\psi(n_i,\bar{d})$. Then X is ω-representable in $T \cup \{\exists\bar{y}\bar{z}\phi\}$; take $\theta(n_i)$ in (2) to be $\exists\bar{y}\bar{z}(\phi \wedge \psi(n_i,\bar{y}))$. But by assumption X is not almost ω-representable in T, and so this case can't arise. □

Exercises for 5.1:

1. (a) Show that Theorem 5.1.2 fails for any first-order language L with uncountably many constants. [Let T say: the elements a_n $(n < \omega)$ are pairwise distinct, and the elements b_i $(i < \omega_1)$ are pairwise distinct. Let $\Phi(x)$ say: x is distinct from all the a_n.] (b) Let L be a first-order language of cardinality λ. Assume the theory T in L has a model. Let $\Phi(\bar{x}) = \{\phi_i : i < \lambda\}$ be a set of formulas of L with the property: for every set $\Theta(\bar{x})$ of formulas of L, if $|\Theta| < \lambda$ and $T \cup \Theta$ is satisfiable in some L-structure, then there is $i < \lambda$ such that $T \cup \Theta \cup \{\neg\phi_i\}$ is satisfiable in some L-structure. Show that T has a model which omits Φ.

2. Let (+) be the following strengthening of Theorem 5.1.2: *For* L, T *as in the theorem, let* α *be any ordinal* $< 2^\omega$, *and for each* $i < \alpha$ *let* $\Phi_i(\bar{x})$ *be a set of formulas of* L *(where the tuple* $\bar{x}$ *may vary with* i*), such that no set* Φ_i *has a support. Then* T *has a model which omits all the* Φ_i $(i < \alpha)$. Show that both (+) and its negation are consistent with ZFC $\cup \{2^\omega > \omega_1\}$. [For consistency use Martin's axiom. For the negation, take a model of ZFC in which $2^\omega > \omega_1$ and there is a family $\{f_i : i < \omega_1\}$ of functions $f: \omega \to \omega$ such that for every $g: \omega \to \omega$ there is $i < \omega_1$ such that $\forall m\ g(m) < f_i(m)$. (Cf. Hechler (1974).) Take the language with 1-ary relation symbols P_{mn} $(m, n < \omega)$, and the theory T with axioms $\forall x(P_{mn}(x) \to P_{mn'}(x))$ (for all $m < \omega$ and all $n' < n < \omega$) and $\forall x(P_{mn}(x) \to P_{m'n}(x))$ (for all $m < m' < \omega$ and all $n < \omega$). The type of an element a of a model A is determined by the function $t_a: \omega \to \omega+1$ where $t_a(m) = \sup\{n+1 : A \models P_{mn}(a)\}$. Put $\Phi_m = \{P_{rs}(x) : r > m \text{ and } s < \omega\}$, $\Psi_i = \{\neg P_{mn}(x) : f_i(m) = n\}$, and try to omit all the Φ_m and Ψ_i simultaneously.]

3. Let T be a theory in a countable first-order language L, and suppose that for some $n < \omega$, S_n is uncountable. (a) Show that there is a family of continuum many countable models of T, none of which is elementarily embeddable in any other model in the family. [Disregard the resemblance to Theorem 5.1.3. Prerequisites for this exercise are the Halpern-Läuchli-Laver-Pincus theorem (Pincus & Halpern (1981)) and a double brandy.] (b) Show that S_1 can be uncountable even when case (b) holds in Theorem 5.1.6.

A model A *of a first-order theory* T *is said to be* *<u>(elementarily) prime</u>* *iff* A *is elementarily embeddable in every model of* T.

4. Let T be a complete theory in a countable first-order language.

(a) If A is a finite or countable model of T, show that A is atomic if and only if A is an elementarily prime model of T. (b) Show that in Theorem 5.1.6, case (b) holds if and only if T has a prime model.

5. Let L be a first-order language of cardinality ω_1 and let T be a complete theory in L. Show that if principal types are dense for T, then T has an atomic model. [Write L as the union of a continuous chain $(L_i : i < \omega_1)$ of countable first-order languages. Write T_i for the set of all consequences of T in L_i. There is a closed unbounded set $X \subseteq \omega_1$ such that for each $i \in X$, (i) if $\phi(\bar{x})$ is a formula of L_i which doesn't isolate a complete type for T, then ϕ doesn't isolate a complete type for T_i, and (ii) for each formula $\psi(\bar{x})$ of L_i such that $\exists\bar{x}\psi \in T_i$, there is a formula $\phi(\bar{x})$ of L_i which isolates a complete type for T_i containing ψ. List X in increasing order as $i_0, i_1, \ldots$, and inductively construct a sequence $(A_i : i < \omega_1)$ using (ii), as follows: A_0 is a finite or countable atomic model of T_{i_0}. A_1 is a finite or countable atomic model of T_{i_1}, and by Exercise 4(a) there is an elementary embedding $A_0 \to A_1|L_{i_1}$, so we can assume $A_1|L_{i_1}$ is an elementary extension of A_0. Etc. Take A to be the "union" of the sequence. Use (i) to show that A is atomic. It is not known if this result holds for languages of cardinality $> \omega_1$.]

6. Show that in Theorem 5.1.7 the following equivalent clause can be added: *(e) For each* $n < \omega$, *there are only finitely many pairwise non-equivalent (modulo* T*) formulas* $\phi(x_1,\ldots,x_n)$ *of* L. Deduce that if L is a countable language and A a countable L-structure such that Th(A) is ω-categorical, then there is a function f: $\omega \to \omega$ such that for each n and each substructure B of A generated by at most n elements, B has at most f(n) elements; show that f depends only on Th(A).

7. Show that if T is an ω-categorical theory in a countable first-order language, then T has an ω-categorical model-companion. [Cf. the discussion after Theorem 4.2.7 above.]

8. Let L be a countable language and A a countable L-structure such that Th(A) is ω-categorical. Let B be a countable L-structure such that every finitely generated substructure of B is embeddable in A. Show that B is embeddable in A.

9. Let L be a countable first-order language with no function symbols. Let L_1, L_2 be languages got from L by removing some relation symbols from the signature. Let W be a countable set of new constants. Define a

condition to be a finite set p of sentences such that p can be written as $p_1 \cup p_2$ with p_i in $L_i(W)$, in such a way that there is no sentence θ of $L_1(W) \cap L_2(W)$ for which $p_1 \vdash \theta$ and $p_2 \vdash \neg\theta$. Let N be the set of all conditions. Show that N is a notion of forcing for L(W). Deduce that if ϕ, ψ are sentences of L_1, L_2 respectively and $\phi \vdash \psi$, then there is a sentence θ of $L_1 \cap L_2$ such that $\phi \vdash \theta$ and $\theta \vdash \psi$.

10. Prove a completeness theorem for $L_{\omega_1\omega}$ as follows. Take a natural deduction calculus for the first-order language L, and add the rules

$$\frac{\phi_i \ (i \in I)}{\bigwedge\{\phi_i : i \in I\}}, \qquad \frac{\bigwedge\{\phi_i : i \in I\}}{\phi_j} \text{ for each } j \in I$$

(and corresponding rules for $\bigvee$). Let F be a countable fragment of $L_{\omega_1\omega}$ (cf. section 1.2), and W a countable set of new constants; let F(W) be the set of formulas got by allowing substitution of constants from W into formulas in F. Define a notion of forcing for F(W) to be a set N of finite sets p of sentences of F(W), such that N satisfies (1)-(12) of section 2.1 for F(W), (1) and (2) of section 2.3 for F(W), and the further conditions: (i) if $\bigwedge\{\phi_i : i \in I\} \cup p \in N$ then for each $i \in I$, $p \cup \{\phi_i\} \in N$; (ii) if $\neg\bigwedge\{\phi_i : i \in I\} \cup p \in N$ then for some $i \in I$, $p \cup \{\neg\phi_i\} \in N$ (and corresponding clauses for $\bigvee$). (a) Show that if F is a countable fragment of $L_{\omega_1\omega}$, N is a notion of forcing for F(W) and $p \in N$, then p has a model. (b) Let F be a countable fragment of $L_{\omega_1\omega}$, and let N be the set of all finite sets p of sentences of F(W) such that (i) at most finitely many constants from W occur in p and (ii) p is consistent in the natural deduction system described above. Show that N is a notion of forcing for F(W). (c) Show that every sentence of $L_{\omega_1\omega}$ lies in some countable fragment of $L_{\omega_1\omega}$. (d) Show that for every sentence ϕ of $L_{\omega_1\omega}$, ϕ is provable in the natural deduction calculus if and only if ϕ is true in every L-structure. [A pedantic point: the natural deduction calculus may need to use extra variables besides those in the formulas of F.]

11. An L-structure A is said to be weakly atomic-compact iff for every set Φ of atomic formulas of L, with arbitrarily many variables, if each finite subset of Φ is satisfiable in A then the whole of Φ is simultaneously satisfiable in A. Let L be a countable first-order language and let Ω be the set of all infinitary rules $\langle\Phi,\theta\rangle$ of the following form: for some formula $\chi(x,y)$ of L, Φ is the set of all

sentences $\exists x_0 \ldots x_n \bigwedge\{\chi(x_i, x_j) : i < j \leqslant n\}$ $(n < \omega)$, and θ is the sentence $\exists x \chi(x,x)$. Let T be a theory in L. Show that T has a weakly atomic-compact model if and only if T is Ω-consistent.

12. (a) Let L be a recursive first-order language, Ω a uniformly recursive set of infinitary rules for L, and T a Π^1_1 theory in L. Show that the set of sentences ϕ of L such that $T \vdash_\Omega \phi$ is a Π^1_1 set. (b) In the context of Theorem 5.1.9, suppose L is a recursive first-order language (and in particular the map $i \mapsto n_i$ is recursive), and T is a Π^1_1 theory. Show that if $X \subseteq \omega$ is internal for every ω-model of T then X is a hyperarithmetical set.

13. (a) Let L be a countable first-order language and let T, U be theories in L (both of them with models). Let $\Phi(\bar{x})$ be a set of formulas of L. Show that there are a model A of T and a model B of U such that for all $\bar{a}$ in A and $\bar{b}$ in B, if $\bar{a}$ satisfies in A the same formulas of Φ as $\bar{b}$ satisfies in B, then there is some formula $\psi(\bar{x})$ of L such that $T \cup \{\exists\bar{x}\psi\}$ has a model, and for all $\phi \in \Phi$, if $A \models \phi(\bar{a})$ then $T \vdash \forall\bar{x}(\psi \rightarrow \phi)$, while if $A \models \neg\phi(\bar{a})$ then $T \vdash \forall\bar{x}(\psi \rightarrow \neg\phi)$. (b) Let L, T be as in the first sentence of Exercise 12(b). Show that there exist ω-models A, B of T such that for every $X \subseteq \omega$, if X is internal for both A and B then X is hyperarithmetical. (c) Show that if P is any non-empty Σ^1_1 set of functions $f: \omega \rightarrow \omega$, then P contains two functions f and g, both of them recursive in Π^1_1, such that any function hyperarithmetical in f and in g is hyperarithmetical. [Use the proof of (b), taking T to be a theory containing enough second-order arithmetic and the statement that P is non-empty. To show that f and g can be chosen recursive in Π^1_1, one must go back to the construction of the models and show that it can be made recursive in a complete Π^1_1 set.]

5.2 SET-THEORETICAL INTERRUPTION: UNBOUNDED SUBSETS

Most people have their appendices somewhere near their middles, and I wouldn't wish to be thought an exception. But in any case there is a good reason for pausing now to survey some set theory. At this halfway point in our journey we leave the level plains of ω to climb up into the uncountable, and there is a sharp change in the weather.

The problem is not that uncountable cardinals are complicated - they aren't. Rather it is that we are building up uncountable structures by approximations of smaller size, and this involves climbing up to uncountable cardinals through the ordinals below them. There are just too many different paths from ω to ω_1, and it matters which route we take. In a similar context Georg Kreisel has sometimes quoted the cigarette advertisement: *It's not how long you make it, it's how you make it long.*

Putting it at its crudest, climbing up to a limit ordinal α means finding an unbounded subset of α, i.e. a set $X \subseteq \alpha$ such that for every $i < \alpha$ there is $j \in X$ such that $i \leqslant j$. The rest of this section collects together the most useful facts about unbounded subsets. I omit most of the proofs: they can be found in textbooks of set theory. The background assumptions throughout are ZFC, i.e. Zermelo-Fraenkel set theory with choice. Sometimes I make the slightly stronger assumption that ZFC has a model.

Regular cardinals

The cofinality of a limit ordinal α, $cf(\alpha)$, is the smallest ordinal β such that α has an unbounded subset of order-type β. One can show that $cf(\alpha)$ is always a cardinal. Infinite cardinals of the form $cf(\alpha)$ are called regular; in fact every regular cardinal is its own cofinality. Infinite cardinals which are not regular are called singular. Equivalently, an infinite cardinal λ is singular iff λ can be written as $\Sigma_{i<\mu}\kappa_i$ where μ and the κ_i are cardinals less than λ; the smallest μ such that λ can be written in this form is in fact $cf(\lambda)$. (Recall how we form cardinal sums: $\Sigma_{i<\mu}\kappa_i$ is the cardinality of a set which is the disjoint union of sets X_i $(i < \mu)$, where each X_i has cardinality κ_i.)

The cardinal next greater than λ is written λ^+ and called the successor of λ. Cardinals of form λ^+ are called successor cardinals. Every infinite successor cardinal is regular.

For any cardinals λ and κ, we write λ^κ for the number of maps $f: \kappa \to \lambda$. Then 2^κ is equal to the number of subsets of a κ-element set. Also if λ is infinite and $\kappa \leqslant \lambda$, then λ^κ is the number of κ-element subsets of λ. We write $\lambda^{<\kappa}$ for $\Sigma\{\lambda^\mu : \mu < \kappa\}$. We call a cardinal λ strongly inaccessible iff λ is a regular cardinal and $2^\mu < \lambda$ for all $\mu < \lambda$. Thus for example ω is strongly inaccessible.

ZFC has not very much to say about the sizes of cardinals λ^κ. The continuum hypothesis, CH for short, says that $2^\omega = \omega_1$. The generalised continuum hypothesis, GCH, asserts that for every infinite cardinal λ, $2^\lambda = \lambda^+$. The CH and GCH are not deducible from ZFC. But they are consistent with ZFC, in the sense that we shall never reach a contradiction by reasoning from the assumptions ZFC + GCH. (That is, unless ZFC by itself already leads to a contradiction, which I assume it doesn't.)

LEMMA 5.2.1. *Let λ be an infinite cardinal.*

(a) $\lambda^{<\lambda} = \lambda$ iff λ is regular and $2^{<\lambda} = \lambda$.

(b) If λ is strongly inaccessible then $\lambda^{<\lambda} = \lambda$.

(c) If λ is regular and the GCH holds then $\lambda^{<\lambda} = \lambda$.

Proof. Cf. Jech (1978) section 6. □

Closed unbounded sets

Let λ be an uncountable regular cardinal. If X is a subset of λ, a limit point of X below λ is a limit ordinal $\delta < \lambda$ such that $X \cap \delta$ is unbounded in δ. We call X closed iff it contains all its limit points below λ. Subsets of λ which are both closed and unbounded are called clubs; some people call them cubs.

A chain of sets is a sequence $(X_i : i < \gamma)$ of sets such that $X_i \subseteq X_j$ whenever $i \leqslant j < \gamma$. The chain is continuous iff for each limit ordinal $\delta < \gamma$, $X_\delta = \bigcup_{i<\delta} X_i$. Clubs often arise in the following two ways:

LEMMA 5.2.2. *Let λ be an uncountable regular cardinal.*

(a) Suppose n is a positive integer and f is an n-ary function from λ to λ. Let Y be the set of all ordinals $\alpha < \lambda$ such that if $i_1, \ldots, i_n < \alpha$ then $f(i_1,\ldots,i_n) < \alpha$. Then Y is a club in λ.

(b) Suppose $(X_i : i < \lambda)$ is a continuous chain of subsets of λ, each set X_i has cardinality $< \lambda$, and $\lambda = \bigcup_{i<\lambda} X_i$. Let Y be the set of all ordinals $\alpha < \lambda$ such that $X_\alpha = \alpha$. Then Y is a club in λ.

Proof. (a) For each subset X of λ, let X' be $X \cup \{f(i_1,\ldots,i_n) : i_1,\ldots,i_n \in X\}$, and let $\alpha(X)$ be the least ordinal α such that $X' \subseteq \alpha$. Since λ is regular, if X has cardinality $< \lambda$ then $\alpha(X) < \lambda$. Now for any ordinal $\beta < \lambda$, define inductively $\beta_0 = \beta$, $\beta_{n+1} = \alpha(\beta_n)$ and $\beta_\omega = \bigcup_{n<\omega} \beta_n$. Since λ has uncountable cofinality, $\beta_\omega < \beta$. Clearly then $\beta_\omega \in Y$ and $\beta \leqslant \beta_\omega$. This shows that Y is unbounded in λ. I leave it to the reader to check that Y is closed. The proof of (b) is similar. □

For example, taking $f(i) = i+1$ in (a), we deduce that the set of limit ordinals $< \lambda$ is a club in λ. A more trivial example of a club in λ is the set of all ordinals strictly between α and λ, where α is a fixed ordinal $< \lambda$.

The next lemma describes some ways of getting new clubs from old ones. Suppose that for each $i < \lambda$, Y_i is a club in λ. Then we write $\Delta_{i<\lambda} Y_i$ for the set $\{\alpha : \text{for all } i < \alpha, \alpha \in Y_i\}$; this set is called the <u>diagonal intersection</u> of the sets Y_i.

LEMMA 5.2.3. *Let λ be an uncountable regular cardinal.*

(a) Suppose μ is a cardinal $< \lambda$, and for each $i < \mu$, Y_i is a club in λ. Then the intersection $\bigcap_{i<\mu} Y_i$ is a club in λ.

(b) Suppose that for each $i < \lambda$, Y_i is a club in λ. Then $\Delta_{i<\lambda} Y_i$ is a club in λ.

<u>Proof</u>. Cf. Jech (1978) section 7. □

<u>Stationary sets</u>

Let λ be an uncountable regular cardinal. We set up a kind of measure on the subsets of λ. A set $X \subseteq \lambda$ will be called <u>fat</u> iff X contains some club. (Here and below, "club" means club in λ.) The set X will be called <u>thin</u> iff $\lambda \setminus X$ is fat. By Lemma 5.2.3(a) the fat sets form a λ-complete filter on the boolean algebra of subsets of λ. Dually, the thin sets form a λ-complete ideal on this algebra. No set is both fat and thin. The filter of fat sets is commonly called the <u>club filter</u>.

A subset S of λ which is not thin is said to be <u>stationary</u>. In other words, S is stationary iff S intersects every club. It follows that every stationary set is unbounded in λ.

By Lemma 5.2.3(a), every club intersects every club, so that all clubs are stationary. But there are plenty of stationary sets which are not clubs. For example:

LEMMA 5.2.4. *Let λ be an uncountable regular cardinal.*

(a) If S is a stationary subset of λ and Y is a club, then $S \cap Y$ is stationary.

(b) For every regular cardinal $\mu < \lambda$, the set of all ordinals $< \lambda$ of cofinality μ is stationary.

(c) Every stationary subset of λ can be partitioned into

λ pairwise disjoint stationary subsets of λ.

Proof. (c) is in Jech (1978) section 35; (a) and (b) are exercises. □

Lemma 5.2.3(b) can be rephrased in terms of stationary sets. The result is a very useful proposition known as Fődor's lemma, which deserves to be better known among mathematicians in general. (See the index for its uses in this book. Group theorists might like its application by Faber (1978) in a proof that generalised soluble groups contain large abelian subgroups.)

A function f is said to be regressive iff $f(i) < i$ for every ordinal i in the domain of f.

LEMMA 5.2.5 (Fődor's Lemma). *Let λ be an uncountable regular cardinal, S a stationary subset of λ and $f: S \to \lambda$ a regressive function. Then there is a stationary subset U of λ, $U \subseteq S$, such that f is constant on U.*

Proof. Cf. Jech (1978) section 7. □

Diamonds

Once again let λ be a regular uncountable cardinal. The principle known as $\diamondsuit_\lambda$ is a kind of workshop manual for constructions of length λ. It lists the tasks we shall meet during the construction, and it tells us when we should tackle each task. The reader will think I am joking: how could one workshop manual serve for all possible tasks of length λ? There is no joke, but the catch is that the manual will generally make a lot of mistakes. It will be right a stationary number of times, and for many constructions this is good enough.

$\diamondsuit_\lambda$ is not deducible from ZFC, but it is consistent with ZFC (in the same sense as the GCH was consistent with ZFC).

To state the principle precisely: $\diamondsuit_\lambda$ says that there is a family $(S_i : i < \lambda)$ of sets, $S_i \subseteq i$, such that for every set $X \subseteq \lambda$,

(1) the set $\{i < \lambda : X \cap i = S_i\}$ is stationary.

The family $(S_i : i < \lambda)$ is sometimes known as the diamond trap. It "traps" every set $X \subseteq \lambda$. We write $\diamondsuit$ for $\diamondsuit_{\omega_1}$.

For an illustration I borrow a result from Exercise 17 of section 6.1 below. If A is a group and X a set of elements of A, then we write $C_A(X)$ for the set of all elements of A which commute with every element of X. By the exercise just mentioned, if B is a countable e.c. group and for each $n < \omega$, X_n is a set of elements of B which is not contained in any finitely generated subgroup of B, then B has a proper extension A which is a countable e.c. group back-and-forth equivalent to B, and such that for each $n < \omega$, $C_B(X_n) = C_A(X_n)$ and X_n is not contained in any finitely generated subgroup of A.

THEOREM 5.2.6. *Assume* $\lozenge$. *Then for every countable e.c. group* B *there is an e.c. group* A *of cardinality* ω_1 *such that* $B \subseteq A$, B *is back-and-forth equivalent to* A, *and* A *contains no uncountable abelian subgroup.*

Proof. By $\lozenge$ there is a family $(S_i : i < \omega_1)$ of sets such that each S_i is a subset of i and (1) holds for every set $X \subseteq \omega_1$.

We shall construct a strictly increasing continuous chain $(A_i : i < \omega_1)$ of countable e.c. groups, by induction on i. The required group A will be the union of the chain. As the construction proceeds, we shall define for each group A_i a family F_i of at most countably many subsets of A_i, none of which are contained in any finitely generated subgroups of A_i. The sets F_i will form a chain. Also we identify the elements of A_0 with the ordinals $< \omega$, the elements of A_1 with the ordinals $< \omega+\omega$, and so on until eventually the elements of A are precisely the ordinals $< \omega_1$.

To start the construction we put $A_0 = B$ and $F_0 = \emptyset$. At limit ordinals δ we put $A_\delta = \bigcup_{i<\delta} A_i$. When A_i and F_i have been chosen, we use the result quoted above from section 6.1 to choose a proper extension A_{i+1} of A_i which is a countable e.c. group back-and-forth equivalent to A_i, such that for all sets $X \in F_i$, $C_{A_i}(X) = C_{A_{i+1}}(X)$ and X is not contained in any finitely generated subgroup of A_{i+1}. Then we put $F_{i+1} = F_i$.

It remains to choose F_δ when δ is a limit ordinal. Here we examine S_δ as provided by the diamond trap. If S_δ is a subset of A_δ which is not contained in any finitely generated subgroup of A_δ, then we put $F_\delta = \{S_\delta\} \cup \bigcup_{i<\delta} F_i$. Otherwise we put $F_\delta = \bigcup_{i<\delta} F_i$.

Let A be the union of the groups A_i $(i < \omega_1)$. Then A is an e.c. group, extends B and is back-and-forth equivalent to B. (For the survival of the necessary properties at limit ordinals, cf. Exercise 4(a) of section 3.2 and the Example after Lemma 4.2.3.) It remains to show that A has no uncountable abelian subgroups.

Suppose H is an abelian subgroup of A. Let U be the set of ordinals $i < \omega_1$ such that $H \cap A_i$ lies inside some finitely generated subgroup of A_i. Let C be the set of ordinals $i < \omega_1$ such that i is the set of elements of A_i; by Lemma 5.2.2(b), C is a club in ω_1. Now we consider two cases.

<u>Case 1</u>: U is stationary. Then by Lemma 5.2.4(a), $U \cap C$ is stationary. Moreover by the construction, every element of C is a limit ordinal. Hence for each $i \in U \cap C$ there is some ordinal $h(i) < i$ such that $H \cap A_i$ is generated by a finite set of elements $\leqslant h(i)$. The function h is regressive and defined on a stationary set; so by Födor's lemma (Lemma 5.2.5) there is a stationary set $V \subseteq U \cap C$ such that h is constant on V. Let α be the constant value of h on V, and choose β so that $\alpha \in A_\beta$. Since V is unbounded in ω_1, it follows that $H \cap A_i \subseteq A_\beta$ for arbitrarily large $i < \omega_1$, and so $H \subseteq A_\beta$. Thus in this case H is countable.

<u>Case 2</u>: U is not stationary. Then there is some club Z which is disjoint from U. By Lemma 5.2.3(a), $C \cap Z$ is a club. By $\diamondsuit$ there is a stationary set of ordinals i such that $H \cap i = S_i$, and so there is such an ordinal i in $C \cap Z$. Since $i \in C$, $H \cap A_i = S_i$; since $i \in Z$, $H \cap A_i$ is not in any finitely generated subgroup of A_i. So $S_i \in F_i$, and hence by construction there were no elements added after A_i which commute with the whole of S_i. But H is abelian and contains S_i. It follows that $H \subseteq A_i$, so that again H is countable. □

As mentioned above, the principles $\diamondsuit_\lambda$ don't follow from ZFC but they are consistent with it. More precisely:

LEMMA 5.2.7. *(a) For each uncountable cardinal λ, if $\diamondsuit_\lambda$ holds then $\lambda^{<\lambda} = \lambda$. So if $\diamondsuit_\lambda$ holds for every regular uncountable cardinal λ, then the GCH holds.*

(b) If the GCH holds, then $\diamondsuit_{\lambda^+}$ holds for every regular uncountable cardinal λ.

(c) In Gödel's constructible universe, $\diamondsuit_\lambda$ holds for every regular uncountable cardinal λ.

<u>Proof</u>. (a) is an exercise. (b) is from Gregory (1976). For (c) cf. Devlin (1984). □

Exercises for 5.2.

1. Let G and H be groups of cardinality λ, where λ is an uncountable regular cardinal. Let the set of elements of G (resp. H) be $\{g_i : i < \lambda\}$ (resp. $\{h_i : i < \lambda\}$). Show: (a) The set $\{\alpha < \lambda : \{g_i : i < \alpha\}$ forms a subgroup of G$\}$ is a club in λ. (b) If f: G $\to$ H is an isomorphism, then the set $\{\alpha < \lambda : f$ maps $\{g_i : i < \alpha\}$ onto $\{h_i : i < \alpha\}\}$ is a club in λ.

2. Let λ be a regular uncountable cardinal and C a club in λ. Show that the set of limit points of C below λ is also a club in λ.

3. (Δ-system lemma) Let λ be an infinite cardinal and X a set of finite sets with $|X| = \lambda^+$. Show that there exist a set $Y \subseteq X$ and a finite set Δ such that $|Y| = \lambda^+$, and for all x, y $\in$ Y, if $x \neq y$ then $x \cap y = \Delta$. [Assume $\bigcup X = \lambda^+$. List X as $\{x_i : i < \lambda^+\}$. Let S be the set of limit ordinals $< \lambda^+$ of cofinality ω, and for each $i \in S$ let f(i) be $\max\{\alpha : \alpha \in x_i$ and $\alpha < i\}$ (= 0 if the set is empty). Use Födor's lemma on f.]

4. Let $(X_i : i < \omega_2)$ be a family of subsets of ω_2 such that each X_i is at most countable. Show that there is a set $Y \subseteq \omega_2$ of cardinality ω_2 such that if $i,j \in Y$ and $i \neq j$ then $i \notin X_j$. [Cf. the proof of Exercise 3.]

5. Prove (a), (b) in Lemma 5.2.4.

6. Prove (a) in Lemma 5.2.7. [If $\mu < \lambda$ then the diamond trap predicts each subset of μ somewhere before λ.]

7. Let λ be a regular uncountable cardinal and assume $\Diamond_\lambda$. Show that there is a family of maps $((f_i: i \to i) : i < \lambda)$ such that for each function g: $\lambda \to \lambda$ the set $\{i < \lambda : g|i = f_i\}$ is stationary.

The next exercise uses some notions in the model theory of arithmetic.

8. Assume $\Diamond$ and the following theorem of Kaufmann: For each countable recursively saturated model B of first-order Peano arithmetic (PA) and each set X of elements of B, if X is not definable (i.e. first-order definable with parameters) in B, then there is a countable recursively saturated proper elementary end-extension A of B such that for every set Y definable in A, $Y \cap \mathrm{dom}(B) \neq X$. Deduce that there exists a recursively saturated model C of PA which has cardinality ω_1, such that every proper initial segment of C is countable, and for every set $X \subseteq \mathrm{dom}(C)$, if for each c in C, $X \cap \{b : b <^C c\}$ is coded by an element of C, then X is definable in C.

The <u>*Mahlo operation*</u> M *on classes of cardinals is defined by:* M(X) *is the class of all uncountable regular cardinals* λ *such that* $X \cap \lambda$ *is stationary in* λ. *If* I *is the class of strongly inaccessible cardinals, we define a cardinal to be* n-<u>*inaccessible*</u> *iff it is in* $M^n(I)$. *Thus* 0-*inaccessible means strongly inaccessible.* 1-*inaccessible cardinals are also called* <u>*Mahlo cardinals*</u>.

9. Show: (a) Every (n+1)-inaccessible cardinal is n-inaccessible. (b) If λ is a Mahlo cardinal then λ is the λ-th strongly inaccessible cardinal. (c) Let λ be an uncountable cardinal and $n < \omega$. Let K be the set of (n+2)-element subsets of λ. Show that λ is n-inaccessible if and only if for every family $(C_i : i < \lambda)$ of partitions of K, if each C_i has cardinality $< \lambda$ then there is a set $X \subseteq \lambda$ of size n+5 such that for all $i \in X$, the set of all (n+2)-element subsets of $X \setminus \{i+1\}$ lies in one class of the partition C_i.

5.3 *SATURATION*

This section will try to adapt forcing to languages and structures of uncountable cardinality - not always very successfully.

Let me begin in strict analogy with section 3.2. We consider a first-order language L and a class K of L-structures. Let A be a structure in K and λ an infinite cardinal. We say that A is <u>λ-existentially compact</u> in K iff:

(1) For every set $\Phi(\bar{x},\bar{y})$ of quantifier-free formulas of L with $|\Phi| < \lambda$, and every sequence $\bar{a}$ of elements of A, if there is a structure B in K such that $B \supseteq A$ and $B \models \exists\bar{y}\bigwedge\Phi(\bar{a},\bar{y})$, then already $A \models \exists\bar{y}\bigwedge\Phi(\bar{a},\bar{y})$.

Note that $\bar{x}$ and $\bar{y}$ are not required to be finite sequences, but since λ is infinite and $|\Phi| < \lambda$, we can assume that they both have length less than λ. Note also that A is e.c. in K if and only if A is ω-existentially compact in K. In the light of this I abbreviate λ-existentially compact to λ-<u>e.c.</u>, with an easy conscience.

It is immediate from the definition that if $\omega \leq \mu < \lambda$ and A is λ-e.c. in K then A is μ-e.c. in K.

EXAMPLE 1. Let K be the class of fields and λ an uncountable cardinal. If A is λ-e.c. in K, then A is an e.c. field and hence it is algebraically closed. Also we can use Φ in (1) to say that for any sequence $\bar{a}$ of fewer than λ elements of A, there is an element distinct from all elements in $\bar{a}$. Thus A is an algebraically closed field of cardinality at least λ. Conversely it is not hard to show that every algebraically closed field of cardinality at least λ is λ-e.c. in K.

EXAMPLE 2. Let K be the class of groups and λ an uncountable cardinal. Let A be λ-e.c. in K. Then A is an e.c. group. If g is an element of infinite order in A and X is a subset of ω, then by Fact 3.3.3 there is a group $B \supseteq A$ containing elements a, b such that

$$\text{(2)} \qquad \text{for all } n < \omega: \quad n \in X \;\Leftrightarrow\; g^n = [[a,b^{2n+1}],a].$$

We can write (2) as $B \models \bigwedge\Phi(g,a,b)$ for a certain countable set Φ of quantifier-free formulas. Since A is λ-e.c., it follows that A already contains elements a, b satisfying (2). But X was any subset of ω, and hence A must have cardinality at least 2^ω. Also since we can encode into A both first-order arithmetic and arbitrary subsets of ω, it follows that complete second-order arithmetic is 1-1 reducible to Th(A) (cf. Exercise 14 for the verification of this).

As in section 3.2, we say that the class K is <u>inductive</u> iff for every chain of structures in K, the union of the chain is also in K. By the lemma on preservation (Lemma 1.2.4(c)), the class of all models of a $\forall_2$ first-order theory is inductive.

THEOREM 5.3.1. *Let* L *be a first-order language,* K *an inductive class of* L*-structures and* λ *an infinite cardinal. Then for every structure* A *in* K *there is a* λ*-e.c. structure in* K *which extends* A.

<u>Proof</u>. Just as Theorem 3.2.1, except that now one must list sets Φ_i of formulas rather than single formulas ϕ_i. □

For large λ, being λ-e.c. is rather a strong requirement. One should expect that the various λ-e.c. models of a theory are not too different from each other. We shall turn this expectation into some theorems, using the back-and-forth machinery of section 4.2. It is technically convenient to start with a slightly stronger notion than λ-e.c., using the idea of a type over a sequence of elements.

Let L be a first-order language, T a theory in L, A a model of T and $\bar{a}$ a sequence of elements of A. We define an <u>∃-type</u> (of T) <u>over</u> $\bar{a}$ <u>in</u> A to be a set $\Phi(\bar{a},\bar{x})$ of $\exists_1$ formulas such that (i) $\Phi(\bar{y},\bar{x})$ is a set of formulas of L and $\bar{x}$ is a tuple of variables (so that $\Phi(\bar{a},\bar{x})$ lies in $L(\bar{a})$, the language got from L by adding $\bar{a}$ as parameters), and (ii) for every finite subset $\Psi(\bar{y},\bar{x})$ of Φ, $T \cup \text{diag}(A) \cup \{\exists\bar{x}\bigwedge\Psi(\bar{a},\bar{x})\}$ has a model. We say that a tuple $\bar{b}$ from A <u>realises</u> $\Phi(\bar{a},\bar{x})$ iff $A \models \bigwedge\Phi(\bar{a},\bar{b})$. If $\Phi(\bar{a},\bar{x})$ is not realised by any tuple in A, we say that A <u>omits</u> $\Phi(\bar{a},\bar{x})$.

In terms of these notions we can paraphrase λ-existential compactness as follows:

LEMMA 5.3.2. *Let* L *be a first-order language,* T *a theory in* L*,* K *the class of all models of* T*,* A *a model of* T *and* λ *an infinite cardinal. Then the following are equivalent:*

(a) A *is* λ*-e.c. in* K.

(b) For every ∃*-type* $\Phi(\bar{a},\bar{x})$ *of* T *in* A*, if* $|\Phi| < \lambda$ *then* $\Phi(\bar{a},\bar{x})$ *is realised in* A.

(c) For every ∃*-type* $\Phi(\bar{a},x)$ *of* T *in* A *with just one variable* x *free, if* $|\Phi| < \lambda$ *then* $\Phi(\bar{a},x)$ *is realised in* A.

<u>Proof</u>. Exercise. For (c) ⇒ (a), extend the sequence $\bar{a}$ one element at a time. □

Now let L be a first-order language, T a theory in L, A a model of T and λ a cardinal. We say that A is λ-<u>existentially saturated</u> for T, or more briefly a λ-<u>e.s.</u> model of T, iff:

(3) For every sequence $\bar{a}$ of fewer than λ elements of A, all ∃-types $\Phi(\bar{a},x)$ of T in A are realised in A.

From the definition and Lemma 5.3.2 it is immediate that if λ is infinite and A is a λ-e.s. model of T then A is a λ-e.c. model of T too. Also if $|L| < \lambda$ and A is λ-e.c. then A is λ-e.s. So the two notions disagree only when the language has cardinality at least λ.

In section 4.2 we introduced the relations $\equiv_0$, $\Rrightarrow_1$ and $\equiv_1$

between structures. Adding another to the list, we write $A \preccurlyeq_{\infty\omega} B$ to mean that $A \subseteq B$ and for every tuple $\bar{a}$ from A, $(A,\bar{a})$ is back-and-forth equivalent to $(B,\bar{a})$. By Karp's theorem (Lemma 4.2.3) this is equivalent to saying that $A \subseteq B$ and for every formula $\phi(\bar{x})$ of $L_{\infty\omega}$ with finitely many free variables and every tuple $\bar{a}$ from A, $A \models \phi(\bar{a})$ iff $B \models \phi(\bar{a})$. So it implies that B is an elementary extension of A.

THEOREM 5.3.3. *Let* L *be a first-order language,* T *a* $\forall_2$ *theory in* L, *and let* A, B *be* ω-*e.s. models of* T.

(a) *If* $\bar{a}$, $\bar{b}$ *are tuples in* A, B *respectively and* $(A,\bar{a}) \Rrightarrow_1 (B,\bar{b})$ *then* $(A,\bar{a})$ *is back-and-forth equivalent to* $(B,\bar{b})$.

(b) *If* $A \subseteq B$ *then* $A \preccurlyeq_{\infty\omega} B$.

Proof. (a) Since A is an ω-e.s. model of T, A is ω-e.c. and hence an existentially closed model of T. So by Lemma 4.1.2, $\exists\text{-tp}_A(\bar{a})$ is a maximal element of $S_n^{\exists}$, where n is the length of $\bar{a}$. Assuming $(A,\bar{a}) \Rrightarrow_1 (B,\bar{b})$, it follows that $(A,\bar{a}) \equiv_1 (B,\bar{b})$.

Define I to be the set of all pairs of tuples $(\bar{c},\bar{d})$ such that $(A,\bar{a},\bar{c}) \equiv_1 (B,\bar{b},\bar{d})$. We claim that I is a back-and-forth system from $(A,\bar{a})$ to $(B,\bar{b})$, as defined by (5)-(8) in section 4.2. Clause (5) there said that if $(\bar{c},\bar{d}) \in I$ then $(A,\bar{a},\bar{c}) \equiv_0 (B,\bar{b},\bar{d})$, while clause (6) said that $(\emptyset,\emptyset) \in I$. Both of these are clearly true. By symmetry it remains only to prove:

(4) If $(\bar{c},\bar{d}) \in I$ and c' is any element of A, then there is an element d' of B such that $(\bar{c}c',\bar{d}d') \in I$.

Let $\Phi(\bar{y},\bar{z},x)$ be the set of all $\exists_1$ formulas $\phi(\bar{y},\bar{z},x)$ of L such that $A \models \phi(\bar{a},\bar{c},c')$. If Ψ is any finite subset of Φ, then clearly $A \models \exists x \bigwedge\Psi(\bar{a},\bar{c},x)$. Since $(A,\bar{a},\bar{c}) \equiv_1 (B,\bar{b},\bar{d})$, it follows that $B \models \exists x \bigwedge\Psi(\bar{b},\bar{d},x)$. Hence $\Phi(\bar{b},\bar{d},x)$ is an $\exists$-type of T in B. Since $\bar{b}$ and $\bar{d}$ are finite and B is ω-e.s., it follows that $B \models \bigwedge\Phi(\bar{b},\bar{d},d')$ for some element d' of B. We deduce that $(A,\bar{a},\bar{c},c') \Rrightarrow_1 (B,\bar{b},\bar{d},d')$, and another application of Lemma 4.1.2 allows us to strengthen $\Rrightarrow_1$ to $\equiv_1$. This proves (4), and so I is a back-and-forth system from $(A,\bar{a})$ to $(B,\bar{b})$ as claimed.

(b) follows from (a) and the lemma on preservation (Lemma 1.2.4(b)). □

If A and B in the theorem are λ-e.s. for some larger cardinal λ, then we get stronger conclusions; cf. Exercise 3.

Clause (a) in the theorem has the following consequence. Let L be a first-order language, T a theory in L and $\phi(\bar{x})$ a formula of L. Then there is a set RES(ϕ) of maximal $\exists$-types of T such that for every ω-e.s. model A of T and every tuple $\bar{a}$ from A,

(5) $\quad A \models \phi(\bar{a}) \quad$ iff $\quad$ for some $\Phi \in \mathrm{RES}(\phi)$, $A \models \bigwedge\Phi(\bar{a})$.

The sets RES(ϕ) look like a kind of analogue of the resultants of section 3.1, though to the best of my knowledge nobody has yet succeeded in making them yield useful algebraic information. Abraham Robinson discovered them in the course of investigating forcing with infinite conditions. His studies led him to the class of structures described by the next theorem.

THEOREM 5.3.4. *Let* L *be a first-order language,* T *a* $\forall_2$ *theory in* L *and* A *an* L*-structure. Then the following are equivalent:*

(a) A *is an elementary substructure of an* ω*-e.s. model of* T.

(b) A *is an e.c. model of* T, *and for every formula* $\phi(\bar{x})$ *of* L *and every* $\Phi(\bar{x}) \in \mathrm{RES}(\phi)$, $A \models \forall\bar{x}(\bigwedge\Phi \rightarrow \phi)$.

(c) For every formula $\phi(\bar{x})$ *of* L, $A \models \forall\bar{x}(\phi \leftrightarrow \bigvee_{\Phi\in\mathrm{RES}(\phi)} \bigwedge\Phi)$.

<u>Proof</u>. (a) $\Rightarrow$ (b): An elementary substructure of an e.c. model of T is again an e.c. model of T, cf. Exercise 4 of section 3.2.

(b) $\Rightarrow$ (c): Suppose (b) holds but (c) fails. Then for some formula $\phi(\bar{x})$ of L and some tuple $\bar{a}$ from A, $A \models \phi(\bar{a})$ but $A \models \neg\bigwedge\Phi(\bar{a})$ for all $\Phi \in \mathrm{RES}(\phi)$. Let Ψ be the $\exists$-type of $\bar{a}$ in A. Since A is an e.c. model of T, Lemma 4.1.2 tells us that Ψ is a maximal $\exists$-type. Hence if we extend A to an ω-e.s. model B of T (as we can by Theorem 5.3.1), Ψ will still be the $\exists$-type of $\bar{a}$ in B. It follows that Ψ must be in either RES(ϕ) or RES($\neg\phi$). We have seen that it is not in RES(ϕ); so it must be in RES($\neg\phi$), and hence by (b), $A \models \neg\phi(\bar{a})$. Contradiction.

(c) $\Rightarrow$ (a): Assume (c), and use Theorem 5.3.1 to extend A to an ω-e.s. model B of T. We shall show that $A \preccurlyeq B$. Let $\phi(\bar{x})$ be any formula of L and $\bar{a}$ a tuple from A. If $A \models \phi(\bar{a})$, then by (c) there is some $\Phi \in \mathrm{RES}(\phi)$ such that $A \models \bigwedge\Phi(\bar{a})$. Since Φ consists of $\exists_1$ formulas, the lemma on preservation (Lemma 1.2.4(b)) tells us that $B \models \bigwedge\Phi(\bar{a})$. By (5) it follows that $B \models \phi(\bar{a})$, as required. □

An L-structure A which satisfies the equivalent conditions of Theorem 5.3.4 is called an infinite-generic model of T. The set of all sentences of L which are true in every infinite-generic model of T is called the infinite forcing companion of T, in symbols T^F.

Let me list a few easy facts about infinite-generic models and T^F. First, the proof of (c) ⇒ (a) in the theorem above shows in fact that if A is an infinite-generic model of T, B is an ω-e.s. model of T and $A \subseteq B$, then $A \preccurlyeq B$. It follows that if A and C are infinite-generic models of T and $A \subseteq C$ then $A \preccurlyeq C$. For by Theorem 5.3.1 we can extend C to an ω-e.s. model B, and then $A \preccurlyeq B$ and $C \preccurlyeq B$, from which it follows that $A \preccurlyeq C$.

Next, suppose T has the joint embedding property. Then for any two ω-e.s. models A, B of T there is a model C of T into which both A and B are embeddable, and by Theorem 5.3.1 we can assume that C is ω-e.s. too. Then $A \preccurlyeq C$ and $B \preccurlyeq C$, so that $A \equiv B$. It follows that if T has the joint embedding property then T^F is a complete theory. (Cf. Theorem 3.4.7 for the analogous fact about T^f.)

Third, T^F may depend on properties of the set-theoretic universe. For example we saw above that if A is an ω_1-e.c. group then second-order arithmetic is reducible to Th(A). Since the theory T of groups has the joint embedding property, and ω_1-e.c. groups are ω-e.s., it follows that second-order arithmetic is reducible to T^F. The reducing function is set-theoretically absolute - it is an explicitly given recursive map on formulas (cf. Exercise 14). But in second-order arithmetic we can write down a sentence which is true if and only if the continuum hypothesis holds. (By contrast the finite forcing companion T^f of the theory of groups is set-theoretically absolute.)

Abraham Robinson (1971 c) maintained that the infinite-generic models of a theory have "a strong claim to be regarded as a proper explication of the notion of algebraic closedness". I report this without having any idea what he meant. To my eye the infinite-generics are simply a well-behaved class within the class of e.c. models. One should add that they belong at the "fat" end of the e.c. models. (Contrast Exercise 12 below with Theorem 4.3.4.)

As mentioned earlier, Robinson himself produced infinite-generic models by a kind of forcing called infinite forcing. Exercises

18 and 19 give a few details. But at this point I prefer to turn to the full first-order analogue of Robinson's infinite forcing. A version of it appears in some powerful unpublished work of Shelah.

For the rest of this section, λ is a fixed uncountable cardinal. We aim to build structures of cardinality λ by compiling chains $(p_i)_{i<\lambda}$ of conditions. Each condition should have cardinality less than λ. To avoid going over the top at limit ordinals below λ, we shall assume that λ is regular.

Let L be a first-order language of cardinality at most λ. We take a set W consisting of λ new constants, and we form L(W) by adding the constants in W to L. These added constants are called the witnesses.

Let T be a theory in L; we always assume T has models. We define a condition to be a set p of fewer than λ sentences of L(W) such that $T \cup p$ has a model. We write N for the set of conditions.

LEMMA 5.3.5. N *is a notion of forcing for* L(W), *and* N *has the properties (6)-(8) of section 2.3. The union of a chain of fewer than* λ *members of* N *is again a member of* N.

Proof. The case-by-case verification is left to the reader. The second sentence, which gives (13) of section 2.1, follows from the compactness theorem and the regularity of λ. □

The construction of a model by forcing will now take λ steps, and this implies some recasting of the definitions in section 2.3. If $\bar{p}$ is a chain $(p_i)_{i<\lambda}$ of conditions, then we write $p_{<i}$ for $\bigcup_{j<i} p_j$. Thus $p_{<i+1}$ is p_i and $p_{<0}$ is the empty set. We write $\bigcup \bar{p}$ for $\bigcup_{i<\lambda} p_i$.

There is a least =-closed set U in L(W) which contains all the atomic sentences in $\bigcup \bar{p}$; we write $A^+(\bar{p})$ for the canonical model of U, and $A(\bar{p})$ for the L-reduct of $A^+(\bar{p})$. As before, we abbreviate these to A^+ and A, and we refer to either as the compiled structure.

Let X be a subset of λ such that

(6) X and $\lambda \setminus X$ are both unbounded subsets of λ, and $0 \notin X$.

Let P be a property. Players $\forall$ and $\exists$ play the game G(P;X) as one would expect. There are λ steps which are used to build up a chain $(p_i)_{i<\lambda}$ of conditions, one condition at a time. At the i-th step, if $i \in X$ then player $\exists$ decides what p_i shall be, while player $\forall$ chooses p_i if $i \notin X$. At the end of the game, player $\exists$ wins iff $\bigcup \bar{p}$ (or the compiled structure $A^+(\bar{p})$ if P is a property of structures) has property P. We say that the property P is enforceable iff for every set X satisfying (6), player $\exists$ has a winning strategy for G(P;X).

We say that a condition q forces a property P iff for every game G(P;X) where X is as in (6), if the players are about to choose p_i and $q \subseteq p_{<i}$, then player $\exists$ is in winning position.

If ϕ is a sentence and P is the property "$A^+(\bar{p})$ is a model of ϕ", then we write G(ϕ;X) for G(P;X). We say that ϕ is enforceable (or that q forces ϕ) to mean that P is enforceable (or that q forces P).

We shall refer to the forcing which has just been defined as λ-forcing. It is surely the most immediate generalisation of the forcing of section 5.1 to cardinality λ. One would reasonably expect λ-forcing to behave like the earlier finite forcing. But it doesn't. Three important differences come to light at once.

(A) There are sets X, Y both satisfying (6), and a property P such that player $\exists$ has a winning strategy for G(P;X) but not for G(P;Y). I give an example at Exercise 20 below. The nub of the matter is that a player who commands the limit-ordinal steps gains an advantage over his rival.

Hence it is no longer true that all "standard" games G(P;-), for a fixed property P, are equivalent. (Contrast the discussion before Lemma 2.3.1; I am taking (6) to define "standard".) This is just a nuisance. We get round it in Lemma 5.3.6(a) below by characterising enforceability in terms of a "standard" game G(P;X) which is as hard as possible for player $\exists$ to win; if she can win this she can win them all.

We define an ordinal α to be odd iff α is of form $\delta + n$ where n is an odd natural number and δ is either 0 or a limit ordinal. We write odds for the set of odd ordinals $< \lambda$. Ordinals which are not odd are even.

LEMMA 5.3.6. *Consider λ-forcing.*

(a) For any property P, P *is enforceable iff player $\exists$ has a winning strategy for* G(P;odds).

(b) For any property P *and condition* q, q *forces* P *iff in each play of* G(P;odds), *if player* $\forall$ *has just chosen* p_0 *so that* $q \subseteq p_0$, *then player* $\exists$ *is in winning position.*

(c) (Conjunction lemma) *If* P *is the conjunction of properties* P_i $(i < \lambda)$, *and* q *forces each* P_i, *then* q *forces* P.

(d) If q *forces* P *and* $p \supseteq q$ *then* p *forces* P.

(e) If for every $q \supseteq p$ *there is* $r \supseteq q$ *such that* r *forces* P, *then* p *forces* P.

Proof. (a) Left to right is because the odd ordinals satisfy (6). For right to left, let X be a set as in (6) and let σ be a winning strategy for player $\exists$ in G(P;odds). She should play G(P;X) as follows. First she should regard any block of her own moves as one single move repeated. Second, if she has to choose p_i and player $\forall$ has a block of moves immediately before the choice of p_i, then she should play as if player $\forall$ has just made one move, viz. $p_{<i}$. Third, if δ is a limit ordinal in X, then she should play as if player $\forall$ had a move immediately before the δ-th step, and he chose $p_{<\delta}$. In this adjusted game the total number of steps is still λ since λ is regular. Thus she reduces G(P;X) to the form G(P;odds), and she can use her strategy σ to win.

Quoting (a) where necessary, parts (b)-(e) can be proved like their analogues in section 2.3. □

(B) The number of sequences of witnesses of length less than λ is $\lambda^{<\lambda}$, and this number may be greater than λ. Some arguments in earlier sections rested on the fact that there were just ω n-tuples of witnesses, so that by the conjunction lemma, if it was enforceable that a particular n-tuple of witnesses had a property, it would be enforceable that all n-tuples had this same property. After ω has been raised to λ, these arguments only survive if we assume that $\lambda^{<\lambda} = \lambda$. To illustrate the difference, here first is a lemma which doesn't need this cardinality assumption.

LEMMA 5.3.7. *Consider* λ*-forcing with a theory* T.

(a) The following property is enforceable: "The compiled structure A^+ *is a model of* $T \cup \bigcup \bar{p}$, *and each element of* A^+ *is of form* c^{A^+} *for* λ *different witnesses* c".

(b) Every condition forces every logically true sentence.

(c) If a condition q *forces a sentence* ϕ, *and* $\phi \vdash \psi$, *then* q *forces* ψ.

Proof. (a) is proved just as Lemma 2.3.1, and (b), (c) are trivial. □

By way of contrast, here is a result which does need the cardinality assumption. We determine which conditions force which sentences of $L(W)_{\lambda^+\lambda}$. (Recall that for a first-order language L, $L_{\lambda^+\lambda}$ is the infinitary language got from L by allowing (i) conjunctions and disjunctions of sets of at most λ formulas, and (ii) quantifiers $\forall\bar{x}$, $\exists\bar{x}$ where $\bar{x}$ is any sequence of variables of length $< \lambda$.)

THEOREM 5.3.8. *Assume* $\lambda^{<\lambda} = \lambda$ *and consider* λ*-forcing with a theory* T *in a first-order language* L *of cardinality at most* λ. *Let* p *be a condition and* ϕ *a sentence of* $L(W)_{\lambda^+\lambda}$ *in which fewer than* λ *witnesses occur.*

(a) *Suppose* ϕ *is atomic, or more generally that* ϕ *is of form* $\forall\bar{x}\bigvee\Phi$ *where* $\Phi(\bar{x})$ *is a set of formulas of* L(W). *Then* p *forces* ϕ *iff* $T \cup p \vdash \phi$.

(b) *If* ϕ *is of form* $\bigwedge_{i<\lambda}\phi_i$, *then* p *forces* ϕ *iff* p *forces every* ϕ_i $(i < \lambda)$.

(c) *If* ϕ *is of form* $\forall\bar{x}\psi$, *and* $\bar{d}$ *is a sequence of distinct witnesses not occurring in* p *or in* ψ, *then* p *forces* ϕ *iff* p *forces* $\psi(\bar{d})$.

(d) p *forces* $\neg\phi$ *iff there is no condition* $q \supseteq p$ *which forces* ϕ.

Proof. The proofs are as Theorem 2.3.4, Exercise 2 of section 2.3, and Theorem 3.4.2. The proof of right to left in (c) requires that the number of sequences of witnesses of the same length as $\bar{x}$ is at most λ. □

(C) We can restrict ourselves without loss to conditions which are the elementary diagrams of structures. (In the case of Robinson forcing one should say diagrams rather than elementary diagrams. As a matter of fact Robinson himself used structures as conditions in his infinite forcing; cf. Exercise 19.)

Since the language L may have cardinality λ, one has to be a little careful here. By expressing the signature of L as the union of a continuous chain of signatures, each of which has cardinality $< \lambda$, we can decompose L into a continuous chain $(L_i)_{i<\lambda}$ of languages, so that $L = \bigcup_{i<\lambda} L_i$ and each L_i is a first-order language of cardinality $< \lambda$. We list the witnesses as a sequence $\bar{c} = (c_i : i < \lambda)$, so that $\bar{c}|\alpha$ means $(c_i : i < \alpha)$.

We say that a condition p is <u>full</u> iff for some ordinal $i < \lambda$, p is the elementary diagram of an L_i-structure, using the witnesses $\bar{c}|i$ to name the elements. This is equivalent to saying that p is a maximal consistent theory in $L_i(\bar{c}|i)$, and for any sentence $\exists x\psi(x)$ in p there is some $j < i$ such that $\psi(c_j) \in p$.

LEMMA 5.3.9. *Consider λ-forcing for an uncountable regular cardinal λ.*

(a) Every condition can be extended to a full condition.

(b) Suppose P *is a property, and write* $G_{full}(P;odds)$ *for the game which is like* G(P;odds) *except that both players are required to play only full conditions. Then player* ∃ *has a winning strategy for* $G_{full}(P;odds)$ *iff she has a winning strategy for* G(P;odds).

<u>Proof</u>. An exercise. Part (b) is proved by the usual device of making player ∃ play the other game in private, and use the results to guide her choices in the public game. □

Finally to put the string round the parcel, we define the classical first-order analogues of λ-e.c. models, and we show that λ-forcing is a way of getting them.

Let L be a first-order language, A an L-structure and $\bar{a}$ a sequence of elements of A. A <u>type over</u> $\bar{a}$ <u>in</u> A is a set of form $\Phi(\bar{a},\bar{x})$ where $\Phi(\bar{y},\bar{x})$ is a set of formulas of L, $\bar{x}$ is a tuple of variables and for every finite subset Ψ of Φ, $A \models \exists\bar{x} \bigwedge\Psi(\bar{a},\bar{x})$. We say the type $\Phi(\bar{a},\bar{x})$ is <u>realised by</u> $\bar{b}$ <u>in</u> A iff $A \models \bigwedge\Phi(\bar{a},\bar{b})$. We call $\bar{a}$ the <u>parameters</u> of the type.

We say that A is λ-<u>compact</u> iff every type of cardinality $< \lambda$ over parameters in A is realised in A. We say that A is λ-<u>saturated</u> iff every type over fewer than λ parameters in A is realised in A.

In both definitions it makes no difference if we restrict to types in just one free variable x; cf. Lemma 5.3.2. Also just as in the existential case, λ-saturated implies λ-compact, and the converse holds when $\lambda > |L|$.

Observe that <u>every</u> structure is ω-compact, so that Theorem 5.3.10 below is no improvement on Lemma 5.3.7(a) unless λ is uncountable.

THEOREM 5.3.10. *Suppose* $\lambda^{<\lambda} = \lambda$. *Let* L *be a first-order language of cardinality at most* λ, *and* T *a theory in* L. *Then the property "The compiled structure* A *is a* λ*-compact model of* T*" is enforceable (in* λ*-forcing with* T*).*

<u>Proof</u>. The proof that follows is not the most direct, but since we have Theorem 5.3.8 we may as well use it.

Because $\lambda^{<\lambda} = \lambda$, there are just λ pairs $\langle\Phi,\bar{d}\rangle$ consisting of a set $\Phi(\bar{y},x)$ of formulas of L with $|\Phi| < \lambda$ and a sequence $\bar{d}$ of fewer than λ witnesses. By the conjunction lemma (Lemma 5.3.6(c)) and Lemma 5.3.7(a), it suffices to show that for each such pair $\langle\Phi,\bar{d}\rangle$ the following is enforceable:

(7) Either $A^+(\bar{p}) \models \exists x \bigwedge\Phi(\bar{d},x)$, or for some finite $\Psi \subseteq \Phi$, $A^+(\bar{p}) \models \neg\exists x \bigwedge\Psi(\bar{d},x)$.

If P is property (7), let players $\forall$ and $\exists$ play G(P;odds), and let player $\forall$ open with p_0. There are two cases.

<u>Case 1</u>: $T \cup p_0 \nvdash \forall x \neg \bigwedge\Phi(\bar{d},x)$. Then by Lemma 5.3.7(c) and Theorem 5.3.8(a), p_0 doesn't force $\neg\exists x \bigwedge\Phi(\bar{d},x)$. So by Theorem 5.3.8(d) there is a condition $p_1 \supseteq p_0$ which does force $\exists x \bigwedge\Phi(\bar{d},x)$. Player $\exists$ can reply with p_1 and be sure of winning.

<u>Case 2</u>: $T \cup p_0 \vdash \forall x \neg \bigwedge\Phi(\bar{d},x)$. Then by the lemma on constants (Lemma 1.2.2) and the compactness theorem there is a finite subset Ψ of Φ such that $T \cup p_0 \vdash \neg\exists x \bigwedge\Psi(\bar{d},x)$. So player $\exists$ wins by playing so that $A^+ \models T \cup p_0$, as she can do by Lemma 5.3.7(a). □

Exercises for 5.3.

1. Let L be a first-order language, λ a cardinal $\geqslant |L|$ and T a $\forall_2$ theory in L. If A is a model of T of cardinality $\leqslant 2^\lambda$, show that A can be extended to a λ^+-e.s. model of T of cardinality $\leqslant 2^\lambda$.

2. (a) Prove Lemma 5.3.2. (b) State and prove the analogous lemma with "λ-saturated" in place of "λ-e.c. in K". (c) Why is it necessary to mention the theory T or class K in the definitions of λ-e.c. and λ-e.s., but not in the definitions of λ-compact and λ-saturated?

The next two exercises are dedicated to those readers who agree with Bertrand Russell that "It is a principle, in all formal reasoning, to generalise to the utmost". In these exercises L *is a first-order language,* K *a class of* L*-structures which is closed under isomorphic copies, and* λ *is an infinite cardinal. A structure* A *in* K *is said to be* λ-*__e.s.__ iff for every set* $\Phi(\bar{y},x)$ *of* $\exists_1$ *formulas of* L *with* $\bar{y}$ *of length* $< \lambda$, *and every sequence* $\bar{a}$ *of elements of* A, *if there is a structure* B *in* K *such that* $A \subseteq B$ *and* $B \models \exists x \bigwedge \Phi(\bar{a},x)$, *then already* $A \models \exists x \bigwedge \Phi(\bar{a},x)$.

3. Assume (i) K is inductive, (ii) $\lambda^{<\lambda} = \lambda > |L|$, (iii) for every structure A in K and set $X \subseteq \mathrm{dom}(A)$, if $|X| \leqslant \lambda$ then there is $B \subseteq A$, $X \subseteq \mathrm{dom}(B)$, $|B| \leqslant \lambda$, $B \in K$. Show that every structure in K of cardinality $\leqslant \lambda$ can be extended to a structure which is λ-e.s. in K and has cardinality $\leqslant \lambda$.

4. Assume that K has the following property:

(*) If A, $B \in K$ and $\bar{a}$, $\bar{b}$ are sequences of elements of A, B respectively, both of length $< \lambda$, such that $(A,\bar{a}) \Rrightarrow_1 (B,\bar{b})$, then for every element c in A there exist $B' \supseteq B$, $B' \in K$ and d in B' such that $(A,\bar{a},c) \Rrightarrow_1 (B',\bar{b},d)$.

Show: (a) If A is λ-e.s. in K, B is a structure in K of cardinality $\leqslant \lambda$ and $B \Rrightarrow_1 A$, then B is embeddable in A. (b) If A and B are λ-e.s. in K, both A and B have cardinality $\leqslant \lambda$, and $A \equiv_1 B$, then $A \cong B$. [If $\bar{b}$ has length $< \lambda$ and $(B,\bar{b}) \Rrightarrow_1 (A,\bar{a})$, argue as in (a) to embed B in A and use the fact that B is λ-e.s. to deduce $(B,\bar{b}) \equiv_1 (A,\bar{a})$.] (c) If A is λ-e.s. in K and has cardinality $\leqslant \lambda$, and $\bar{a}$, $\bar{b}$ are sequences of length $< \lambda$ in A such that $(A,\bar{a}) \equiv_1 (A,\bar{b})$, then there is an automorphism f of A such that $f\bar{a} = \bar{b}$.

5. Let K be the class of all models of a first-order theory T. (a) Show that K has property (*) of Exercise 4. (b) Show that a structure A is λ-e.s. in K in the sense described before Exercise 3 if and only if A is a λ-e.s. model of T.

A model A *of a theory* T *is said to be* <u>*homogeneous-universal*</u> *iff (i)* (homogeneous) *if* $B \subseteq A$, $B \models T$ *and* $|B| < |A|$ *then every embedding of* B *into* A *extends to an automorphism of* A, *and (ii)* (universal) *every model of* T *of cardinality* $\leqslant |A|$ *is embeddable in* A.

6. Let L be a first-order language, λ a cardinal $> |L|$ and T a $\forall_2$ theory in L. Show: (a) If T has a homogeneous-universal model of cardinality λ then T has the amalgamation and joint embedding properties (cf. section 3.2). (b) If T has the amalgamation and joint embedding properties, then a model of T of cardinality λ is homogeneous-universal if and only if it is λ-e.s.

7. Let L be a first-order language, T a $\forall_2$ theory in L, and K the class of infinite-generic models of T. Show that K is the unique class of L-structures such that (i) every model of T can be extended to a structure $\in$ K, (ii) every structure in K is a model of T, (iii) if A, B $\in$ K and $A \subseteq B$ then $A \preccurlyeq B$, and (iv) if B $\in$ K and $A \preccurlyeq B$ then A $\in$ K.

8. Show that the union of a chain of infinite-generic models of T is again an infinite-generic model of T.

9. Let L be a countable first-order language and T a $\forall_2$ theory in L. (a) Show that the class of ω-e.s. models of T is axiomatisable by a set of sentences of $L_{\omega_1\omega}$. (b) Show that the class of infinite-generic models of T is axiomatisable by a set of sentences of $L_{\omega_1\omega}$.

10. Let L be a first-order language, T a $\forall_2$ theory in L. Show that $T_\forall = (T^F)_\forall \subseteq T^F = (T_\forall)^F = (T^F)^F$.

11. For a theory T as in Exercise 10, show that the following are equivalent: (a) T is companionable. (b) T^F is a model-companion of T. (c) Every model of T^F is infinite-generic.

12. Let L be a first-order language and T a $\forall_2$ theory in L which has the joint embedding property. Show that if ϕ is an $\exists_3$ sentence of L and $A \models \phi$ for some e.c. model A of T, then $B \models \phi$ for every infinite-generic model B of T. [Use Corollary 3.2.5. Cf. Theorem 4.3.4 for a kind of dual.]

13. Let R be a ring with 1 and let T be the theory of left R-modules. Show that every e.c. model of T is infinite-generic. [Cf. Exercise 8 of section 4.3.]

14. (a) Improve Exercise 8 of section 3.3 by showing that for every sentence ϕ of second-order arithmetic (allowing quantification over sets

of natural numbers), there is a first-order sentence ϕ^* of the language of groups such that for every ω-e.s. group G, $G \models \phi^*$ iff ϕ is true in the natural numbers. Show that the map $\phi \mapsto \phi^*$ can be chosen recursive. [For an arbitrary element g of infinite order, represent the number n by g^n. Represent a set $X \subseteq \omega$ by a pair of elements a, b as in (2).] (b) Show that the class of infinite-generic groups is not axiomatisable by a sentence of $L_{\omega_1\omega}$, where L is the language of groups. [By a theorem of Lopez-Escobar (cf. chapter 10 of Keisler (1971)) it suffices to find a formula $\phi(x,y,\bar{z})$ of L such that (i) for every infinite-generic group G and any tuple $\bar{g}$ from G, $\{(a,b) : G \models \phi(a,b,\bar{g})\}$ is a well-ordering, and (ii) for every countable ordinal α there are an infinite-generic group G and $\bar{g}$ in G such that the well-ordering in (i) has length $> \alpha$. Use part (a).]

15. Show that for every $k \geq 2$, Exercise 14 works also for the class of nilpotent groups of class k.

16. Let L be a countable first-order language and T a $\forall_2$ theory in L with the joint embedding property. Show that if T has a countable ω-e.s. model then T has an $\exists$-atomic model. [Cf. Exercises 3, 5 of section 4.2.]

17. Complete the proof of Lemma 5.3.9.

18. Define <u>Robinson</u> λ-<u>forcing</u> as λ-forcing, but with the following changes: T is a $\forall_2$ theory in L, and a <u>condition</u> is a set p of fewer than λ basic sentences of L(W) such that $T \cup p$ has a model. (a) Show that if is a regular cardinal $\geq 2^{|L|}$, then it is enforceable that the compiled structure is an ω-e.s. model of T. (b) Show that if $2^{|L|} \leq \lambda \leq \mu$ and λ, μ are regular cardinals, p is a condition for Robinson λ-forcing and ϕ is a sentence of L(W), then the following are equivalent: (i) p forces ϕ (in Robinson λ-forcing), (ii) for every μ-e.s. model B of T, if the witnesses occurring in p and ϕ are read as names of any elements of B so that $B \models \bigwedge p$, then $B \models \phi$. (So the cardinal λ is essentially irrelevant.)

19. Let L be a first-order language, T a $\forall_2$ theory in L and K the class of all substructures of models of T. For all B in K and all sentences ϕ of L with parameters added for elements of B, we define: (i) If ϕ is atomic then $B \Vdash \phi$ iff $B \models \phi$; (ii) $B \Vdash \phi \wedge \psi$ iff $B \Vdash \phi$ and $B \Vdash \psi$; (iii) $B \Vdash \forall x\psi(x)$ iff for all $C \in K$ with $B \subseteq C$ and all elements c of C, $C \Vdash \psi(c)$; (iv) $B \Vdash \neg\phi$ iff there is no $C \in K$, $C \supseteq B$, such that $C \Vdash \phi$. (Treat $\vee$, $\rightarrow$, $\leftrightarrow$, $\exists$ as defined symbols.) Show: (a) If $B \in K$, ϕ is a sentence of L with parameters for elements of B, and λ is a regular

cardinal $\geqslant 2^{|L|}$, then $B \Vdash \phi$ iff diag(B) forces ϕ in Robinson λ-forcing (with the parameters used as witnesses). (b) B is an infinite-generic model of T iff for every sentence ϕ with parameters from B, $B \models \phi \Leftrightarrow B \Vdash \phi$.

20. Let λ be a regular uncountable cardinal and L the first-order language whose signature consists of λ constants b_i $(i < \lambda)$ and one 1-ary relation symbol R. Let T be the empty theory. Let P be the property "$\{i < \lambda : A \models R(b_i)\}$ is a club in λ". Let X be the set of all limit ordinals $< \lambda$, and Y the set of all successor ordinals $< \lambda$. Show that player $\exists$ has a winning strategy for G(P;X) in Robinson λ-forcing, but not for G(P;Y).

21. Let λ be a regular cardinal, A and B λ-saturated structures, and $\bar{a}$, $\bar{b}$ sequences of length $< \lambda$ from A, B respectively, such that $(A,\bar{a}) \equiv (B,\bar{b})$. Show that for every formula $\phi(\bar{x})$ of $L_{\infty\lambda}$, $A \models \phi(\bar{a})$ iff $B \models \phi(\bar{b})$. [Combine Exercise 4 above with a suitable generalisation of Karp's theorem, Lemma 4.2.3.]

22. Let λ be a regular cardinal and L a first-order language of cardinality $\leqslant \lambda$. Let B be a λ-compact L-structure of cardinality λ, and let T be eldiag(B) (in some language L(B) of cardinality λ). In λ-forcing with T, show that it is enforceable that the compiled structure A is isomorphic to B.

23. Let λ be a regular cardinal, and let L_1, L_2 be first-order languages of cardinality $\leqslant \lambda$ such that $L_3 = L_1 \cap L_2$ has cardinality $< \lambda$. Let T_1, T_2 be theories in L_1, L_2 respectively such that $T_1 \cap T_2$ is a complete theory in L_3, and each of T_1, T_2 has a model. Consider games G_i of length λ for L_i-structures, with conditions for λ-forcing defined in terms of T_i, for $i = 1,2$. Let P_i be a property of L_i-structures, for $i = 1,2$. Suppose player $\exists$ has a winning strategy σ for $G_2(P_2;\text{odds})$. Show that player $\exists$ can play $G_1(P_1;\text{odds})$ in such a way that if $\bar{p}^1$ is the resulting play, then there is a play $\bar{p}^2$ of $G_2(P_2;\text{odds})$ in which player $\exists$ uses σ, such that $L_3(W) \cap \bigcup \bar{p}^1 = L_3(W) \cap \bigcup \bar{p}^2$.

REFERENCES FOR CHAPTER 5

5.1: Keisler (1977) contains an excellent survey of the topics of this section. The Henkin-Orey theorem (Theorem 5.1.8) was proved in Henkin (1957) and Orey (1956). The omitting types theorem followed soon afterwards, in Engeler (1959), Ryll-Nardzewski (1959) and

Svenonius (1959); all three authors omitted a single type, and used this to characterise ω-categorical theories. In (1961) Vaught observed that one can omit countably many non-principal types at once, proved Lemma 5.1.4 and Exercise 4, and showed that if principal types are dense for a countable theory T then T has atomic models. Theorem 5.1.9 is from Grzegorczyk et al. (1961), which is a very direct extension of the topological completeness proof given by Rasiowa & Sikorski (1950). Theorem 5.1.3 is from Shelah (1978 a) Theorem IV.5.16.

Exercise 2 answers a question of Henryk Kotlarski, who gave this application of Martin's axiom (unpublished). Exercise 3(a): Shelah (1978 a) section VIII.1. Knight (1978) proves a version of Exercise 5 and attributes it to Harrington and Shelah; cf. Shelah (1975 a; 1983 b). Exercise 7: Pouzet (1972 b) and Saracino (1973). Exercise 8: Macpherson (1983). Exercise 9: this model-theoretic proof of Craig's interpolation theorem is due to Henkin (1963) and Smullyan (1963). Exercise 10: Karp (1964). Makkai (1969) combined the ideas of Smullyan (1963) and Exercise 10 to prove various preservation and interpolation theorems for $L_{\omega_1\omega}$; see also Keisler (1971). Exercise 11: Fuhrken & Taylor (1971). Exercise 12: Gandy et al. (1960). Exercise 13: Grilliot (1972).

5.2: The principles $\Diamond_\lambda$ were discovered by Ronald Jensen in about 1969, and Lemma 5.2.7(c) is due to him; cf. Devlin (1984). One of the deepest incursions of diamonds into algebra is their use by Shelah (1975 b) to show that in the constructible universe all Whitehead groups are free. Lemma 5.2.7(b) is strengthened in Shelah (1980 a).

Theorem 5.2.6 is from Macintyre (1976). Shelah (1977) showed that with a different proof, not using forcing, one can weaken the assumption $\Diamond$ to $2^\omega = \omega_1$. (The crucial idea in Shelah's approach is to define each A_{i+1} as the union of a countable chain $(A^n_{i+1} : n < \omega)$ of subgroups in such a way that each $A_i \cap A^n_{i+1}$ lies within a finitely generated subgroup of A_i. This has the effect that if a subset X of A_i is not in any finitely generated subgroup of A_i, then it is not in any finitely generated subgroup of A_{i+1} either.) Shelah (198- b) extends Macintyre's results in other directions using $\Diamond$. Macintyre's paper contains analogous results for locally finite groups and for skew fields. For locally finite groups Hickin (1978) proved a stronger theorem from ZFC alone.

Macintyre's result is by no means the only one which was

proved first from $\diamondsuit$ and then from a weaker assumption. Shelah in particular has made a habit of "eliminating diamonds", to quote the title of his (1983 d). See section 6.3 below for an example in boolean algebras where the assumption $\diamondsuit$ was weakened to $2^\omega = \omega_1$, and Shelah (1978 d) for remarks on removing $\diamondsuit$ altogether.

Exercise 3: Erdős & Rado (1960). Exercise 4: cf. Hajnal (1961) for a stronger result. Exercise 8: Kaufmann (1977), cf. the references to section 6.2. Exercise 9(c): Schmerl (1974).

5.3: The shovel-it-all-in idea behind λ-saturated structures goes back to Fraissé (1953) (cf. Exercise 12 of section 4.2) and Jónsson (1956; 1960). Morley & Vaught (1962) discovered λ-saturated structures as an application of Jónsson's results. Robinson's (1971 b) discovery of infinite forcing and infinite-generics (cf. Exercise 19 for a version of his definition of forcing) seems to have been independent of these papers. In 1970 Ed Fisher pointed out that both infinite-generic models and Jónsson's homogeneous-universal structures can be seen as parts of the theory of λ-e.s. structures; in spirit at least, Theorems 5.3.3, 5.3.4 and Exercises 3-6 are from Fisher (1970). Wood (1972) treats infinite-generics from a similar point of view; cf. Cherlin (1972; 1976) for yet another way of reaching the infinite-generics, and Hirschfeld & Wheeler (1975) for a general study of them. I should mention that Robinson's school referred to the ω-e.s. models as existentially universal models. (Speaking of terminology, a saturated structure is a structure A which is $|A|$-saturated. I had no occasion to discuss them; but cf. Exercise 4(b).) Shelah's λ-forcing appears in Shelah (1983 d; 198- c).

Russell's opinion (not mine) about generalisation is in Chapter 18 of Russell (1919). Exercises 3, 4 generalise Jónsson (1956; 1960). Shelah (198- a) loosens up Jónsson's conditions in a different way; cf. Makowsky (198-) for an exposition. Exercise 6(b): Boffa (1972 a) gave a weak form of the implication from left to right. Exercise 7: Robinson (1971 c). Exercises 8, 10: Robinson (1971 b). Exercise 9: Wood (1972). Exercise 11: Eklof & Sabbagh (1970). Exercise 13: Sabbagh (1971). Exercise 14(a) is due to Macintyre (unpublished), cf. Ziegler (1980); (b) is from Macintyre (1975). Exercise 15: cf. Hodges (1980). Exercise 16: Simmons (1975 b). Exercise 23 is adapted from Robinson's proof of his consistency lemma, cf. Chang & Keisler (1973) section 3.1.

6 PROPER EXTENSIONS

The cistern contains, the fountain overflows.

William Blake, Proverbs of Hell (c. 1790)

So far we have started with a theory T and built a model A of T. What happens if we start instead with a countable structure B and build an extension A of B?

According to the diagram lemma (Lemma 1.2.3), an extension of B is essentially the same thing as a model of diag(B). So one's first instinct is to replace T by diag(B) and then proceed as before. Unfortunately it is often enforceable that A = B (cf. Exercise 5(b) of section 3.4), and so we make no progress.

In this chapter I shall describe some rather more sophisticated machinery, which does three things. First, it makes A "the same sort of structure" as B - for example if B is a group, it will be enforceable that A is a group. Second, A will always contain elements not in B (in other words, it will be a proper extension of B). We ensure this by introducing a new constant ∞, the overflow constant, and then legislating that anything we can say about the element named ∞ is already true of a "large number" of elements of B. The notion of a "large number" has to obey certain axioms, but otherwise we can choose it to suit the problem. And third, if only a "small number" of elements of B have some property, then A will have no new elements with that property. Small stays small.

The machinery looks ad hoc, but this and the next two chapters will show that it has many applications. Section 6.2 below applies it to models of arithmetic, and section 6.3 to boolean algebras.

6.1 *LARGENESS PROPERTIES*

Let us fix some notation. L is a countable first-order language and B is an L-structure of cardinality ω. We define L' to be the first-order language got by adding to L the following constants: (i) the elements of B, (ii) a countable set W of new constants, known as the <u>witnesses</u>, and (iii) the <u>overflow constant</u> ∞. We write L(B) for L with only the constants (i) added.

A <u>largeness property</u> (over B) is a set Q of formulas $\phi(x)$ of L(B) with the following four properties. To express these properties we write $Qx\phi$ for "$\phi(x) \in Q$". The axioms should be thought of as being true in B:

(Axiom 1) $\forall x(\phi \to \psi) \to (Qx\phi \to Qx\psi)$.

(Axiom 2) $Qx(\phi \vee \psi) \to Qx\phi \vee Qx\psi$.

(Axiom 3) $\neg Qx\ x=b$ (for each element b of B).

(Axiom 4) $Qx\ x=x$.

More generally let Φ be a set of formulas $\phi(x)$ of L(B). We say Q is a <u>largeness property for</u> Φ iff Q is as above, but with ϕ, ψ in Axioms 1 and 2 restricted to come from Φ.

We shall always understand $Qy\phi(y)$ to mean the same thing as $Qx\phi(x)$. Although this seems a triviality, it should be stated as another axiom, since it is not deducible from Axioms 1-4:

(Axiom 0) $Qx\phi(x) \leftrightarrow Qy\phi(y)$ (for any two variables x, y).

Most of our examples will also satisfy another axiom, Axiom 5, which will be introduced later in this section.

EXAMPLE 1. Let $Qx\phi$ mean "There are infinitely many elements x such that $\phi(x)$". Then Q is a largeness property. Axiom 1 says that every set containing an infinite set is infinite. Axiom 2 says that if an infinite set is the union of two other sets, then at least one of these other sets is infinite. Axiom 3 says that no singleton set is infinite. Axiom 4 says that B has infinitely many elements. (Recall that we assumed B has cardinality ω.)

EXAMPLE 2. Let A be a proper elementary extension of B, and let $Qx\phi$ mean "There is an element of A which satisfies $\phi(x)$ and is not in B". Then Q is a largeness property over B.

EXAMPLE 3. Let B be a group, and let $Qx\phi$ mean "There is no finitely generated subgroup G of B such that every element satisfying $\phi(x)$ is in G". Axioms 1-3 are clearly true. To make Axiom 4 true we must assume that B is not finitely generated.

EXAMPLE 4. Let B be an infinite ordered group and let $Qx\phi$ mean "There is no y such that for all x, $\phi(x)$ implies $-y \leqslant x \leqslant y$". Then Q is a largeness property over B. In this example $Qx\phi$ happens to be expressible by a first-order formula.

Until further notice, let Q be a largeness property over B <u>for the set of $\exists_1$ first-order formulas</u>. As the examples above show, many largeness properties are actually largeness properties for all first-order formulas. We shall get slightly sharper results by restricting to $\exists_1$ formulas.

Consider a finite set p of basic sentences of L'. We can write the conjunction $\bigwedge p$ as $\phi(\infty,\bar{c})$ where $\phi(x,\bar{y})$ is a formula of L(B) and $\bar{c}$ is a tuple of distinct witnesses. We shall call p a <u>condition</u> iff

(1) $B \models Qx\exists\bar{y}\phi(x,\bar{y})$ (i.e. the formula $\exists\bar{y}\phi(x,\bar{y})$ is in Q).

Observe that by Axiom 1, the property (1) depends only on p and not on the choice of expression $\phi(\infty,\bar{c})$ for $\bigwedge p$. We write N for the set of conditions.

LEMMA 6.1.1. *(a)* N *is not empty.*

(b) *If* p *is a condition then* p *has a model.*

(c) *If* p *is a condition and* ψ *is a basic sentence of* L' *such that* $p \vdash \psi$*, then* $p \cup \{\psi\}$ *is a condition.*

(d) *If* p *is a condition and* t *is a closed term of* L'*, then for any witness* d *which doesn't occur in* p *or* t*,* $p \cup \{t=d\}$ *is a condition.*

(e) *If* p *is a condition and* ψ *is an atomic sentence of* L',

then either $p \cup \{\psi\}$ *or* $p \cup \{\neg\psi\}$ *is a condition.*

<u>Proof</u>. (a) By Axiom 4, $\{\infty=\infty\}$ is a condition.

(b) Suppose p has no model. Then writing $\bigwedge p$ in the form $\phi(\infty,\bar{c})$ as above, we have $\vdash \forall x \neg\exists\bar{y}\phi(x,\bar{y})$. Hence $\vdash \forall x(\exists\bar{y}\phi(x,\bar{y}) \rightarrow x=b)$, where b is any element of B. By Axiom 3, $B \models \neg Qx(x=b)$. So by Axiom 1, $B \models \neg Qx\exists\bar{y}\phi(x,\bar{y})$, and hence p is not a condition.

(c) is proved like (b), using Axiom 1.

(d) Write $\bigwedge p$ as $\phi(\infty,\bar{c})$ and suppose t is $s(\infty,\bar{c})$ for some term $s(x,\bar{y})$ of L(B). Then $B \models Qx\exists\bar{y}\phi$, and hence by Axiom 1, $B \models Qx\exists\bar{y}z(\phi \wedge s(x,\bar{y})=z)$. It follows that $p \cup \{t=d\}$ is a condition, where d is any witness not occurring in p or t.

(e) Write $\bigwedge p$ as $\phi(\infty,\bar{c})$ and ψ as $\theta(\infty,\bar{c})$. Then since p is a condition, $B \models Qx\exists\bar{y}\phi$, and so $B \models Qx\exists\bar{y}((\phi \wedge \theta) \vee (\phi \wedge \neg\theta))$ by Axiom 1. By Axiom 2 and the corresponding property of $\exists\bar{y}$, it follows that either $B \models Qx\exists\bar{y}(\phi \wedge \theta)$ or $B \models Qx\exists\bar{y}(\phi \wedge \neg\theta)$. Hence either $p \cup \{\psi\}$ or $p \cup \{\neg\psi\}$ is a condition. □

In section 2.3 we defined notions of forcing for a countable first-order language L(W) with an infinite set W of witnesses. The language L' is just such a language, and in fact we have:

THEOREM 6.1.2. N *is a notion of forcing for* L', *and it satisfies (6)-(8) of section 2.3.*

<u>Proof</u>. This is easy to verify from Lemma 6.1.1 and direct inspection. □

Accordingly we can use N to form construction sequences $\bar{p}$ and L'-structures $A^+(\bar{p})$ as in section 2.3. So it makes sense to talk of enforceable properties and forcing.

THEOREM 6.1.3. *The following is enforceable: "The compiled structure* $A^+(\bar{p})$ *is a model of* $\bigcup\bar{p}$ *and of every* $\forall_2$ *sentence of* L(B) *which is true in* B; *every element of* $A^+(\bar{p})$ *is of form* $c^{A^+(\bar{p})}$ *for infinitely many witnesses* c; $\bigcup\bar{p}$ *contains* $diag(A^+(\bar{p}))$; *and the element* $\infty^{A^+(\bar{p})}$ *is not equal to* $b^{A^+(\bar{p})}$ *for any element* b *of* B".

<u>Proof</u>. We use the conjunction lemma (Lemma 2.3.3(e)) to split this property into manageable pieces. By Lemma 2.3.1 it is enforceable

that A^+ is a model of $\bigcup \bar{p}$ in which every element is named by infinitely many witnesses. By Lemma 6.1.1(e) it is enforceable that for every atomic sentence ϕ of L', either ϕ or $\neg\phi$ is in $\bigcup \bar{p}$. Together these properties imply that $\bigcup \bar{p}$ is the diagram of A^+ in L'.

Every $\forall_2$ sentence θ of L(B) can be brought to the form $\forall \bar{z}(\exists \bar{w}\psi_1 \vee \ldots \vee \exists \bar{w}\psi_n)$, where each of $\psi_1, \ldots, \psi_n$ is a conjunction of basic formulas. Suppose θ is true in B; we have to show that θ is enforceable. Since it is enforceable that all elements are named by infinitely many witnesses, it suffices to enforce the sentence

(2) $$\exists \bar{w}\psi_1(\bar{c},\bar{w}) \vee \ldots \vee \exists \bar{w}\psi_n(\bar{c},\bar{w})$$

for each tuple $\bar{c}$ of distinct witnesses. Writing θ' for the sentence (2), let players $\forall$ and $\exists$ play the game $G(\theta';\text{odds})$, and suppose player $\forall$'s first move is p_0. Write $\bigwedge p_0$ as $\phi(\infty,\bar{c},\bar{d})$ where $\bar{d}$ lists the distinct witnesses occurring in p_0 but not in $\bar{c}$. Since p_0 is a condition, $B \models Qx\exists \bar{z}\bar{y}\phi(x,\bar{z},\bar{y})$. Then by Axiom 1, $B \models Qx\exists \bar{z}\bar{y}((\phi \wedge \exists \bar{w}\psi_1) \vee \ldots \vee (\phi \wedge \exists \bar{w}\psi_n))$, and so by Axiom 2,

(3) $$B \models Qx\exists \bar{z}\bar{y}\bar{w}(\phi \wedge \psi_1) \vee \ldots \vee Qx\exists \bar{z}\bar{y}\bar{w}(\phi \wedge \psi_n).$$

Hence for some i ($1 \leqslant i \leqslant n$) there is a condition p_1 consisting of p_0 and the conjuncts of $\psi_i(\bar{c},\bar{e})$ where $\bar{e}$ is a tuple of distinct new witnesses. If player $\exists$ replies with p_1 and then plays to make $A^+ \models \bigcup \bar{p}$, she wins $G(\theta';\text{odds})$. Thus θ is enforceable.

Finally if b is any element of B, then by Lemma 6.1.1(e) it is enforceable that at least one of the sentences $\infty=b$, $\infty\neq b$ is in $\bigcup \bar{p}$. If $\infty=b$ were in $\bigcup \bar{p}$, we could infer (via Axiom 1 again) that $B \models Qx\ x=b$, contradicting Axiom 3. Hence it is enforceable that $\infty\neq b$ is in $\bigcup \bar{p}$, and since it is also enforceable that $A \models \bigcup \bar{p}$, we can enforce that ∞ and b name different elements of A^+. □

One immediate corollary of the theorem is that it is enforceable that A^+ is a model of diag(B). So by the diagram lemma (Lemma 1.2.3) it is enforceable that A is up to isomorphism an extension of B. Thus we can assume in future that $B \subseteq A$, and in fact that $b^{A^+} = b$ for every element b of B.

Without confusion we can refer to the element ∞^{A^+} simply as ∞. This element is called the <u>overflow element</u>. By the last part of the theorem, it is enforceable that ∞ is not in B. So we can assume in future that A is a proper extension of B.

For the rest of this section the slogan is that small is beautiful. We study how to keep small sets from growing. It was Keisler who first worked out how to do this, and accordingly the following axiom is sometimes known as <u>Keisler's Axiom 4</u> (though it was written down earlier by Craig and Fuhrken):

$$\text{(Axiom 5)}\quad Qx\exists y\phi \rightarrow \exists yQx\phi \vee Qy\exists x\phi.$$

Axiom 5 is a kind of regularity axiom. Write Y_b for $\{a : B \models \phi(a,b)\}$. Axiom 5 says that if $\bigcup\{Y_b : b \text{ is in } B\}$ is large, then either Y_b is large for some b, or $\{b : Y_b \neq \emptyset\}$ is large. Since ω is a regular cardinal, it follows that Axiom 5 holds in Example 1 above for all formulas ϕ of L(B). We shall see many other examples where Axiom 5 holds, and a few where it doesn't. (See for example Exercises 1, 10, 13, 14, 15.)

THEOREM 6.1.4. *Let Q be a largeness property for $\exists_1$ formulas, satisfying Axiom 5 for $\exists_1$ formulas. Then it is enforceable that for each quantifier-free formula $\psi(x)$ of* L(B), *if* $B \models \neg Qx\psi$ *then every element of* $A^+(\bar{p})$ *satisfying ψ is already in* B.

<u>Proof</u>. Thanks to Axiom 1 and disjunctive normal form, the content of the theorem is not altered if we replace "quantifier-free" by "conjunction of basic formulas". So let $\psi(x)$ be a conjunction $\theta_0(x)\wedge\ldots\wedge\theta_n(x)$ of basic formulas of L(B). By Theorem 6.1.3 it suffices to show that for each witness d the following property P is enforceable:

(4) Either $\bigcup\bar{p} \vdash \neg\psi(d)$, or there is an element b of B such that $(d=b) \in \bigcup\bar{p}$.

We play G(P;odds). Suppose player $\forall$ offers p_0, and $\bigwedge p_0$ is $\phi(\infty,\bar{c},d)$. (There is no loss in including d in the formula.) Then $B \models Qx\exists\bar{y}z\phi(x,\bar{y},z)$. Now there are two cases to consider.

<u>Case 1</u>: $B \models Qx\exists\bar{y}z(\phi \wedge \neg\psi(z))$. Then by Axiom 2 there is some

i such that $B \models Qx\exists\bar{y}z(\phi \wedge \neg\theta_i(z))$. Player $\exists$ can win by choosing p_1 to be $p_0 \cup \{\chi\}$, where χ is a basic sentence equivalent to $\neg\theta_i(d)$.

Case 2: $B \models \neg Qx\exists\bar{y}z(\phi \wedge \neg\psi(z))$. Then by Axiom 2, $B \models Qx\exists z\exists\bar{y}(\phi \wedge \psi(z))$. But by assumption $B \models \neg Qz\psi(z)$, and hence (Axiom 1) $B \models \neg Qz\exists x\exists\bar{y}(\phi \wedge \psi(z))$. It follows by Axiom 5 that $B \models \exists zQx\exists\bar{y}(\phi \wedge \psi(z))$. With a little rearrangement this implies $B \models \exists wQx\exists\bar{y}z(\phi \wedge z=w)$. So for some element b of B, $B \models Qx\exists\bar{y}z(\phi \wedge z=b)$. This shows that player $\exists$ can win by choosing p_1 to be $p_0 \cup \{d=b\}$.

Hence in both cases player $\exists$ can be sure of winning. □

EXAMPLE 5. Let T be a theory in a first-order language L with a relation symbol <, and let T contain sentences that say "< linearly orders all the elements, and there is no last element". Let B be a model of T. We say that an extension C of B is an end extension iff for every element c of C and every element b of B, if $c <^C b$ then c is already in B. In short, A is an end extension of B iff all new elements are added at the end.

If B is a model of T and $\phi(x)$ is a formula of L(B), we define $Qx\phi$ to mean that cofinally many elements in B satisfy ϕ, i.e. $\forall y\exists x(y<x \wedge \phi(x))$. This is called the cofinal interpretation of Q. It is easy to check that Q is a largeness property over B.

Consider the case where L and B are countable and Axiom 5 holds for all $\exists_1$ formulas of L(B). For each element b of B we have $B \models \neg Qx(x<b)$, by choice of Q. Then Theorem 6.1.4 says that it is enforceable that in the compiled extension A, no new elements get added to the left of b. So it is enforceable that A is an end extension of B. (It is also enforceable that A is linearly ordered without last element, since this can be said with a $\forall_2$ sentence; cf. Theorem 6.1.3. But it need not be enforceable that Axiom 5 holds in A, so in general there is no hope of iterating this construction.)

Let me describe two applications of Example 5. Both of them will apply the spirit of Example 5 rather than the letter of it, because in both cases we shall generalise from $\exists_1$ to a larger class of formulas.

The first application is a version of Vaught's two cardinal theorem. In this application we replace "$\exists_1$" by "first-order". One effect of this change is that now A will be an elementary extension of B, and so A will satisfy Axiom 5 too. Hence we can iterate to form longer and longer end extensions of B.

To be precise, let L be a countable first-order language, B a countable L-structure and Q a largeness property over B. Redefine a condition to be a finite set p of sentences of L' (not necessarily basic), such that if $\bigwedge p$ is $\phi(\infty,\bar{c})$ then $B \models Qx\exists\bar{y}\phi(x,\bar{y})$. Redefine N to be the set of all conditions in this new sense. Then Lemma 6.1.1 holds with the words "basic" and "atomic" struck out; Theorem 6.1.2 holds unaltered; Theorem 6.1.3 holds with "$\forall_2$" deleted and diag($A^+(\bar{p})$) replaced by eldiag($A^+(\bar{p})$), so that $A(\bar{p})$ is an elementary extension of B; and Theorem 6.1.4 remains true with "$\exists_1$" replaced by "first-order", and "quantifier-free" struck out. The proofs are essentially as before (though in the proofs of Theorems 6.1.3 and 6.1.4 we can now avoid having to use disjunctive normal forms). We can call these revised theorems the full first-order versions of Lemma 6.1.1-Theorem 6.1.4.

Recall from section 1.2 that $\phi(A)$ means the set of elements a of A such that $A \models \phi(a)$.

THEOREM 6.1.5 (Two-Cardinal Theorem). *Let* L *be a countable first-order language,* λ *an infinite cardinal and* C *an* L*-structure of cardinality* λ^+. *Then there exists an* L*-structure* D *of cardinality* ω_1, *such that* $C \equiv D$ *and for every formula* $\psi(x)$ *of* L, *if* $|\psi(C)| \leqslant \lambda$ *then* $\psi(D)$ *is countable or finite.*

Proof. Adding a symbol to L and C if necessary, we can suppose that the elements of C are ordered by a relation $<^C$ in order-type λ^+. As in Example 5 above, write $Qx\phi$ for $\forall y\exists x(x<y \wedge \phi(x))$. Since λ^+ is a regular cardinal, Q is a largeness property over C satisfying Axiom 5 for all first-order formulas. Let B be any countable elementary substructure of C. Since the formulas $Qx\phi$ are actually first-order, Q is also a largeness property over B satisfying Axiom 5. Also Th(B) is a theory T as in Example 5.

By the full first-order versions of Theorems 6.1.3 and 6.1.4, it follows that B has a countable elementary end extension A_0. Now if $\psi(x)$ is any formula of L such that $|\psi(C)| \leqslant \lambda$, then since λ^+ is regular, $\psi(C)$ is not unbounded in C, and hence $C \models \neg Qx\psi$. Therefore $B \models \neg Qx\psi$, and so by Theorem 6.1.4 (again in the full first-order version) we can choose

A_0 so that $\psi(A_0) = \psi(B)$. In fact there are at most countably many such formulas ψ of L, so that by the conjunction lemma we can make a choice of A_0 which works for them all.

Again since the formulas $Qx\phi$ are first-order and $B \preccurlyeq A_0$, Q is a largeness property over A_0 satisfying Axiom 5. So we can iterate. Taking unions at limit ordinals (cf. the elementary chain lemma, Lemma 1.2.7), we build a chain $(A_i : i < \omega_1)$ of countable L-structures such that for each $i < \omega_1$, A_{i+1} is a proper elementary end extension of A_i, and for each formula $\psi(x)$ of L, if $|\psi(C)| \leqslant \lambda$ then $\psi(A_{i+1}) = \psi(A_i)$. To deduce the theorem, take D to be the union of the chain. □

The construction of D in the proof of the theorem above is actually a two-dimensional task. Each structure A_{i+1} is built up in <u>finite</u> pieces, and each piece ties A_{i+1} to a <u>finite</u> number of elements of A_i. Compare with picture (7) in section 1.1, and contrast with the one-dimensional attitudes of section 5.3 above.

The next application of Example 5 is useful for studying models of arithmetic. There is a set-theoretic version too; cf. Exercise 13.

Let L be a countable first-order language whose symbols include a 2-ary relation symbol <. We write $(\forall x{<}y)\phi$ for $\forall x(x{<}y \rightarrow \phi)$, and $(\exists x{<}y)\phi$ for $\exists x(x{<}y \wedge \phi)$. The quantifiers $(\forall x{<}y)$ and $(\exists x{<}y)$ are said to be <u>bounded</u>, and a formula of L is said to be Δ_0 iff all quantifiers occurring in it are bounded. For each $n < \omega$, a formula of L is said to be Π_n (resp. Σ_n) iff it has the form $\forall\bar{x}_1\exists\bar{x}_2\forall\bar{x}_3\ldots\bar{x}_n\psi$ (resp. $\exists\bar{x}_1\forall\bar{x}_2\exists\bar{x}_3\ldots\bar{x}_n\psi$) with ψ a Δ_0 formula. We allow any of the alternating quantifier blocks $\forall\bar{x}_i$ or $\exists\bar{x}_i$ to be empty, so that for example a Σ_n formula is also a Σ_{n+1} formula and a Π_{n+1} formula. The Π_0 formulas and the Σ_0 formulas are both the same as the Δ_0 formulas.

Assume now that T is a theory in L which implies "< linearly orders everything and there is no last element". Let B be a countable model of T and let Q have the cofinal interpretation (as in Example 5). Let n be a fixed positive integer. Then Q is a largeness property for the class of Σ_n formulas of L(B). Redefine a <u>condition</u> to be a finite set p of Π_{n-1} sentences and Σ_{n-1} sentences of L' such that if $\bigwedge p$ is $\phi(\infty,\bar{c})$ then $B \models Qx\exists\bar{y}\phi(x,\bar{y})$. (Note that $\exists\bar{y}\phi$ here is logically equivalent to a Σ_n

formula.) Redefine N to be the set of all conditions in this new sense. Then Lemma 6.1.1(a)-(d) holds with "basic" replaced by "Π_{n-1} or Σ_{n-1}"; (e) holds with "atomic" replaced by "Δ_0". Theorem 6.1.2 holds without alteration. Theorem 6.1.3 holds with "$\forall_2$" replaced by "Π_{n+1}". Theorem 6.1.4 holds with "$\exists_1$" and "quantifier-free formula" replaced by "Σ_n" and "boolean combination of Π_{n-1} formulas" respectively. This is all easily verified. We call these revised theorems the Σ_n versions of Lemma 6.1.1-Theorem 6.1.4.

Write $Ax5(\Sigma_m)$ for Axiom 5 for Σ_m formulas. $Ax5(\Sigma_m)$ is a set of sentences of L(B). It turns out that $Ax5(\Sigma_m)$ for a positive integer m is equivalent to another axiom that number theorists of the logical sort have sometimes studied. If Φ is a class of formulas, the Φ collection schema is the axiom schema

$$(\mathrm{Coll}(\Phi)) \quad \forall\bar{z}u((\forall x<u)\exists\bar{y}\phi \rightarrow \exists w(\forall x<u)(\exists\bar{y}<w)\phi),$$
where $\phi(x,\bar{y},\bar{z},u)$ is any formula of L in Φ,

writing $(\exists y_1 \ldots y_k<w)$ for $(\exists y_1<w)\ldots(\exists y_k<w)$.

LEMMA 6.1.6. *Let* T, B, Q, n *be as above. Then the following are equivalent:*

(a) $B \models Ax5(\Sigma_n)$.
(b) $B \models \mathrm{Coll}(\Pi_n)$.

Proof. (a) $\Rightarrow$ (b): This is the harder direction. We go by induction on n, starting at n = 0, and at the same time we show by induction on n that if $B \models Ax5(\Sigma_n)$ then

(5) For every $k \leqslant n+1$ and every Σ_k formula $\phi(x,y,\bar{z})$ of L(B) there is a Σ_k formula $\psi(y,\bar{z})$ of L(B) such that $B \models \forall y\bar{z}(\psi \leftrightarrow (\forall x<y)\phi)$.

Note that for any n, if $B \models \mathrm{Coll}(\Pi_n)$ then one can deduce (5) using the induction assumption that (5) holds for all $k \leqslant n$.

Assume $B \models Ax5(\Sigma_n)$ and suppose ϕ is a Π_n formula of L(B) such that for some element b of B,

(6) $B \models \neg\exists w(\forall x<b)(\exists\bar{y}<w)\phi$.

There is a Σ_n formula ψ which is logically equivalent to $\neg\phi$. By (6) and Exercise 2(a) we have

(7) $$B \models Qw\exists x(x<b \wedge (\forall\bar{y}<w)\psi).$$

By induction hypothesis (5) when $n > 0$, or a triviality when $n = 0$, the formula $x<b \wedge (\forall\bar{y}<w)\psi$ is equivalent to a Σ_n formula, so that by Ax5(Σ_n) we deduce

(8) $$\text{either } B \models \exists xQw(x<b \wedge (\forall\bar{y}<w)\psi), \text{ or } B \models Qx\exists w(x<b \wedge (\forall\bar{y}<w)\psi).$$

Paraphrasing (8),

(9) $$\text{either } B \models \neg(\forall x<b)\exists\bar{y}\phi, \qquad \text{or } B \models Qx(x<b \wedge \exists w(\forall\bar{y}<w)\psi).$$

The second option in (9) contradicts the definition of Q, and so the first option holds. Comparing with (6), this proves $B \models \mathrm{Coll}(\Pi_n)$.

(b) ⇒ (a): Assume $B \models \mathrm{Coll}(\Pi_n)$, and suppose $\phi(x,y)$ is a Σ_n formula of L(B). Since $n \geqslant 1$, there is a Π_n formula $\psi(y,z)$ which is equivalent modulo T to $\forall x(\phi(x,y) \to x < z)$. Reasoning inside B, if $\neg\exists yQx\phi$ then $\forall y\exists z\psi$. If also $\neg Qy\exists x\phi$ then there is some element b such that for all x and y, $\phi(x,y)$ implies $y < b$. Then $(\forall y<b)\exists z\psi$, and so by Coll($\Pi_n$) there is an element c such that $(\forall y<b)(\exists z<c)\psi$. This implies that $\forall x(\forall y<b)(\phi \to x < c)$ and hence (by choice of b) $\neg Qx\exists y\phi$. Ax5(Σ_n) is proved. □

When T and L are as in Example 5 and A, B are models of T, we say that A is a Σ_n extension of B iff $B \subseteq A$ and for every Σ_n formula $\phi(\bar{x})$ of L and every tuple $\bar{b}$ from B, $B \models \phi(\bar{b})$ iff $A \models \phi(\bar{b})$.

THEOREM 6.1.7. *Let* T *be a theory in a countable first-order language* L, *and suppose* T *implies "< linearly orders all the elements and there is no last element". Let* B *be a countable model of* T *and* n *a positive integer. Then the following are equivalent:*

(a) Coll(Π_n) *holds in* B.

(b) B *has a proper* Σ_{n+1} *end extension.*

Proof. (a) ⇒ (b) is by the easy direction in the lemma,

coupled with the Σ_n version of the earlier results of this section.

(b) ⇒ (a): Let A be a proper Σ_{n+1} end extension of B, and suppose $\phi(x,y)$ is a Σ_n formula of L(B). Assuming $B \models Qx\exists y\phi \wedge \neg Qy\exists x\phi$, we deduce $B \models \exists yQx\phi$ as follows. By assumption $B \models \forall w\exists xy(w < x \wedge \phi)$ and there is an element b such that $B \models \forall xy(\phi \to y < b)$. Since these are equivalent modulo T to Π_{n+1} formulas and hence to the negations of Σ_{n+1} formulas, and A is a Σ_{n+1} extension of B, it follows that

$$(10) \qquad A \models \forall w\exists xy(w < x \wedge \phi) \wedge \forall xy(\phi \to y < b).$$

Since A is a proper end extension of B, it follows from the first formula in (10) that $A \models \exists y\phi(a,y)$ for some element a which is in A but not in B. Hence, using the second formula in (10), there is an element c of A such that $A \models \phi(a,c) \wedge c < b$. But A is an end extension of B, and so c is in B. It follows that for every element d of B, $A \models \exists x(d < x \wedge \phi(x,c))$ and hence $B \models \exists x(d < x \wedge \phi(x,c))$. In short, $B \models Qx\phi(x,c)$ and so $B \models \exists yQx\phi$. This proves that $Ax5(\Sigma_n)$ holds in B, and we deduce (a) by the lemma. □

What has this to do with models of arithmetic? More than meets the eye. The axioms $Coll(\Pi_n)$ are in fact a form of the induction axiom for Peano arithmetic. See Exercise 11 for a precise statement.

Exercises for 6.1.

1. Show that in each of (a)-(d), Q is a largeness property over B. Show also that Axiom 5 is satisfied in cases (a), (c), and in case (b) when κ is regular. Give a counterexample to Axiom 5 in case (b) when κ is singular. (a) A is an uncountable elementary extension of B, and $Qx\phi$ means "There are uncountably many elements x in A such that $\phi(x)$". (b) κ is an infinite cardinal, B has cardinality $\geq \kappa$, and $Qx\phi$ means "There are at least κ elements x in B such that $\phi(x)$". (c) B is a model of ZFC and is a regular cardinal in the sense of B; $B \models Qx\phi$ holds iff $B \models$ "There are at least κ elements x such that $\phi(x)$". (d) Th(B) is not ω-stable, and $Qx\phi$ holds iff ϕ has Morley rank ∞. (ϕ has Morley rank ∞ iff there are an elementary extension A of B, formulas $\psi_i(x,\bar{y}_i)$ of L and tuples $\bar{a}_i$ in A such that for every finite sequence s of 0's and 1's, $A \models \exists x(\phi \wedge \bigwedge_{i<lh(s)}\psi_i(x,\bar{a}_i)^{s(i)})$, where ψ^0 is $\neg\psi$ and ψ^1 is ψ. Th(B) is not ω-stable iff $Qx\ x=x$ holds.)

2. Let Q be a largeness property over B. (a) Show that $\forall x\phi$ implies $Qx\phi$, and that $Qx\phi$ implies $\exists x\phi$. (b) Show that $\exists xQy\phi$ implies $Qy\exists x\phi$, and that $Qy\forall z\phi$ implies $\forall zQy\phi$. Find counterexamples to the converses. (c) Show that the following holds:

$$\text{(Axiom 3')} \quad \forall yz\neg Qx(x=y \vee x=z).$$

(d) Show that if Axiom 5 holds, then $Qx\exists y_1\ldots y_n\phi$ implies $Qy_1\exists y_2\ldots y_n x\phi \vee \exists y_1 Qy_2\exists y_3\ldots y_n x\phi \vee \ldots \vee \exists y_1\ldots y_{n-1}Qy_n\exists x\phi \vee \exists y_1\ldots y_n Qx\phi$.

3. Let L be a first-order language, B an L-structure, and suppose that for every formula $\phi(x)$ of L(B), $Qx\phi$ is defined in such a way that Axioms 0, 1, 3' (cf. Exercise 2 above), 4 and 5 all hold in B. Show that Axioms 2 and 3 also hold in B.

4. If Q is a largeness property, define $Q^o x\phi$ to mean $\neg Qx\neg\phi$. Show that the following axioms hold: (i) $\forall x(\phi \to \psi) \to (Q^o x\phi \to Q^o x\psi)$. (ii) $Q^o x\phi \wedge Q^o x\psi \to Q^o x(\phi \wedge \psi)$. (iii) $Q^o x(x\neq b)$ for each element b. (iv) $\neg Q^o x(x\neq x)$. (If Q is only a largeness property for $\exists_1$ formulas, then ϕ, ψ in (i) and (ii) are $\forall_1$ formulas.)

5. Let Q be a largeness property over B for $\exists_1$ formulas, $\phi(x,\bar{y})$ a quantifier-free formula of L(B) such that $B \models Qx\exists\bar{y}\phi(x,\bar{y})$, and P an enforceable property. Show that there is an extension A of B such that $A \models \exists\bar{y}\phi(\infty,\bar{y})$ and A satisfies P.

6. Forcing with the notion of forcing N of Lemma 6.1.1, show that if p is a condition and $\psi(\infty,\bar{c})$ is a $\forall_1$ sentence of L(B), then p forces $\psi(\infty,\bar{c})$ iff $B \models Q^o x\forall\bar{y}\bar{z}(\phi(x,\bar{y},\bar{z}) \to \psi(x,\bar{y}))$ where $\bigwedge p$ is $\phi(\infty,\bar{c},\bar{d})$ and Q^o is as in Exercise 4.

The next exercise supplies the Omitting Types Theorem for forcing as in Lemma 6.1.1-Theorem 6.1.4.

7. (a) Let L be a countable first-order language, B a countable L-structure and Q a largeness property for $\exists_1$ formulas over B. Let $\psi_i(\bar{z})$ $(i < \omega)$ be $\forall_1$ formulas of L(B). Show that the following are equivalent: (i) $\neg\exists\bar{z}\bigwedge_{i<\omega}\psi_i(\bar{z})$ is enforceable. (ii) For each $\exists_1$ formula $\phi(x,\bar{z})$ of L(B), if $B \models Qx\exists\bar{z}\phi$ then there is $i < \omega$ such that $B \models Qx\exists\bar{z}(\phi \wedge \neg\psi_i)$.
(b) State and verify the full first-order version of (a).

8. Verify the statements made in the text about the versions of Lemma

6.1.1-Theorem 6.1.4 which hold (a) for the full first-order version of forcing as in this section, and (b) for the Σ_n version.

The next exercise gives a local version of Theorem 6.1.4. It can be used to keep a single small set small in cases where Axiom 5 fails in general.

9. Let L be a countable first-order language, B a countable L-structure and Q a largeness property over B for $\exists_1$ formulas. Let $\psi(x)$ be a formula of L(B) such that $B \models \neg Qx\psi$, and suppose that for every $\exists_1$ formula $\phi(x,y)$ of L(B),

$$B \models Qx\exists y(\psi(y) \wedge \phi) \rightarrow \exists y(\psi(y) \wedge Qx\phi).$$

Show that it is enforceable that $\psi(A) = \psi(B)$.

10. (a) In the context of Example 5, show that Axiom 5 for $\exists_1$ formulas is equivalent to the following: For all $\exists_1$ formulas $\phi(x,y)$ of L(B), $B \models \forall z(Qy(\exists x{<}z)\phi \rightarrow (\exists x{<}z)Qy\phi)$. (b) Let B be a model of ZFC and κ an infinite cardinal of B. Define $B \models Qx\phi$ to mean $B \models$ "For every set y of cardinality $\leqslant \kappa$ there is a set x of cardinality $\leqslant \kappa$ such that $y \subseteq x$ and $\phi(x)$". Show that Q is a largeness property over B, and that for every formula $\phi(x,y)$ of L(B), $B \models Qx(\exists y{\in}\kappa)\phi \rightarrow (\exists y{\in}\kappa)Qx\phi$.

For the next two exercises, L *is the first-order language of arithmetic with* <, *and* T *is the set of* $\forall_2$ *consequences of first-order Peano arithmetic including the definition of* <. *For any* $n < \omega$, *we write* $I\Sigma_n$ *for the induction schema* $\forall\bar{z}(\phi(0,\bar{z}) \wedge \forall x(\phi(x,\bar{z})\rightarrow\phi(x+1,\bar{z})) \rightarrow \forall x\phi(x,\bar{z}))$ *where* ϕ *ranges over* Σ_n *formulas of* L.

11. (a) Show that for every $n < \omega$, $\mathrm{Coll}(\Pi_n)$ is equivalent to $\mathrm{Coll}(\Sigma_{n+1})$. (b) Show that for every $n < \omega$, $T \cup I\Sigma_{n+1}$ implies $\mathrm{Coll}(\Pi_n)$, and $T \cup I\Sigma_0 \cup \mathrm{Coll}(\Pi_n)$ implies $I\Sigma_n$. (c) Show that for every $n < \omega$, $T \cup I\Sigma_n$ implies $\forall\bar{z}(\exists y\phi(y,\bar{z}) \rightarrow \exists y(\phi(y,\bar{z}) \wedge (\forall x{<}y)\neg\phi(x,\bar{z})))$ for all Σ_n formulas $\phi(y,\bar{z})$ of L. (d) Show that if B is a model of $T \cup I\Sigma_0$ which has a proper Σ_1 end-extension, then $B \models \mathrm{Coll}(\Pi_1)$ (and hence if B is countable, B has a proper Σ_2 end-extension too).

12. (a) Show that for all $n > 0$, if B is a countable model of $T \cup I\Sigma_n$ then it is enforceable (in Σ_n forcing) that for every element a of A there is a Σ_n formula $\phi(x,y)$ of L(B) such that $A \models \phi(\infty,a) \wedge \forall xyz(\phi(x,y)\wedge\phi(x,z) \rightarrow y=z)$. [Use (c) of Exercise 11.]
(b) Show that in (a), ϕ can be chosen to be $\psi(x,y,b)$ where ψ is in L and

is independent of a. [Use a universal Σ_n formula.] (e) For any $n > 0$, let T' be the set of all Π_{n+2} consequences of first-order Peano arithmetic. With ψ as in (b), let χ be the sentence $\exists u x \forall y (\exists z < u) \psi(x,y,z)$. Show that $T' \vdash \neg\chi$, but if B is a countable model of T' then it is enforceable (in Σ_n forcing) that $A \models \chi$, so that A is not a Σ_{n+2} extension of B. (d) Show that for every $n < \omega$, first-order Peano arithmetic is not equivalent to the set of its Π_n consequences.

For the next two exercises, L *is the language of set theory. We define* Π_n, Σ_n *formulas,* Σ_n *end extensions and* $\mathrm{Coll}(\Pi_n)$ *exactly as for arithmetic, but with* $\in$ *in place of* $<$.

13. Let T consist of the set-theory axioms Extensionality, Pair, Union, Foundation and Δ_0 Separation. We write $Qx\phi(x)$ for $\neg\exists y \forall x(\phi(x) \rightarrow x \in y)$. Let B be a model of T. (a) Show that Q is a largeness property over B. (b) Show that for all $n \geqslant 1$, if $B \models \mathrm{Coll}(\Pi_n)$ then $B \models \mathrm{Ax5}(\Sigma_n)$, and hence if B is also countable then B has a proper Σ_{n+1} end extension.
(c) Suppose there is a function F defined on the ordinals of B, with a Σ_n definition in B, such that $B \models \forall x \exists \alpha (x \in F(\alpha)) \wedge \forall \alpha\beta(\alpha < \beta \rightarrow F(\alpha) \subseteq F(\beta))$. Show that for all $n \geqslant 1$, if B has a proper Σ_{n+1} end extension then $B \models \mathrm{Ax5}(\Sigma_n)$, and if $B \models \mathrm{Ax5}(\Sigma_n)$ then $B \models \mathrm{Coll}(\Pi_n)$. [In (7) put $B \models Q\alpha \exists x z(x \in b \wedge z = F(\alpha) \wedge (\forall \bar{y} \in z)\psi)$.]

14. Let B be a countable model of ZFC and λ an uncountable regular cardinal in the sense of B. (a) For each formula $\phi(x)$ of L(B), interpret $Qx\phi$ as "The set $\{x < \lambda : \phi(x)\}$ is stationary in λ" (in the sense of B). Show that Q is a largeness property over B. Show also that it is enforceable (in full first-order forcing) that the ordinals of B below λ are a proper initial segment of the ordinals of A below λ, and ∞ is the first ordinal of A which is not in B. [Use Lemma 5.2.3(b).] (b) For each formula $\phi(x)$ of L(B), interpret $Qx\phi$ as "The set $\{x < \lambda : \phi(x)\}$ is unbounded in λ". Show that Q is a largeness property over B. Show also that it is enforceable (in full first-order forcing) that the ordinals of B below λ are a proper initial segment of the ordinals of A below λ, and that there is no first ordinal of A which is not in B. [In both parts Axiom 5 fails, but Exercise 9 shows how to stop the ordinals below λ from growing.]

For the next three exercises we return to forcing in the sense of Lemma 6.1.1-Theorem 6.1.4.

15. In a countable first-order language L, let T be a $\forall_2$ theory and B a countable model of T. Let $Qx\phi$ mean that there is a model D of T such that $B \subseteq D$ and $D \models \phi(d)$ for some element d not in B. Show: (a) If B is an e.c. model of T and $\phi(x)$ is an $\exists_1$ formula of L(B), then $B \models Qx\phi$ if and only if there are infinitely many elements b of B such that $B \models \phi(b)$. (b) It is enforceable that the compiled extension A of B is an e.c. model of T. [Use Corollary 3.2.4. If $B \models Qx\exists\bar{y}\phi$ but $B \models \neg Qx\exists\bar{y}(\phi \wedge \psi)$, then for some elements $b_1, \ldots, b_n$ of B, $T \cup \text{diag}(B) \vdash \forall x\bar{y}(\phi \wedge x \neq b_1 \wedge \ldots \wedge x \neq b_n \rightarrow \neg\psi)$.]

16. Let B be an e.c. group and L the first-order language of groups. For each $\exists_1$ formula $\phi(x)$ of L(B), let $Qx\phi$ mean that there is no finitely generated subgroup G of B such that $\phi(B) \subseteq G$. (a) Show that the following are equivalent, for any primitive formula $\phi(x,\bar{y})$ of L(B), listing as $\bar{b}$ the parameters from B which occur in ϕ: (i) $B \models Qx\exists\bar{y}\phi(x,\bar{y})$. (ii) There is some $g \notin \langle\bar{b}\rangle_B$ such that $B \models \exists\bar{y}\phi(g,\bar{y})$. (iii) $B \models \exists zx\bar{y}(\phi(x,\bar{y}) \wedge z \in C(\bar{b}) \wedge z \notin C(x))$ (where $C(\bar{b})$ is the centraliser of the elements $\bar{b}$). (b) Deduce that Q is a largeness property over B satisfying Axiom 5, for $\exists_1$ formulas. (c) Show that for every $\exists_1$ formula $\phi(x)$ of L(B), $Qx\phi$ holds if and only if $\phi(B)$ is finite.

17. Let B be a countable e.c. group and Q as in Exercise 16. We use Q to force over B. Show: (a) It is enforceable that the compiled structure A is an e.c. group whose finitely generated subgroups are all embeddable in B, and hence that A is back-and-forth equivalent to B. (b) Let X be a set of elements of B which is not contained in any finitely generated subgroup of B; then it is enforceable that $C_A(X) = C_B(X)$ and X is not contained in any finitely generated subgroup of A. (c) It is enforceable that A is generated by B and ∞. [The proof I have of (c) uses small cancellation theory, cf. Lyndon & Schupp (1977), but there may be an easier way.]

The next exercise draws out an easy consequence of Exercise 17. See Theorem 5.2.6 above for a subtler corollary.

18. Show that every countable e.c. group is back-and-forth equivalent to some uncountable e.c. group.

6.2 DEFINABLE ULTRAPOWERS

Travelling around the Welsh borderlands one sometimes comes across pre-Cambrian outcrops. Just by looking, you can't see that they are more than a hundred million years older than the surrounding Silurian rocks. But if you know that they are, you can feel it somehow. The construction to be explained in this section gives me something of the same feeling. The device is some thirty years older than any of the other forcing constructions described in the book. As usual, the man thirty years ahead of his time was Thoralf Skolem.

As in the previous section, let L be a first-order language and B an L-structure. Define L(B) to be L together with constants for the elements of B, just as before. Define L' to be L(B) together with just one new constant, the overflow constant ∞, <u>and no witnesses</u>. Let Q be a largeness property over B <u>for quantifier-free formulas</u>. Define a <u>condition</u> to be a finite set p of basic sentences of L' such that if $\bigwedge p$ is $\phi(\infty)$ for some formula $\phi(x)$ of L(B), then $B \models Qx\phi$.

The absence of witnesses damages some of our theorems. But the reader can verify that the following survive, assuming B is countable.

Lemma 6.1.1 remains true except for clause (d). Theorem 6.1.3 remains true except that "$\forall_2$" becomes "$\forall_1$", and the statement that every element is named by infinitely many witnesses has to be weakened to "every element of $A^+(\bar{p})$ is of form $t^{A^+(\bar{p})}$ for some closed term t of L'". Theorem 6.1.4 remains true without alteration; as stated it requires Q to be defined on $\exists_1$ formulas, but we can recast Axiom 5 to remove this feature (cf. Exercise 1). Call all this the <u>witness-free version</u> of constructing extensions by forcing.

In the witness-free version it is enforceable that A satisfies all the $\forall_1$ sentences of L(B) which are true in B. Suppose now that T is a theory with Skolem functions. (Cf. Chang & Keisler (1973) section 3.3 or Sacks (1972) Chapter 11. For present purposes we can define a theory with Skolem functions to be a theory T such that (i) T is equivalent to a $\forall_1$ theory and (ii) every formula $\phi(\bar{x})$ is equivalent modulo T to a quantifier-free formula $\psi(\bar{x})$.) Suppose also that B is a model of T. Then it is enforceable that A is also a model of T, and hence that $B \preccurlyeq A$ (since quantifier-free formulas are preserved in embeddings).

In a state of nature it seems that there are very few theories with Skolem functions. All the examples that one meets were constructed artificially. But here is one of the less unnatural examples.

We shall say that a theory T in a first-order language L is <u>Peano-like</u> iff L has at least the symbols $+$, $\cdot$, 0, 1, $<$, and T contains the familiar Peano axioms for first-order arithmetic, including the

<u>induction schema</u>

(1) $\forall\bar{z}(\phi(0,\bar{z}) \wedge \forall x(\phi(x,\bar{z}) \rightarrow \phi(x+1,\bar{z})) \rightarrow \forall x\phi(x,\bar{z}))$
for all formulas $\phi(x,\bar{z})$ of L.

Then T implies that for all $\bar{z}$, if there is an element x such that $\phi(x,\bar{z})$, then there is a unique first such element x. We can add to L function symbols $F_\phi(\bar{z})$, and put into T sentences which say that for all $\bar{z}$, $F_\phi(\bar{z})$ is the first element x such that $\phi(x,\bar{z})$, or 0 if there is no such element. The resulting theory T^* says nothing that wasn't already expressed in T, since the functions F_ϕ are explicitly definable. But T^* is a Peano-like theory with Skolem functions.

Thus the witness-free version can be used for constructing proper elementary extensions of models of Peano-like theories. But before developing this, let me turn to Skolem's picture.

Let L, B and Q be as in the second paragraph of this section. For any formula $\phi(x)$ of L(B), we write $\phi(B)$ for the set of all elements of B which satisfy ϕ. We write $\mathscr{B}(B)$ for the boolean algebra of subsets of dom(B) of form $\phi(B)$ with ϕ quantifier-free. We call the set $\phi(B)$ <u>small</u> iff $B \models \neg Qx\phi$; by Axiom 1 this depends only on $\phi(B)$ and not on the choice of ϕ. We write $\mathscr{I}(B)$ for the set of small sets $\phi(B) \in \mathscr{B}(B)$. By Axioms 1-4, $\mathscr{I}(B)$ is a proper ideal in $\mathscr{B}(B)$ containing all finite sets.

Write $\mathscr{T}$ for the set of all terms t(x) of L(B) with at most the variable x. Let $\mathscr{U}$ be an ultrafilter in $\mathscr{B}(B)$. We define a relation $\sim_{\mathscr{U}}$ on $\mathscr{T}$ by:

(2) $s \sim_{\mathscr{U}} t \quad \text{iff} \quad \{b : s^B(b) = t^B(b)\} \in \mathscr{U}.$

Note that the set in (2) is $\phi(B)$ where $\phi(x)$ is the formula $s(x)=t(x)$. Since $\mathscr{U}$ is a filter, $\sim_{\mathscr{U}}$ is an equivalence relation on $\mathscr{T}$. Write $t^\sim$ for the equivalence class of t.

We define an L(B)-structure $B_{\mathscr{U}}$ with domain $\{t^\sim : t \in \mathscr{T}\}$ by putting:

(3) $(t_1^\sim,\ldots,t_n^\sim) \in R^{B_{\mathscr{U}}} \quad \text{iff} \quad \{b : B \models R(t_1(b),\ldots,t_n(b))\} \in \mathscr{U},$

(4) $F^{B_{\mathscr{U}}}(\tilde{t}_1,\ldots,\tilde{t}_n) = \tilde{s}$ iff

$\{b : B \models F(t_1(b),\ldots,t_n(b))=s(b)\} \in \mathscr{U}$.

Since $\mathscr{U}$ is a filter, (3) and (4) are sound definitions. (Cf. Chang & Keisler (1973) section 4.1.) The L-reduct $B_{\mathscr{U}}|L$ is called the definable ultrapower of B by $\mathscr{U}$, written $\Pi_{\mathscr{U}}B$.

LEMMA 6.2.1. *For all quantifier-free formulas* $\phi(x_1,\ldots,x_n)$ *of* L(B) *and all terms* $t_1, \ldots, t_n \in \mathscr{U}$,

(5) $B_{\mathscr{U}} \models \phi(\tilde{t}_1,\ldots,\tilde{t}_n)$ iff $\{b : B \models \phi(t_1(b),\ldots,t_n(b))\} \in \mathscr{U}$.

Proof. Exactly as Łoś's theorem, Theorem 4.1.9 in Chang & Keisler (1973). The first step is to show that for each term $s(x_1,\ldots,x_n)$ of L(B), $s^{B_{\mathscr{U}}}(\tilde{t}_1,\ldots,\tilde{t}_n) = s(t_1,\ldots,t_n)^{\sim}$. □

It follows from Lemma 6.2.1 that B is naturally embedded in $\Pi_{\mathscr{U}}B$ by the map taking each element b to $\tilde{b}$, so that we can regard $\Pi_{\mathscr{U}}B$ as an extension of B. Also it follows from Lemma 6.2.1 that any $\forall_1$ sentence of L(B) which is true in B is true in $\Pi_{\mathscr{U}}B$ too. Hence if B is a model of a theory with Skolem functions, then $B \preccurlyeq \Pi_{\mathscr{U}}B$.

LEMMA 6.2.2. $\Pi_{\mathscr{U}}B$ *is generated by* $\tilde{x}$ *and the elements of* B. *If* $\mathscr{U}$ *is disjoint from* $\mathscr{I}$(B) *then* $\tilde{x}$ *is not in* B.

Proof. The first sentence follows at once from the construction. For the second, suppose $B_{\mathscr{U}} \models \tilde{x}=b$. Then by Lemma 6.2.1, if $\mathscr{U}$ is disjoint from $\mathscr{I}$(B) we have $B \models Qx(x=b)$, which contradicts Axiom 3 for largeness properties. □

The reader has probably guessed that the definable ultrapower construction and the witness-free game construction are one and the same thing. Indeed they are when B is countable. Let $\bar{p}$ be a play in the witness-free game over B. Define $\mathscr{F}(\bar{p})$ to be the set of all sets which are of form $\phi(B)$ for some quantifier-free $\phi(x)$ such that $\phi(\infty) \in \bigcup\bar{p}$. By the definition of conditions, $\mathscr{F}(\bar{p})$ is disjoint from $\mathscr{I}$(B). By Lemma 6.1.3 it is enforceable that

(6) for quantifier-free $\phi(x)$ in L(B), $\phi(B) \in \mathcal{F}(\bar{p}) \Leftrightarrow A^+ \models \phi(\infty)$.

(6) implies that $\mathcal{F}(\bar{p})$ is an ultrafilter in $\mathcal{B}(B)$.

THEOREM 6.2.3. *Let* L, B *and* Q *be as in the second paragraph of this section, and suppose* B *is countable.*

(a) *Let* $\bar{p}$ *be a play of a witness-free game to compile* $A^+(\bar{p})$ *over* B, *and suppose (6) holds. Then* $A^+ \cong B_{\mathcal{F}(\bar{p})}$ *by an isomorphism fixing all elements of* B *and taking* ∞ *to* $x^\sim$.

(b) *If* $\mathcal{U}$ *is any ultrafilter in* $\mathcal{B}(B)$ *which is disjoint from* $\mathcal{I}(B)$, *then there is a possible play* $\bar{p}$ *such that* $\mathcal{F}(\bar{p}) = \mathcal{U}$.

Proof. Left to the reader. □

Curiously the definable ultrapower construction seems not to have been studied for its own sake. Apart from occasional implicit uses (such as the argument of Shelah reported in the next section), all the published applications of it seem to be devoted to the special case for which Skolem first introduced it, namely models of arithmetic and their close relatives. This special case has one very peculiar feature: B doesn't have to be countable. This is because a Peano-like theory treats bounded sets of elements just as if they were finite, so that its models behave in many ways just as if they were countable. The trick is to internalise the argument within the model.

The next lemma illustrates this. Here and to the end of the section, we take Qx to mean "For an unbounded set of elements x". Over a model of a Peano-like theory, Q is a largeness property; this was verified in Example 5 of the previous section. It is not hard to deduce Axiom 5 from the induction schema.

LEMMA 6.2.4. *Let* T *be a Peano-like theory,* B *a model of* T *and* $\chi(x)$ *a formula of* L(B) *such that* $B \models Qx\chi$. *Then for every formula* $\phi(x,\bar{y})$ *of* L(B) *there is a formula* $\theta(x)$ *of* L(B) *such that*

(a) $B \models Qx\theta \wedge \forall x(\theta \rightarrow \chi)$, *and*

(b) *for every tuple* $\bar{a}$ *from* B, *either* $B \models \phi(b,\bar{a})$ *for all large enough* $b \in \theta(B)$, *or* $B \models \neg\phi(b,\bar{a})$ *for all large enough* $b \in \theta(B)$.

Proof. If B consisted of just the natural numbers we could

reason as follows. The tuples $\bar{a}$ can be listed as $\bar{a}_0$, $\bar{a}_1$, ... Since $\chi(B)$ is unbounded, it has an unbounded overlap either with $\phi(B,\bar{a}_0)$ or with $\neg\phi(B,\bar{a}_0)$. If the former, define θ_0 to be $\chi \wedge \phi(x,\bar{a}_0)$; otherwise define θ_0 to be $\chi \wedge \neg\phi(x,\bar{a}_0)$. Similarly define θ_1 to be either $\theta_0 \wedge \phi(x,\bar{a}_1)$ or $\theta_0 \wedge \neg\phi(x,\bar{a}_1)$, ensuring that $\theta_1(B)$ is unbounded. Etc. Finally define θ so that $\theta(B)$ consists of the first element of $\theta_0(B)$, the second element of $\theta_1(B)$, the third element of $\theta_2(B)$, etc. etc.

Since T is Peano-like, this entire argument can be encoded inside B so that the induction covers all tuples $\bar{a}$. □

Let me apply this to prove a theorem of Haim Gaifman (which is generally and incorrectly known as the MacDowell-Specker theorem). If $B \preccurlyeq A$ and a is an element of A, we say that a has <u>definable type over</u> B iff for every formula $\phi(x,\bar{y})$ of L(B) there is a formula $\psi(\bar{y})$ of L(B) such that for every tuple $\bar{b}$ in B, $A \models \phi(a,\bar{b})$ iff $B \models \psi(\bar{b})$. Note that if A is generated by a and the elements of B, and a has definable type over B, then

(7) for every formula $\phi(x)$ of L(A), $\phi(A) \cap \mathrm{dom}(B)$ is of form $\psi(B)$ for some $\psi(x)$ in L(B).

Elementary extensions A of B with property (7) are called <u>conservative</u>.

THEOREM 6.1.5. *Let* T *be a Peano-like theory in a countable first-order language* L, *and let* B *be a model of* T. *Then* B *has a conservative proper elementary end extension.*

<u>Proof</u>. Without loss we assume T has Skolem functions, so that we can construct the end extension as a definable ultrapower of B. It will be convenient to use the language of games. The two players choose a sequence $(p_i)_{i<\omega}$ of conditions. But now we define a <u>condition</u> to be a set of form $\phi(B)$ which is in $\mathscr{B}(B)$ but not in $\mathscr{I}(B)$. The conditions must be chosen so that $p_0 \supseteq p_1 \supseteq \dots$. At the end of the game, when the sequence $\bar{p}$ of conditions has been chosen, we define $\mathscr{F}(\bar{p})$ to be the filter in $\mathscr{B}(B)$ generated by the sets of form $\{b : B \models \phi(b) \wedge c<b\}$ with $\phi(B) \in \{p_i : i < \omega\}$ and c in B. Since a final segment of an unbounded set is again unbounded, $\mathscr{F}(\bar{p})$ is disjoint from $\mathscr{I}(B)$.

Let $\phi(x,\bar{y})$ be a formula of L(B). We claim that the following is enforceable:

(8) There is a formula $\theta(x)$ of $L(B)$ such that $\theta(B) \in \mathcal{F}(\bar{p})$, and for every tuple $\bar{a}$ from B, either $\theta(B) \cap \phi(B,\bar{a})$ or $\theta(B) \cap \neg\phi(B,\bar{a})$ is a final segment of $\theta(B)$.

This follows at once from Lemma 6.2.4.

Since L is countable, the conjunction lemma implies that (8) can be made true simultaneously for all formulas $\phi(x,\bar{y})$ of $L(B)$. Then $\mathcal{F}(\bar{p})$ is an ultrafilter $\mathcal{U}$ which is disjoint from $\mathcal{I}(B)$. Hence $\Pi_{\mathcal{U}}B$ is a proper elementary extension of B.

We show that $x^{\sim}$ has definable type over B. Let $\phi(x,\bar{y})$ be any formula of $L(B)$, and take $\theta(x)$ as in (8). Then for every tuple $\bar{a}$ from B, $\Pi_{\mathcal{U}}B \models \phi(x^{\sim},\bar{a})$ iff $\phi(B,\bar{a}) \in \mathcal{U}$ iff $B \models \exists z \forall x(z<x \wedge \theta(x) \to \phi(x,\bar{a}))$. Thus $\Pi_{\mathcal{U}}B$ is a conservative extension of B.

This fact implies that $\Pi_{\mathcal{U}}B$ is an end extension of B too. For let $t(x,\bar{c})^{\sim}$ be any element of $\Pi_{\mathcal{U}}B$ and suppose there is an element d of B such that $\Pi_{\mathcal{U}}B \models t(x,\bar{c})^{\sim} < d$. Then the set of all elements of B which are greater than $t(x,\bar{c})^{\sim}$ is non-empty and definable in B, so that by the induction schema it has a least element b. We infer that $t(x,\bar{c})^{\sim}+1 = b$, so that $t(x,\bar{c})^{\sim}$ is in B. □

A linearly ordered set is said to be κ-like iff it has cardinality κ and every proper initial segment has cardinality less than κ. (For example, ω-like means of order-type ω. But there are many different ω_1-like orderings.) Since models of a Peano-like theory have a distinguished ordering, we can meaningfully describe them as κ-like or not κ-like.

COROLLARY 6.2.6. *Let* B *be a model of a Peano-like theory in a countable language. Then for every cardinal* κ *greater than the cardinality of* B, B *has a* κ*-like conservative elementary end extension.*

Proof. Keep iterating Theorem 6.2.5, taking unions at limit ordinals. It is clear that the conservative property survives. □

If B is a model of a Peano-like theory, then by a class of B we mean a set $X \subseteq \mathrm{dom}(B)$ such that for every element b of B, the set of elements of X which are $<^B$ b is in $\mathscr{B}(B)$. The reason for the name is as follows. Suppose C is a model of second-order arithmetic; then C has two sorts of element, namely numbers and classes of numbers. If B is the first-order part of C, whose elements are just the numbers, then the classes in C will all be classes of B in the sense just defined.

Conversely if this model C is a model of the comprehension axiom for first-order formulas, then any set in $\mathscr{B}(B)$ must be a class of C. So if every class of B is in $\mathscr{B}(B)$, then C is the unique expansion of B to a model of second-order arithmetic with first-order comprehension. We shall say that B is rather classless iff every class of B is in $\mathscr{B}(B)$.

COROLLARY 6.2.7. *Let* T *be a Peano-like theory in a countable language, and suppose* T *has a model. Then for every cardinal* κ *of uncountable cofinality,* T *has a rather classless* κ*-like model.*

Proof. T has a countable model B_0. As in the previous corollary we can build up a continuous chain $(B_i)_{i<\kappa}$ of conservative elementary end extensions, so that the union $A = \bigcup_{i<\kappa} B_i$ is κ-like. It only has to be shown that A is rather classless.

Let X be a class of A. By taking a suitable subchain of $(B_i)_{i<\kappa}$, we can express A as the union of a continuous chain $(A_j)_{j<cf(\kappa)}$ so that A is a conservative elementary end extension of each A_j. For any $j < cf(\kappa)$, let c be an element of A which is not in A_j. Since X is a class of A, the elements of X below c form a set $\in \mathscr{B}(A)$. By conservativeness it follows that $X \cap dom(A_j)$ is of form $\phi_j(A_j, \bar{a}_j)$ for some formula ϕ_j of the language of T and some parameters $\bar{a}_j$ from A_j.

For each limit ordinal $\delta < cf(\kappa)$, $\bar{a}_\delta$ lies in some $A_{f(\delta)}$ with $f(\delta) < \delta$. This defines a regressive function f on the stationary set of all limit ordinals $< cf(\kappa)$ (cf. Lemma 5.2.4(b)). By Fődor's lemma (Lemma 5.2.5), f has a constant value α on some stationary subset S of $cf(\kappa)$. Since the language is countable, there are a formula $\phi(x,\bar{y})$ and an unbounded subset U of $cf(\kappa)$ such that $U \subseteq S$ and ϕ_j is ϕ for all j in U. If j, k are any two elements of U with $\alpha \leqslant j < k$, then $\phi(A_j, \bar{a}_j) = \phi(A_j, \bar{a}_k)$ since both sides of the equation are equal to $X \cap dom(A_j)$. But then also $\phi(A_k, \bar{a}_j) = \phi(A_k, \bar{a}_k)$ since $A_j \preccurlyeq A_k$. Fixing j, it follows that $X \cap dom(A_k) = \phi(A_k, \bar{a}_j)$ for arbitrarily large $k < cf(\kappa)$, and so $X = \phi(A, \bar{a}_j)$. □

Exercises for 6.2.

1. Verify the statements at the beginning of this section about the forms of Lemma 6.1.1 and Theorems 6.1.3 and 6.1.4 which are true for the witness-free construction. Show that in Theorem 6.1.4 for this construction, the assumption that Q obeys Axiom 5 for $\exists_1$ formulas can be weakened to: For all quantifier-free formulas $\phi(x)$ and $\psi(x)$ of L(B) and every term t(x) of L(B), if $B \models Qx(\phi(x) \wedge \psi(t(x)))$ and $B \models \neg Qx\psi(x)$, then there is an element b of B such that $B \models Qx(\phi(x) \wedge t(x)=b)$.

2. Using the language of rings with 1, let B be a countable field and let $Qx\phi$ mean "There are infinitely many elements which satisfy ϕ". Show that Q is a largeness property over B satisfying Axiom 5, but in the witness-free construction over B using Q, it is enforceable that A is the polynomial ring B[x]. *(So with the best will in the world it is hopeless to expect $\forall_2$ sentences to be preserved from* B *to* A *in this construction.)*

3. Verify Theorem 6.2.3.

4. Let L be a first-order language, T a theory in L with Skolem functions functions, A and B models of T and a an element of A, such that $B \preccurlyeq A$ and a has definable type over B. For each formula $\phi(x,\bar{y})$ of L let $d\phi(\bar{y})$ be a formula of L(B) such that if $\bar{b}$ is any tuple from B then $A \models \phi(a,\bar{b})$ iff $B \models d\phi(\bar{b})$. (a) Show that if C is any structure with $B \preccurlyeq C$, then there is an elementary extension D of C generated over C by an element d such that for all formulas $\phi(x,\bar{y})$ of L and all tuples $\bar{c}$ from C, $D \models \phi(d,\bar{c})$ iff $C \models d\phi(\bar{c})$. (b) Let $(D_i : i \leqslant \lambda)$ be a continuous elementary chain with $B \preccurlyeq D_0$ and each D_{i+1} related to D_i as D is to C in (a), with generating element d_i. Show that for every formula $\chi(x_1,\ldots,x_n)$ of L(B) and all $i_1 < \ldots < i_n < \lambda$ and $j_1 < \ldots < j_n < \lambda$, $D_\lambda \models \chi(d_{i_1},\ldots,d_{i_n})$ iff $D_\lambda \models \chi(d_{j_1},\ldots,d_{j_n})$.

5. Let the structure B be the standard model of first-order Peano arithmetic. Show that B has elementary end extensions which are not conservative. [For any set X of primes, use compactness to add an element divisible by just the primes in X.]

6. Show that if L is a countable first-order language, T is a Peano-like theory in L with Skolem functions, and B is a model of T, then B has a conservative proper end extension A such that for all C, if $B \preccurlyeq C \preccurlyeq A$ then either B = C or C = A. [In the notation of the proof of Theorem 6.2.5, we want that if $t^A(x^\sim,\bar{a}) \notin B$ then there is a set $\theta(B) \in \mathcal{U}$ such that the map

$b \mapsto t^B(b,\bar{a})$ is 1-1 on $\theta(B)$. Then $x^\sim$ is definable with parameters $t^A(x^\sim,\bar{a})$ and elements of B.]

7. Let T be a countable Peano-like theory with Skolem functions. If A, B are models of T, we write $B < A$ to mean that A is a proper end extension of B (necessarily elementary). (a) Show that if A is the extension of B constructed in the proof of Theorem 6.2.5, then there is no structure C such that $B < C < A$. (b) Show that if B is a countable model of T and α is any countable order-type, then there is an end extension A of B such that the order-type of $\{C : B < C < A\}$ is that of the non-empty proper initial segments of α. (c) Show that if T has models then it has 2^ω non-isomorphic countable models.

8. Show that if L is a countable first-order language whose symbols include $<$, C is an L-structure such that $\mathrm{dom}(C) = \omega$ and $<^C$ is the usual ordering of ω, and $B \equiv C$, then B has a proper elementary end extension.

One can adapt definable ultrapowers to construct Σ_{n+1} *end extensions, by using partial Skolem functions for* Σ_n *formulas. Some uniformisation principle is needed to find the Skolem functions; on the other hand the structures need not be countable. The next exercise shows this idea at work in set theory. Notation is as for Exercise 13 of section 6.1.* L_α *is the set of all sets constructible before* α.

9. Let α be a limit ordinal and $n \geq 1$. Write $\mathscr{B}$ for the boolean algebra generated by the Π_n-definable subsets of L_α. A Π_n <u>ultrafilter on</u> L_α is a non-principal ultrafilter $\mathscr{U}$ in $\mathscr{B}$ such that for every Π_n formula $\phi(x,y)$ of $L(L_\alpha)$ and element b of L_α, if $\phi(L_\alpha,a) \in \mathscr{U}$ for every $a \in b$ then $\bigcap\{\phi(L_\alpha,a) : a \in b\} \in \mathscr{U}$. Show: (a) If A is any proper Σ_{n+1} end extension of L_α, a is in A but not in L_α, and $\mathscr{U}$ is the set of all sets $\phi(L_\alpha)$ such that $\phi(x)$ is a boolean combination of Π_n formulas of $L(L_\alpha)$ and $A \models \phi(a)$, then $\mathscr{U}$ is a Π_n ultrafilter on L_α. (b) If $\mathscr{U}$ is a Π_n ultrafilter on L_α, then there is a proper Σ_{n+1} end extension A of L_α got as follows: the elements of A are the equivalence classes under $\sim$ of Σ_n-definable partial functions t, where $s \sim t$ iff $\{b : L_\alpha \models s(b)=t(b)\} \in \mathscr{U}$ and $s^\sim \in t^\sim$ iff $\{b : L_\alpha \models s(b) \in t(b)\} \in \mathscr{U}$.

10. Let L be a first-order language, λ an infinite cardinal and B an L-structure of cardinality λ such that every subset of $\mathrm{dom}(B)$ is of form R^B for some 1-ary relation symbol R of L. Let A be a proper elementary extension of B and a an element of A which is not in B. (a) Show that

$\{X \subseteq \mathrm{dom}(A) : \text{for some } R,\ X = R^B \text{ and } A \models R(a)\}$ is a non-principal ultrafilter $\mathcal{U}$ on dom(B). Show also that $\mathcal{U}$ is not λ^+-complete. (b) If $\mathcal{U}$ is the ultrafilter in (a), Th(B) has Skolem functions and A is generated by B and a, show that $A \cong \Pi_{\mathcal{U}} B$. (c) Show that if $\lambda^\omega > \lambda$ and λ is less than the first measurable cardinal, then not every structure of cardinality λ has a proper elementary extension of cardinality λ. [If λ is less than the first measurable and $\mathcal{U}$ is not λ^+-complete, then $\mathcal{U}$ is countably incomplete and hence $|\Pi_{\mathcal{U}} B| \geq \lambda^\omega$.]

11. Let T be a Peano-like theory and λ an uncountable cardinal. Show that if T has a λ-like ω-saturated rather classless model then there exists a λ-Aronszajn tree (cf. section 8.1 below). [If B is such a model, regard any element b of B as coding a set b^p of ordered pairs. The tree consists of the elements b such that $b^p \cap (\omega \times \mathrm{dom}\ B)$ encodes a truth definition ($\langle n,a\rangle \in b^p \Leftrightarrow B \models \phi_n(a)$) for an initial segment of B, for formulas of arithmetic. In the hypothesis, ω-saturation can be weakened to recursive saturation.]

12. Show: (a) Theorem 6.2.5 fails if L is allowed to be uncountable. (b) No model of ZFC has a proper conservative elementary end extension. (c) If V = L then L_{ω_1} has no proper Σ_2 end extension.

6.3 UNCOUNTABLE BOOLEAN ALGEBRAS

This section aims at a single result on boolean algebras, using the witness-free construction. In section 4.1 we played several games in parallel. Now we shall follow the opposite tack and play uncountably many games in series. Chapter 7 will exploit the same idea.

I shall regard boolean algebras as structures in a first-order language L whose symbols are $\wedge$, $\vee$, $\tilde{}$, 0, 1; $\tilde{}$ means complement. The algebra ordering $\leq$ is defined by: $a \leq b$ iff $a \wedge b = a$. A chain is a set of elements which is linearly ordered in this ordering. Elements a, b are comparable iff $a \leq b$ or $b \leq a$. A pie is a set of Pairwise Incomparable Elements. An atom is an element $b \neq 0$ such that $0 \leq a \leq b$ implies $a = 0$ or $a = b$. A boolean algebra is atomless iff it has no atoms.

THEOREM 6.3.1. *(Assuming $2^\omega = \omega_1$) there is an uncountable boolean algebra which has no uncountable chains and no uncountable pies.*

The eye of intuition hardly tells one whether this theorem is going to be easy work. In fact significant results about uncountable boolean algebras tend to be difficult, and this is a significant result. I hope to show that games give a sense of direction to the proof.

The proof falls into four parts. In part I we use games to construct a proper extension of a countable atomless boolean algebra. This part is a straightforward example of the techniques of section 6.1, in the witness-free form introduced in section 6.2.

Part II discusses how to build up a battery of uncountably many extensions as in part I, so that the later extensions are "generic" in relation to the earlier ones. This part will be familiar ground to set theorists who work with iterated forcing. The argument is very general and never mentions boolean algebras. The purpose of the argument, which may not be clear at first, is to arrange in order the tasks which have to be performed during the construction, so that we meet each task in time to carry it out.

Part III constructs the boolean algebra of cardinality ω_1 by amalgamating ω_1 separate countable extensions of a countable algebra; it combines the work of parts I and II. Finally part IV shows that the construction works. The argument involves some heavy combinatorics, but again the drift will be familiar to set theorists. We show that any uncountable set of elements contains an uncountable subset which is in some sense well-behaved.

To open part I of the proof, we take a countable atomless boolean algebra B, fixed for the rest of the argument. Writing L for the first-order language of boolean algebras as described above, we form L(B) by adding the elements of B as constants, and L' by adding to L(B) an overflow constant ∞. For any formula $\phi(x)$ of L(B) we take $Qx\phi$ to mean: there are infinitely many elements x such that $\phi(x)$.

LEMMA 6.3.2. *Q is a largeness property over* B, *satisfying Axioms 0-5 of section 6.1 for all formulas of* L(B). *We have* $B \models Qx\phi(x)$ *iff: there is* a *in* B *such that* $B \models \phi(a)$ *and* a *is not in the subalgebra of* B *generated by the parameters in* ϕ.

<u>Proof</u>. The first sentence is as in section 6.1. For the second, let D be the subalgebra of B generated by the parameters in ϕ.

Left to right is clear since D is finite. For right to left, choose a so that $B \models \phi(a)$ and $a \notin D$. Since B is atomless, there are infinitely many elements a' of B such that for each d in D, $d \lessgtr a'$ iff $d \lessgtr a$. For each such a' there is an automorphism of B fixing D pointwise and taking a to a' (cf. Exercise 1). So $B \models \phi(a')$ and hence $B \models Qx\phi$. □

By Lemma 6.3.2 we can use Q to compile a proper extension $A(\bar{p})$ of B, using the witness-free game introduced at the beginning of section 6.2. In this game, a condition p mentions at most ∞ and finitely many elements $\bar{b}$ of B. Let us call p total iff p completely specifies the boolean algebra generated by $\bar{b}$ and ∞ (e.g. by giving the complete tables for $\wedge$ and $\vee$). Using Axiom 2 one can show that every condition can be extended to a total condition. But actually we can do better than this, as follows.

Let p be a condition. Write $\bigwedge p$ as $\phi(\infty,\bar{b})$, where $\bar{b}$ are all the elements of B mentioned in p. Let D_1 be the subalgebra of B generated by $\bar{b}$. Since $B \models Qx\phi(x,\bar{b})$, Lemma 6.3.2 supplies an element e of B such that $B \models \phi(e,\bar{b})$ and e is not in D_1. Let D_2 be the algebra generated by $\bar{b}$ and e. Let D be a subalgebra of D_2 containing D_1 and maximal with the property that $e \notin D$. It is easy to see that D has exactly one atom a such that $0 < a \wedge e < a$, and that D and e generate D_2. Now we can extend p to a condition q which describes D_2 and names all the elements of D. Thus:

LEMMA 6.3.3. *For every condition* p *there is a condition* $q \supseteq p$ *such that*

(a) q *is total;*

(b) *the elements of* B *mentioned in* q *form a subalgebra* D *of* B*;*

(c) *there is exactly one atom* d *of* D *such that* q *implies* $0 < d \wedge \infty < d$. □

We shall call the atom d in (c) above the split atom of q. When C is any boolean algebra $\supseteq$ B, we say that B is dense in C iff for every $c \neq 0$ in C there is $b \in B$ with $c \geqslant b > 0$.

LEMMA 6.3.4. *It is enforceable that* $A(\bar{p})$ *is an atomless boolean algebra* $\supseteq$ B, A *is generated by* ∞ *and the elements of* B, $A \models \bigcup \bar{p}$, *and* B *is dense in* A.

Proof. By Theorem 6.1.3 in the witness-free version (cf. the beginning of section 6.2), it is enforceable that A is a model of U_p^- extending B and generated by ∞ and the elements of B, and all the $\forall_1$ sentences of L(B) which are true in B are true in A too. These sentences include the axioms defining boolean algebras, so that A is a boolean algebra. To make B dense in A it suffices to put non-zero elements of B below all non-zero elements of form $\infty \wedge b$ or $\tilde{\infty} \wedge b$ with b in B; Lemma 6.3.3 makes this easy for player $\exists$. Since B is atomless and dense in A, A is also atomless. □

Suppose now that A has been constructed as in Lemma 6.3.4. Write I_A, or I for short, for the set $\{b \in B : \infty \wedge b \in B\}$. Then I is an ideal in B.

LEMMA 6.3.5. *It is enforceable that* I *is a maximal ideal of* B. *If* J *is any maximal ideal of* B, *then it is enforceable that* $I \neq J$.

Proof. Suppose player $\forall$ has just played condition q, and let b be an element of B. Extending q if necessary, player $\exists$ can assume that q is as in Lemma 6.3.3, and that b is named in q. Write D for the algebra of elements of B mentioned in q, and d for the split atom of q. If $d \leqslant b$ then $\infty \wedge \tilde{b}$ is in B, and this puts $\tilde{b}$ into I. If $d \leqslant \tilde{b}$ then likewise b goes into I. Hence it is enforceable that for every element b of B, either b or $\tilde{b}$ is in I. To make I a maximal ideal, player $\exists$ need only ensure that $\infty \notin B$, using Theorem 6.1.3 (witness-free version).

The argument above shows that every atom of D except d will lie in I, and so d will never reach I. Now let J be any maximal ideal of B. If J contains d then $I \neq J$ without further ado. Suppose then that J contains all atoms of D except d. Since B is atomless, it has an element g such that $0 < g < d$. Just one of g and $d \wedge \tilde{g}$ lies in J; let it be g. Then player $\exists$ can add to q the sentence "$d \wedge \tilde{g} \wedge \infty = 0$". This will put $d \wedge \tilde{g}$ into I and hence prevent g from reaching I. It will follow that $I \neq J$. □

We turn to part II of the proof. Imagine the following game G of length ω_1: a countable atomless boolean algebra B is given, and players $\forall$ and $\exists$ play the game discussed above, ω_1 times in succession, so that at the α-th game a structure $A_\alpha = A(\bar{p}_\alpha)$ is constructed. Player $\exists$

wins G iff the whole bundle of resulting sets $\bigcup \bar{p}_\alpha$ has a certain property P. The property P is said to be hyperenforceable iff player ∃ has a winning strategy for this game G. Such games of length ω_1 will be called hypergames. Another version of them will appear in Chapter 7.

If for each $\alpha < \omega_1$, Q_α is an enforceable property, then clearly it is hyperenforceable that each A_α has property Q_α $(\alpha < \omega_1)$. Also the conjunction of countably many hyperenforceable properties is hyperenforceable, by the conjunction lemma for the games of length ω. This is even true for an uncountable conjunction of hyperenforceable properites, provided that each of the smaller games is concerned with just countably many of them. In fact we shall consider hypergames where the set of tasks to be performed by player ∃ keeps growing as the game proceeds.

Let N be the notion of forcing which was used to build A from B in part I of the proof. A subset X of N is said to be dense iff for every $p \in N$ there is $q \in X$ such that $p \subseteq q$. More generally if m is a positive integer and $X \subseteq N^m$, then we call X dense iff for every $(p_1,\ldots,p_m) \in N^m$ there is $(q_1,\ldots,q_m) \in X$ with $p_k \subseteq q_k$ for each k $(1 \leq k \leq m)$.

Imagine that as a hypergame is played, just before each A_α is constructed, a countable set S_α of dense subsets of N is named. Either player is allowed to put dense sets into S_α, and the choice of S_α may depend on what happened in the games to construct A_β $(\beta < \alpha)$.

LEMMA 6.3.6. *(a) It is hyperenforceable that for each* $\alpha < \omega_1$ *and each* $X \in S_\alpha$, $\bigcup \bar{p}_\alpha$ *contains some condition* $\in X$.

(b) Let X *be a dense subset of* N^m, *for some positive integer* m. *Then it is hyperenforceable that: whenever* $\alpha_1 < \ldots < \alpha_m < \omega_1$, *there is* $(q_1,\ldots,q_m) \in X$ *such that* $\bigcup \bar{p}_{\alpha_k}$ *contains* q_k $(1 \leq k \leq m)$.

(c) Let m *be a positive integer and* X_β $(\beta < \omega_1)$ *dense subsets of* N^m. *Then it is hyperenforceable that: for each* $\beta < \omega_1$, *whenever* $\beta < \alpha_1 < \ldots < \alpha_m < \omega_1$, *there is* $(q_1,\ldots,q_m) \in X_\beta$ *such that for each* k $(1 \leq k \leq m)$, $\bigcup \bar{p}_{\alpha_k}$ *contains* q_k.

Proof. (a) If X is dense in N, then for each $\alpha < \omega_1$ it is enforceable that $\bigcup \bar{p}_\alpha$ contains some element of X. Since each S_α is countable, (a) follows by the conjunction lemma for countable games.

(b) Suppose $1 \leqslant k \leqslant m$. Given $\alpha_1 < \ldots < \alpha_k < \omega_1$ and $r_{k+1}, \ldots, r_m \in N$, let $X(\alpha_1,\ldots,\alpha_k,r_{k+1},\ldots,r_m)$ be the set

(1) $$\{q_k \in N : \text{for some } q_1 \subseteq \bigcup \bar{p}_{\alpha_1}, \ldots, q_{k-1} \subseteq \bigcup \bar{p}_{\alpha_{k-1}}, q_{k+1} \supseteq r_{k+1}, \ldots, q_m \supseteq r_m, \text{ we have } (q_1,\ldots,q_m) \in X\}.$$

This set is well-defined as soon as $A_{\alpha_{k-1}}$ has been constructed. For a fixed $\alpha = \alpha_k$ there are just countably many possibilities for $\alpha_1, \ldots, \alpha_{k-1}, r_{k+1}, \ldots, r_m$, and for each such sequence the set (1) is well-defined at the beginning of the construction of A_α. So we can assume without loss that every dense set of this form lies in S_α. We can assume this simultaneously for all k ($1 \leqslant k \leqslant m$) and all α_k. Let player ∃ play her winning strategy for (a). We shall show that the conclusion of (b) holds.

We claim that for all k and all $\alpha_1, \ldots, \alpha_k, r_{k+1}, \ldots, r_m$:

(2) $X(\alpha_1,\ldots,\alpha_k,r_{k+1},\ldots,r_m)$ is dense in N;

(3) There are $q_1 \subseteq \bigcup \bar{p}_{\alpha_1}, \ldots, q_k \subseteq \bigcup \bar{p}_{\alpha_k}$ such that for some $q_{k+1} \supseteq r_{k+1}, \ldots, q_m \supseteq r_m$ we have $(q_1,\ldots,q_m) \in X$.

The claim is proved by induction on k, for (2) and (3) simultaneously. When k = 1, (2) is immediate from the fact that X is dense, and (3) follows from player ∃'s strategy. Assuming the claim true for k-1 we prove it for k. Consider $X(\alpha_1,\ldots,\alpha_k,r_{k+1},\ldots,r_m)$ and any $r_k \in N$. By induction hypothesis (3) there is $q_k \supseteq r_k$ such that for some $q_1 \subseteq \bigcup \bar{p}_{\alpha_1}, \ldots, q_{k-1} \subseteq \bigcup \bar{p}_{\alpha_{k-1}}, q_{k+1} \supseteq r_{k+1}, \ldots, q_m \supseteq r_m$ we have $(q_1,\ldots,q_m) \in X$. Hence (2) holds for k. Again (3) follows by player ∃'s strategy. This proves the claim.

(b) follows from (3) when k = m.

(c) Player ∃ can arrange to take care of the sets X_β ($\beta < \alpha$) during the construction of A_α. Although there are uncountably many sets X_β, she has only countably many to deal with at any one time, and she can manage this by the conjunction lemma for countable games. □

Somehow we have to exploit the continuum hypothesis $2^\omega = \omega_1$, so that we can say something about every uncountable set of elements of the final boolean algebra. Recall that with the stronger assumption ◊

(cf. section 5.2) we would have been able to predict each uncountable set during the course of the construction. Here we can't do that. But the next lemma shows that there is a hyperenforceable property which controls all uncountable sets M of a certain special form. In part IV of the proof we shall see why this is enough.

LEMMA 6.3.7. *(Assuming* $2^\omega = \omega_1$*) the following is hyperenforceable:*

For any positive integer m, *let* M *be any set of* m*-tuples* $\{(\alpha_1^\gamma,\ldots,\alpha_m^\gamma) : \gamma < \omega_1\}$ *such that* $\alpha_1^\gamma < \ldots < \alpha_m^\gamma < \omega_1$ *for each* γ, *and if* $\beta < \gamma$ *then* $\alpha_m^\beta < \alpha_1^\gamma$. *Then there is some* $(r_1,\ldots,r_m) \in N^m$ *such that for all* $(q_1,\ldots,q_m) \in N^m$, *if* $q_1 \supseteq r_1, \ldots, q_m \supseteq r_m$ *then there are uncountably many* $\gamma < \omega_1$ *such that each* $\bigcup \bar{p}_{\alpha_k^\gamma}$ *contains* q_k $(1 \leqslant k \leqslant m)$.

Proof. Since a conjunction of countably many hyperenforceable properties is hyperenforceable, we can fix m. Since $2^\omega = \omega_1$ and N is countable, we can list as X_β $(\beta < \omega_1)$ all dense subsets of N^m. With this choice of the X_β, let player ∃ use her strategy which ensures the conclusion of Lemma 6.3.6(c). We claim that this works for the present lemma too.

For let M be as stated, and suppose for contradiction that the set $X = \{(q_1,\ldots,q_m)$: there are only countably many $\gamma < \omega_1$ such that $q_1 \subseteq \bigcup \bar{p}_{\alpha_1^\gamma}, \ldots, q_m \subseteq \bigcup \bar{p}_{\alpha_m^\gamma}\}$ is dense in N^m. Then $X = X_\beta$ for some $\beta < \omega_1$, and hence by player ∃'s strategy, for all but countably many γ there is $(q_1,\ldots,q_k) \in X$ such that $\bigcup \bar{p}_{\alpha_k^\gamma}$ contains q_k $(1 \leqslant k \leqslant m)$. Since N^m is countable, the same $(q_1,\ldots,q_m)$ must serve for uncountably many γ, which contradicts the definition of X. □

Now comes the construction proper, which is part III of the proof. We require the players ∀ and ∃ to play the game of part I over and over again, so that they build boolean algebras A_α $(\alpha < \omega_1)$ extending the countable atomless boolean algebra B. We write ∞_α for the overflow constant used in the construction of A_α, and I_α for the ideal I_{A_α} of A_α. Thus the players play a hypergame as in part II above, and we have:

LEMMA 6.3.8. *The following is hyperenforceable. For each* $\alpha < \omega_1$, A_α *is an atomless boolean algebra generated by* B *and* ∞_α, B *is dense in* A_α, $A_\alpha \models \bigcup \bar{p}_\alpha$, I_α *is a maximal ideal of* B *(and hence* $A_\alpha \neq B$*); for all* $\beta < \alpha < \omega_1$, $I_\beta \neq I_\alpha$*; and the hyperenforceable property of Lemma 6.3.7 holds.*

Proof. Lemmas 6.3.4, 6.3.5 and 6.3.7. □

We assume that player ∃ uses her strategy to ensure the properties in Lemma 6.3.8. There remains the problem of packing the separate algebras A_α into a single uncountable algebra. This is not a very delicate part of the construction, and I use the first method that comes to hand:

LEMMA 6.3.9. *Every atomless boolean algebra is existentially closed in the class of boolean algebras.*

Proof. The class of atomless boolean algebras is axiomatised by a $\forall_2$ first-order theory T, and any two countable models of T are isomorphic. So by Theorem 3.2.10, T is model-complete. It follows by Lemma 3.2.9 and Theorem 3.2.3 that every atomless boolean algebra is e.c. in the class of models of $T_\forall$. By Corollary 3.2.2 and the lemma on preservation (Lemma 1.2.4(a)), the models of $T_\forall$ are just those boolean algebras which can be extended to atomless boolean algebras. I leave it as an exercise to show that every boolean algebra can be extended to an atomless boolean algebra. □

It follows by Theorem 3.2.7 that an atomless boolean algebra is a strong amalgamation base in the class of boolean algebras. Hence we can construct a chain $(B_\alpha : \alpha < \omega_1)$ of boolean algebras as follows, by induction on α. B_0 is B. For each limit ordinal $\delta \leqslant \omega_1$, B_δ is $\bigcup_{\alpha<\delta} B_\alpha$. When B_α has been constructed, we consider the extensions $B \subseteq B_\alpha$ and $B \subseteq A_\alpha$, and we use the strong amalgamation property of B to find a boolean algebra $C \supseteq B_\alpha$ and an embedding $f: A_\alpha \to C$ such that f is the identity on B and $B_\alpha \cap f(A_\alpha) = B$. We define $B_{\alpha+1}$ to be the subalgebra of C generated by B_α and $f(A_\alpha)$. Since $A_\alpha \neq B$, the choice of C implies that $B_{\alpha+1} \neq B_\alpha$. Hence the chain is strictly increasing. Using the maps f, we can assume without loss that each A_α is a subalgebra of B_{ω_1}, and hence that B_{ω_1} is generated by B and the elements ∞_α $(\alpha < \omega_1)$.

Write A^* for B_{ω_1}. Certainly A^* is an uncountable boolean algebra. This concludes the construction.

There remains part IV of the proof, in which we show that A^* has no uncountable chains and no uncountable pies. We start by bringing each element of A^* to a kind of normal form.

LEMMA 6.3.10. *For every element* b *of* A^* *there is a finite boolean algebra* $D \subseteq B$ *such that* b *can be written as*

$$(4) \qquad b = (d_1 \wedge \infty'_{\alpha_1}) \vee \ldots \vee (d_m \wedge \infty'_{\alpha_m}) \vee d$$

where $d_1, \ldots, d_n$ *are the atoms of* D, $m \leqslant n$, $d \in D$, $d \leqslant d_{m+1} \vee \ldots \vee d_n$, *each* ∞'_{α_k} *is either* ∞_{α_k} *or* $\tilde{\infty}_{\alpha_k}$, *and for each* k $(1 \leqslant k \leqslant m)$, $0 < d_k \wedge \infty_{\alpha_k} < d_k$ *and* $\tilde{d}_k \wedge \infty_{\alpha_k} \in B$.

Proof. We can write $b = t(\bar{d}, \infty_{\alpha_1}, \ldots, \infty_{\alpha_m})$ for some boolean term t, some tuple $\bar{d}$ from B and some $\alpha_1 < \ldots < \alpha_m < \omega_1$. If $1 \leqslant h < k \leqslant m$ then $I_{\alpha_h} \neq I_{\alpha_k}$ by construction. So we can assume without loss that $\bar{d}$ contains elements of $I_{\alpha_h} \setminus I_{\alpha_k}$ and $I_{\alpha_k} \setminus I_{\alpha_h}$. Let D be the algebra generated by $\bar{d}$ and let $d_1, \ldots, d_n$ be its atoms. Each set $B \setminus I_{\alpha_k}$ contains just one atom, no two the same atom. So we can number the atoms so that $d_k \notin I_{\alpha_k}$ and $\tilde{d}_k \wedge \infty_{\alpha_k} \in B$ $(1 \leqslant k \leqslant m)$, using the definition of I_{α_k}.

If $h \neq k$ then $d_h \in I_{\alpha_k}$ and hence $d_h \wedge \infty_{\alpha_k} \in B$. Adding finitely many elements to $\bar{d}$ if necessary, we can suppose that

(5) if $h \neq k$ then $d_h \wedge \infty_{\alpha_k}$ lies in D and hence is d_h or 0.

(Adding elements to D changes the atoms of D. It should be checked that there is no regress here.)

Now $b = (d_1 \wedge b) \vee \ldots \vee (d_n \wedge b)$. Consider for example $d_1 \wedge b$. Writing $t(\bar{d}, \infty_{\alpha_1}, \ldots, \infty_{\alpha_m})$ in disjunctive normal form as $t_1 \vee \ldots \vee t_g$, $d_1 \wedge b$ can be written as $(d_1 \wedge t_1) \vee \ldots \vee (d_1 \wedge t_g)$. If one of the terms t_f contains ∞_{α_k} or $\tilde{\infty}_{\alpha_k}$ for some $k \neq 1$, then by (5), $d_1 \wedge \infty_{\alpha_k}$ is either d_1 or 0, and likewise $d_1 \wedge \tilde{\infty}_{\alpha_k}$; hence the term $d_1 \wedge t_f$ can be simplified to remove ∞_{α_k}. Thus each $d_1 \wedge t_f$ reduces to one of the forms d_1, 0, $d_1 \wedge \infty_{\alpha_1}$ or $d_1 \wedge \tilde{\infty}_{\alpha_1}$. The union of terms of any of these four forms again reduces to a term of one of these forms. Similarly with $d_2, \ldots, d_n$.

Put $d = (d_{m+1} \vee \ldots \vee d_n) \wedge b$. If $1 \leqslant k \leqslant m$ and $d_k \wedge b$ is either d_k or 0, renumber so as to incorporate $d_k \wedge b$ in d. For the remaining k $(1 \leqslant k \leqslant m)$ we have $0 < d_k \wedge \infty_{\alpha_k} < d_k$, and $d_k \wedge b$ is either $d_k \wedge \infty_{\alpha_k}$ or $d_k \wedge \tilde{\infty}_{\alpha_k}$.

So we have brought b to the required form (4). □

Now let $J = \{b_\gamma : \gamma < \omega_1\}$ be an uncountable set of elements of A^*. We shall show that J has an uncountable subset which can be written in a simple form.

We start from Lemma 6.3.10, which writes each element b_γ in the form

(6) $$b_\gamma = (d_1^\gamma \wedge \infty'_{\alpha_1^\gamma}) \vee \ldots \vee (d_{m_\gamma}^\gamma \wedge \infty'_{\alpha_{m_\gamma}^\gamma}) \vee d^\gamma.$$

Using the fact that B is countable, we can assume (after replacing J by a suitable uncountable subset and renumbering) that:

(7) m_γ is a fixed integer m for all $\gamma < \omega_1$;

(8) for all $\beta < \gamma < \omega_1$ and all k ($1 \leqslant k \leqslant m$), $\infty'_{\alpha_k^\beta}$ is $\infty_{\alpha_k^\beta}$ iff $\infty'_{\alpha_k^\gamma}$ is $\infty_{\alpha_k^\gamma}$;

(9) $d_1^\gamma, \ldots, d_m^\gamma, d^\gamma$ are fixed elements $d_1, \ldots, d_m$, d of B;

(10) for each k ($1 \leqslant k \leqslant m$), $\tilde{d}_k \wedge \infty_{\alpha_k^\gamma}$ is a fixed element d'_k of B.

By the Δ-system lemma (Exercise 3 of section 5.2), replacing J again by a suitable uncountable subset, we can suppose that

(11) for all $\beta < \gamma < \omega_1$ we have $\alpha_1^\beta = \alpha_1^\gamma, \ldots, \alpha_h^\beta = \alpha_h^\gamma, \alpha_m^\beta < \alpha_{h+1}^\gamma$

where h is a fixed integer. Thus we can write α_k^γ as α_k when $1 \leqslant k \leqslant h$.

Applying Lemma 6.3.7 to the set $M = \{(\alpha_{h+1}^\gamma, \ldots, \alpha_m^\gamma) : \gamma < \omega_1\}$, we find elements $r_{h+1}, \ldots, r_m \in N$ such that

(12) for all $q_{h+1} \supseteq r_{h+1}, \ldots, q_m \supseteq r_m$ in N, there is an uncountable set $X \subseteq \omega_1$ such that for all $\gamma \in X$ we have $q_{h+1} \subseteq \bigcup \bar{p}_{\alpha_{h+1}^\gamma}, \ldots q_m \subseteq \bigcup \bar{p}_{\alpha_m^\gamma}$.

Without loss we can suppose that each r_k is as q in Lemma 6.3.3; let E_k be the algebra of elements mentioned in r_k, and e_k the split atom of r_k.

Put $Y = \{\gamma < \omega_1 : r_k \subseteq \bigcup \bar{p}_{\alpha_k^\gamma}\ (h < k \leqslant m)\}$. By (12), Y is uncountable, and clearly each X occurring in (12) must be a subset of Y.

Hence by one final decimation we can suppose that Y is ω_1, i.e.

(13) for all $\gamma < \omega_1$, $r_{h+1} \subseteq \bigcup \bar{p}_{\alpha^\gamma_{h+1}}$, ..., $r_m \subseteq \bigcup \bar{p}_{\alpha^\gamma_m}$.

Given the preceding lemmas, these are all fairly routine reductions of J. We should stand back a moment and see what they have achieved.

By our various reductions, each element b_γ of J can be written in the form

(14)
$$\begin{aligned} b_\gamma = {} & (d_1 \wedge \infty'_{\alpha^\gamma_1}) \vee \ldots \vee (d_h \wedge \infty'_{\alpha^\gamma_h}) \vee d \\ & \vee (d_{h+1} \wedge \tilde{e}_{h+1} \wedge \infty'_{\alpha^\gamma_{h+1}}) \vee \ldots \vee (d_k \wedge \tilde{e}_k \wedge \infty'_{\alpha^\gamma_k}) \\ & \vee (d_{h+1} \wedge e_{h+1} \wedge \infty'_{\alpha^\gamma_{h+1}}) \vee \ldots \vee (d_k \wedge e_k \wedge \infty'_{\alpha^\gamma_k}). \end{aligned}$$

In this expression for b_γ, the first line is clearly independent of γ. So is the second line, for the following reason. Suppose $h < k \leqslant m$. By part III and (13), $A_{\alpha^\gamma_k} \models r_k$. By choice of r_k and e_k, r_k specifies that $\tilde{e}_k \wedge \infty$ and $\tilde{e}_k \wedge \tilde{\infty}$ are certain fixed elements of B. So by (8) the second line in (14) is independent of γ.

Turning to the third line, we claim that for each k ($h < k \leqslant m$), $d_k \wedge e_k \neq 0$. For otherwise $e_k \wedge \infty'_{\alpha^\gamma_k} = e_k \wedge d_k \wedge \infty'_{\alpha^\gamma_k}$, which is an element d^* of B independent of γ, by (10). Then we get a contradiction from (12) by taking $q_k = r_k \cup \{e_k \wedge \infty \neq d^*\}$ (or $= r_k \cup \{e_k \wedge \tilde{\infty} \neq d^*\}$, according to how k sits in (8)), and $q_{k'} = r_{k'}$ when $k' \neq k$.

Put $z_1 = d_{h+1} \wedge e_{h+1}$, ..., $z_{m-h} = d_m \wedge e_m$ and $z_0 = (z_1 \vee \ldots \vee z_{m-h})^\sim$. Then

(15) $$z_0 \vee \ldots \vee z_{m-h} = 1, \quad \text{and} \quad z_j \wedge z_k = 0 \text{ whenever } 0 \leqslant j < k \leqslant m-h.$$

(15) says that z_0, ..., z_{m-h} form a <u>partition of unity</u>. The first two lines of (14) name an element c of A^* independent of γ, and $c \leqslant z_0$.

In sum, each element b_γ of J can be written as

(16) $$b_\gamma = (z_1 \wedge \infty'_{\alpha^\gamma_{h+1}}) \vee \ldots \vee (z_{m-h} \wedge \infty'_{\alpha^\gamma_m}) \vee c$$

where c is an element $\leqslant z_0$ and independent of γ, and z_0, ..., z_{m-h} are a partition of unity in B, in which z_1, ..., z_{m-h} are non-zero. This is our promised "simple form" for an uncountable subset of the original set J.

The substance of the next lemma is that as γ varies, the elements $z_k \wedge \infty'_{\alpha_{h+k}^{\gamma}}$ in (16) are dense below z_k $(1 \leqslant k \leqslant m-h)$ in a strong sense.

LEMMA 6.3.11. *Let* J *be any uncountable set of elements of* A^*. *Then there are a partition of unity* $z_0, \ldots, z_n$ *lying in* B *with* $n > 0$ *and* $z_1, \ldots, z_n$ *all non-zero, and an element* $c \leqslant z_0$ *in* A^*, *such that:*

> *Whenever* a_k, a'_k *are elements of* B *such that* $a_k < a'_k \leqslant z_k$ $(1 \leqslant k \leqslant n)$, *there is an uncountable subset* H *of* J *such that for all* $b \in H$, $b \wedge z_0 = c$ *and* $a_k \leqslant b \wedge z_k \leqslant a'_k$ $(1 \leqslant k \leqslant n)$.

Proof. Clearly it does no harm to replace J by an uncountable subset of J. So we can assume that J has been normalised as in the discussion leading up to (16). Putting $n = m-h$, let $z_0, \ldots, z_n$ be the partition of unity introduced at (15), and let c be as in (16). Note that $m-h > 0$ by (16).

Suppose now that elements a_k, a'_k of B are given, so that $a_k < a'_k \leqslant z_k$ for all k $(1 \leqslant k \leqslant m-h)$. For each k we shall define a condition $q_{h+k} \supseteq r_{h+k}$. There are two cases, according to which of the possibilities in (8) holds at h+k.

Suppose first that $\infty'_{\alpha_{h+k}^{\gamma}} = \infty_{\alpha_{h+k}^{\gamma}}$ for all $\gamma < \omega_1$. Then put

$$(17) \qquad q_{h+k} = r_{h+k} \cup \{a_k \leqslant z_k \wedge \infty,\ z_k \wedge \infty \leqslant a'_k\}.$$

The choice of z_k implies that q_{h+k} is a condition. When $\infty'_{\alpha_{h+k}^{\gamma}} = \tilde{\infty}_{\alpha_{h+k}^{\gamma}}$ for all $\gamma < \omega_1$, the definition of q_{h+k} is the same except that ∞ is replaced by $\tilde{\infty}$ throughout. From (16) it is clear that in both cases we have

$$(18) \qquad b_\gamma \wedge z_k = z_k \wedge \infty'_{\alpha_{h+k}^{\gamma}}$$

for all k $(1 \leqslant k \leqslant m-h)$ and all $\gamma < \omega_1$.

Now at last we use the full strength of Lemma 6.3.7. With $q_{h+1}, \ldots, q_m$ as just defined, let X be as in (12). Since $A_\alpha \models \bigcup \bar{p}_\alpha$ for each $\alpha < \omega_1$, we infer

(19) $\quad a_k \leqslant z_k \wedge \infty'_{\alpha_{h+k}}\gamma \leqslant a'_k$ for all k $(1 \leqslant k \leqslant m-h)$ and all $\gamma \in X$.

Comparing (18) and (19), we have the lemma with $H = \{b_\gamma : \gamma \in X\}$. □

The proof of Theorem 6.3.1 is now almost immediate.

First suppose for contradiction that J is an uncountable chain in A^*. Find $z_0, \ldots, z_n$ as in Lemma 6.3.11. Since $z_1 \neq 0$ and B is atomless, we can find y in B such that $0 < y < z_1$. Choosing $y = a_1 < a'_1 < z_1$ in the lemma, we find b_1 in J so that $y \leqslant b_1 \wedge z_1 < z_1$. Next, choosing $\tilde{y} \wedge z_1 = a_1 < a'_1 < z_1$ in the lemma, we find b_2 in J so that $\tilde{y} \wedge z_1 \leqslant b_2 \wedge z_1 < z_1$. Since J is a chain, b_1 and b_2 are comparable. Suppose $b_1 \leqslant b_2$. Then $y \leqslant b_1 \wedge z_1 \leqslant b_2 \wedge z_1$ and so $z_1 \leqslant y \vee (y \wedge z_1) \leqslant b_2 \wedge z_1 < z_1$, contradiction. Similarly $b_2 \leqslant b_1$ leads to contradiction.

Next suppose that J is an uncountable pie in A^*. Choose $z_0, \ldots, z_n$ as in Lemma 6.3.11, and for each k $(1 \leqslant k \leqslant n)$ find $0 < y_k < z_k$ in B. Putting $a_k = 0$ and $a'_k = y_k$ in the lemma, we find b_1 in J such that $b_1 \wedge z_k \leqslant y_k$ for each k, and so $b_1 \leqslant c \vee y_1 \vee \ldots \vee y_n$. Likewise putting $a_k = y_k$ and $a'_k = z_k$ in the lemma, we find b_2 in J such that $b_2 \geqslant c \vee y_1 \vee \ldots \vee y_n$. This contradicts the assumption that J is a pie.

□ Theorem 6.3.1

Exercises for 6.3.

1. (a) Let A, A' be boolean algebras, B a finite subalgebra of A, f: B → A' an embedding and C a subalgebra of A generated by B and an element c. Let d be any element of A' such that for each b in B, $b \lessgtr c$ iff $f(b) \lessgtr d$. Show that f can be extended to an embedding g: C → A' such that g(c) = d. (b) If A, A' are countable atomless boolean algebras, B a finite subalgebra of A and f: B → A' an embedding, show that there are 2^ω distinct isomorphisms g: A → A' extending f. (c) Show that up to isomorphism there is just one countable atomless boolean algebra, and it has 2^ω automorphisms.

2. (a) Show that every existentially closed boolean algebra is atomless. [For each boolean algebra A and element a of A one must show that there is a boolean algebra $B \supseteq A$ in which a is not an atom. Use Stone representation to write A as an algebra of subsets of a set X, and split an element of X.]

3. Show that every infinite boolean algebra contains an infinite chain and an infinite pie.

An <u>*antichain*</u> *in a boolean algebra* B *is a set* $X \subseteq B \setminus \{0\}$ *such that if* a, b *are distinct elements of* X *then* $a \wedge b = 0$.

4. Show: (a) Every antichain in a boolean algebra is a pie. (b) If λ is any cardinal and B is the free boolean algebra on generators b_i $(i < \lambda)$ (otherwise known as the algebra of propositions in propositional logic with sentence letters b_i $(i < \lambda)$), then B has no uncountable chains and no uncountable antichains. [Use the Δ-system lemma or Fődor's lemma.]

5. An element of a boolean algebra is said to be <u>countable</u> iff there are at most countably many elements below it. Show that if a boolean algebra B has at least ω_2 countable elements then B has a pie of cardinality ω_2. [In Exercise 4 of section 5.2, for each countable element i take X_i to be the set of elements below i.]

6. Show: (a) If B is an infinite boolean algebra which contains no uncountable pie, then B has a countable dense subalgebra. (b) If B is a boolean algebra with a countable dense subalgebra, then B is isomorphic to some algebra of subsets of ω.

For any linearly ordered set I, *the* <u>*interval algebra*</u> *of* I *is defined to be the smallest boolean algebra of subsets of* I *which contains all the sets of form* $\{i \in I : i \leqslant a\}$ *with* $a \in I$.

7. Show: (a) If B is the interval algebra of some linearly ordered set, and B has uncountable cardinality, then B contains either an uncountable chain or an uncountable pie. (b) If $2^{\omega} = \omega_1$ then there is an interval algebra of cardinality ω_1 with no uncountable pie.

8. Let A^* be the boolean algebra constructed in the proof of Theorem 6.3.1. (a) Show that if $\{b_\gamma : \gamma < \omega_1\}$ is a family of elements of A^* then there are $\beta < \gamma < \omega_1$ such that $b_\gamma < b_\beta$. (b) Deduce that every ideal of A^* is generated by a countable set of elements.

9. Let B be a countable atomless boolean algebra, and let players $\forall$ and $\exists$ play witness-free games (as in part I of the proof of Theorem 6.3.1) to construct a boolean algebra $A \supseteq B$. Show: (a) For each maximal chain X in B, it is enforceable that X is a maximal chain in A. (b) For each maximal pie X in B, it is enforceable that X is a maximal pie in A.

10. (a) Assuming $\Diamond$, use Exercise 9 to show that there exists an

atomless boolean algebra A^* of cardinality ω_1 such that (i) A^* has no uncountable chains or pies, and (ii) every element $\neq 0$ in A^* has uncountably many elements below it. (b) Show that if A is a boolean algebra which has no uncountable pies and satisfies (ii) above, then A has no non-trivial automorphisms. [If f is a non-trivial automorphism of A then there is a non-zero element a of A such that $f(a) \wedge a = 0$. Consider the elements $b \vee (f(a) \wedge f(b)^{\sim})$ with $b \leqslant a$.]

REFERENCES FOR CHAPTER 6

6.1: Keisler (1970) showed how Axioms 0-5 allow one to construct proper elementary extensions of countable structures without adding new elements to small sets; this is the full first-order version of the results up to Theorem 6.1.4. In fact Keisler (1966) had already used a special case of Axiom 5 to prove the two-cardinal theorem, Theorem 6.1.5. (The two-cardinal theorem was originally proved by Vaught (1961) using homogeneous structures.) The notion of a largeness property is from Shelah (1983 d) (he calls it a notion of bigness). Theorem 6.1.7 is from Paris & Kirby (1978). On Lemma 6.1.6 cf. Mills & Paris (1984).

Exercises 2, 3 are from Keisler (1970), as is the idea of Exercise 7. Exercise 10(b): this largeness property occurs in Jensen's proof of the gap-2 two-cardinal theorem in L, cf. Devlin (1973). Exercise 11: the first implication in (b) is from Parsons (1970) and the rest is from Paris & Kirby (1978). Exercise 12: (b,c) Paris (unpublished), (d) Rabin (1962). Exercise 13: Kaufmann (1981 a). Exercise 14: Hutchinson (1976), extending Keisler & Morley (1968). Exercise 17(a,b): Macintyre (1976). Exercise 18 is Theorem 12 of Macintyre (1972 b), though it is superseded now by Theorem 4.1.11 above.

6.2: Skolem (1934; 1955) used a version of definable ultrapowers to show that non-standard countable models of complete first-order arithmetic exist. MacDowell & Specker (1961) internalised his argument and thereby showed that every model of a countable Peano-like theory has a proper elementary end extension. Gaifman talked about definable types and conservative extensions in lectures at UCLA in 1967; before he went into print in Gaifman (1976), several other people rediscovered Theorem 6.2.5, including Phillips (1974) and Schmerl (1973).

Corollary 6.2.7 and Exercise 11 are from Schmerl (1981); cf. Schmerl (1973) for a simpler argument (due to S. Simpson) which gives

Corollary 6.2.7 for regular κ. Not every model of first-order Peano arithmetic (PA) can be expanded to a model of second-order arithmetic. Barwise & Schlipf (1975) showed that every recursively saturated model of PA can be expanded to a model of second-order arithmetic with Δ^1_1 comprehension and restricted induction. Naturally people asked whether PA has rather classless recursively saturated models. Kaufmann (1977) showed that it has such a model of cardinality ω_1; cf. Exercise 8 of section 5.2. He used $\diamondsuit$, but Shelah (1978 d) gave an absoluteness argument which removed this assumption. Shelah (1978 d) also contains a construction of a rather classless ω_1-saturated model of PA. Schmerl (1981) simplified and extended Kaufmann's argument by applying Theorem 6.2.5 to PA with carefully chosen satisfaction classes.

Exercise 4: Gaifman (1967; 1976). More recently the idea of (b) has appeared in stability theory under the name of Morley sequences. Exercises 5, 6: Gaifman (1976). Exercise 7(b,c): Rubin in Shelah (1978 b). Exercise 8: Shelah (1978 b). Exercise 9: Kaufmann (1981 a) and Kranakis (1982). (Cf. Kotlarski (1980) for another view of Skolem ultrapowers in set theory.) Exercise 10: Rabin (1959); cf. Chang & Keisler (1973) Theorem 6.4.5. Exercise 12: (a) Mills (1978), (b) Kaufmann (1983 b), (c) Prikry in Kranakis (1982).

6.3: Exercise 10(a), which implies Theorem 6.3.1 under the assumption of $\diamondsuit$, is due to Baumgartner & Komjáth (1981) and independently to Rubin (1983); Rubin's paper contains many refinements of the result. Shelah (1981) then proved Theorem 6.3.1, assuming only the CH. Note that the relative crudity of part III of the construction prevents us from saying much about how the different countable extensions combine in A^*, so that the conclusions are weaker than Rubin's. Baumgartner (1980) showed that it is consistent with ZFC that every uncountable boolean algebra contains an uncountable pie. Shelah (1983 a; 198- b) contain further constructions of boolean algebras assuming $\diamondsuit$.

Exercise 2 and Lemma 6.3.9: folklore, but cf. Carson (1973) and Lipshitz & Saracino (1973) for a related idea in ring theory. Exercises 5, 6, 9, 10(b): Baumgartner & Komjáth (1981). Exercise 7: (a) Rubin (1983), (b) Bonnet, cf. Bonnet & Shelah (198-); also E. S. Berney (unpublished). Exercise 8: Shelah (1981), cf. Rubin (1983).

7 GENERALISED QUANTIFIERS

Many shall run to and fro, and knowledge shall be increased.
Daniel xii 4

In Chapter 6 we saw that if a countable structure B satisfies certain axioms, then by playing games one can build a proper extension A of B whose properties closely reflect those of B. Sometimes we iterated this procedure, to construct an uncountable chain of structures starting from B. The union of the chain was an uncountable extension of B.

This iteration is not always possible. The axioms that B has to satisfy involve some new quantifiers Qx. In general our games allow us to carry up some first-order properties from B to A, but not properties involving Qx. So A may fail to satisfy the axioms, and then we are snookered.

How did we get round this obstacle in Chapter 6? In Theorem 6.1.5 and Corollary 6.2.6 we used the fact that the relevant quantifier Qx was expressible by ordinary first-order formulas; but in general it isn't. In section 6.3 we evaded the problem: instead of iterating directly, we made uncountably many separate extensions of B and then amalgamated them.

In broad terms it is clear where one should look for a general solution. If we want to transport from B to A some sentences S_0 involving Qx, we have to require that B satisfies some other axioms S_1 got by applying the Qx quantifier to these sentences S_0. If we want to transport up those other axioms S_1 too, B must satisfy yet a third set of axioms S_2 got by applying Qx to the second set. And so on. In short, we must take on board sentences in which there are any number or nested occurrences of Q, just as if Q was ∀ or ∃. We shall have to handle "many" as freely as we do "all" and "some".

Quantifiers like "many" are no newcomers to logic. As a fan of C. S. Peirce I have to mention that already in 1885 he was advocating the quantifier $\frac{2}{3}x$. For example he used it to formalise the sentence "At least $\frac{2}{3}$ of the company have white neckties". Unfortunately the sober influence of Frege prevailed, and for the best part of a hundred years we were stuck with just "all" and "some". But the last twenty years have amply made up for this abstinence, and we now know the model-theoretic properties of dozens of interesting quantifiers.

The main languages I discuss below are of form L(Q), got by adding Q to a first-order language L. I have included a section on the Magidor-Malitz quantifiers, which are a challenging variant of Q. I am sorry there wasn't room for a section on the club quantifier and stationary logic. All these languages obey completeness and compactness theorems, though the proof of this for Magidor-Malitz quantifiers assumes the set-theoretic principle $\Diamond$.

A good deal is known about the behaviour of these quantifiers in particular mathematical theories. See the survey article Baudisch <u>et al</u>. (198-).

7.1 L(Q)

Let L be a countable first-order language. The language L(Q) is formed as follows. Its formulas are exactly the same as those of L, except that it has a new quantifier symbol Q besides $\forall$ and $\exists$. The new symbol Q binds variables in just the same way as $\forall$ and $\exists$ do. We interpret formulas of L(Q) in L-structures by:

(1) $A \models Qx\phi(x)$ iff there are uncountably many elements a of A such that $A \models \phi(a)$.

This is called the ω_1-<u>interpretation</u>. For any infinite cardinal λ there is a λ-<u>interpretation</u> of L(Q), which puts "at least λ" in place of "uncountably many" in (1). We shall use only the ω_1-interpretation in this chapter. Chapter 8 will look at larger λ.

In the ω_1-interpretation, every uncountable L-structure is a model of all the following sentences of L(Q), where $\phi(x,\bar{z})$, $\psi(x,\bar{z})$ and $\chi(x,y,\bar{z})$ are any formulas of L(Q):

(Axiom 0) $\forall\bar{z}(Qx\phi \leftrightarrow Qy\phi(y,\bar{z}))$.

(Axiom 1) $\forall\bar{z}(\forall x(\phi \to \psi) \to (Qx\phi \to Qx\psi))$.

(Axiom 2) $\forall\bar{z}(Qx(\phi \vee \psi) \to Qx\phi \vee Qx\psi)$.

(Axiom 3) $\forall z \neg Qx(x=z)$.

(Axiom 4) $Qx(x=x)$.

(Axiom 5) $\forall\bar{z}(Qx\exists y\chi \to \exists yQx\chi \vee Qy\exists x\chi)$.

The reasons are just the same as in section 6.1 above. The sentences of the forms of Axioms 0-5 are together called the <u>axioms of</u> L(Q) (for the ω_1-interpretation). By the argument of Exercise 3 of section 6.1, Axioms 2 and 3 can together be replaced by one axiom: $\forall yz\neg Qx(x=y \vee x=z)$.

Let T be a theory in L(Q) and ψ a sentence of L(Q). We shall write $T \vdash_{0-5} \psi$ to mean that if each formula beginning with Q is treated as atomic, then ψ can be deduced from T and Axioms 0-5 by first-order logic. Likewise we shall say that T is <u>$\vdash_{0-5}$-consistent</u> iff there is no contradiction ψ such that $T \vdash_{0-5} \psi$. $\vdash_{0-4}$ is understood the same way, but with Axioms 0-4.

I shall sketch a proof of Keisler's remarkable completeness theorem: $T \vdash_{0-5} \psi$ iff every uncountable model of T is also a model of ψ. Left to right follows at once from the choice of the axioms. Just as in the proof of the completeness theorem in first-order logic, right to left reduces to showing that if T is $\vdash_{0-5}$-consistent then it has an uncountable model.

Let T be a $\vdash_{0-5}$-consistent theory. We want an uncountable model of T. The idea will be to construct this model A_{ω_1} as the union of an elementary chain $(A_\alpha : \alpha < \omega_1)$ of countable L-structures. Each structure A_α will have attached to it a labelling I_α of all the L(Q)-sentences $\phi(\bar{a})$ with parameters $\bar{a}$ from A_α, which labels each such sentence as TRUE or FALSE. If $Qx\phi(x)$ is labelled FALSE, then we arrange that $I_{\alpha+1}$ labels $\phi(b)$ FALSE whenever b is in $A_{\alpha+1}$ but not in A_α; so "small" sets never grow. On the other hand we arrange that if $Qx\phi(x)$ is labelled TRUE then for uncountably many different elements b the sentence $\phi(b)$ will be labelled TRUE during the course of the construction. Note that I_α may label a sentence TRUE even though the sentence is false in A_α under the ω_1-interpretation: for example I_α will always label $Qx(x=x)$ TRUE.

The following device for labelling the sentences is neat and turns out to be quite general enough. Let A be an L-structure. Then take I to be any set of subsets of dom(A). We interpret sentences of L(Q) in the pair (A,I) by:

(2) $(A,I) \models Qx\phi(x)$ iff $\{a \text{ in } A : (A,I) \models \phi(a)\} \notin I$.

(Note $\notin$ rather than $\in$. Later this will make I an ideal in a boolean algebra, rather than the complement of an ideal. Cf. $\mathcal{J}(B)$ in section 6.2.) Clause (2) is part of an inductive definition of when a sentence is true in (A,I). The remaining clauses are as one would expect, ignoring I. For example $(A,I) \models \neg\phi$ iff $(A,I) \not\models \phi$.

Pairs of the form (A,I) are called <u>weak</u> L(Q)-<u>structures</u>. The interpretation of sentences of L(Q) in (A,I), using clause (2), is called the <u>weak interpretation</u> of L(Q) (as opposed to the ω_1-interpretation defined earlier). If (A,I) satisfies a theory T in the weak interpretation, we call (A,I) a <u>weak model</u> of T. The <u>cardinality</u> of a weak model (A,I) is the cardinality of A.

There should be no danger of confusion between the weak interpretation and the ω_1-interpretation, because the former uses pairs (A,I) and the latter uses L-structures A. Thus the notation "$(A,I) \models \phi$" will always refer to the weak interpretation. For emphasis we say that A is a <u>standard model</u> of a theory T iff A is a model of T in the ω_1-interpretation.

We say that (A,I) is an <u>elementary extension</u> of (B,J), in symbols $(B,J) \preccurlyeq (A,I)$, iff $B \subseteq A$ and for every formula $\phi(\bar{x})$ of L(Q) and all tuples $\bar{b}$ from B, $(B,J) \models \phi(\bar{b})$ iff $(A,I) \models \phi(\bar{b})$. Likewise an <u>elementary embedding</u> of (B,J) into (A,I) is a map f: dom(B) $\to$ dom(A) which preserves all formulas of L(Q).

An <u>elementary chain</u> of weak L(Q)-structures is a sequence $(A_i, I_i)_{i<\gamma}$ such that $(A_i, I_i) \preccurlyeq (A_j, I_j)$ whenever $i < j < \gamma$. We form the <u>union</u> (A,I) of the chain as follows. A is $\bigcup_{i<\gamma} A_i$, and I is the set $\{X \subseteq \text{dom}(A) : \text{there is } i < \gamma \text{ such that } X \cap \text{dom}(A_j) \in I_j \text{ for all } j \geqslant i\}$.

LEMMA 7.1.1. *Let* (A,I) *be the union of an elementary chain* $(A_i, I_i)_{i<\gamma}$ *of weak* L(Q)-*structures. Then for each* $i < \gamma$, $(A_i, I_i) \preccurlyeq (A,I)$.

<u>Proof</u>. We proceed just as in the proof of the elementary

chain lemma, Lemma 1.2.7. For every formula $\phi(\bar{x})$ of L(Q), every $i < \gamma$ and every tuple $\bar{b}$ from A_i we show

(3) $$(A_i, I_i) \models \phi(\bar{b}) \quad \text{iff} \quad (A, I) \models \phi(\bar{b})$$

by induction on the complexity of ϕ. The interesting clause is where $\phi(\bar{b})$ is $Qx\psi(x,\bar{b})$. We have:

$$\begin{aligned}(A,I) \models Qx\psi(x,\bar{b}) \quad & \text{iff} \quad \{a \text{ in } A : (A,I) \models \psi(a,\bar{b})\} \notin I \\ & \text{iff} \quad \text{for some } j \geqslant i, \\ & \qquad \{a \text{ in } A_j : (A_j, I_j) \models \psi(a,\bar{b})\} \notin I_j \\ & \text{iff} \quad (A_i, I_i) \models Qx\psi(x,\bar{b})\end{aligned}$$

using the definition of I, the induction hypothesis and the fact that the chain is elementary. □

Now we shall define forcing over a fixed countable weak L(Q)-structure (B,J). The definition is a straightforward adaptation of section 6.1. If (B,J) satisfies Axioms 0-4 then we can think of Q as being a largeness property over B.

Let L(Q)' be the language L(Q) with the following new constants added: (i) each element of B, (ii) a countable set W of new constants called <u>witnesses</u>, and (iii) an <u>overflow</u> constant ∞. We write L(Q,B) for L(Q) with just the constants (i) added. If p is a finite set of sentences of L(Q)', we can write $\bigwedge p$ as $\phi(\infty,\bar{c})$ where $\phi(x,\bar{y})$ is a formula of L(Q,B) and $\bar{c}$ lists the distinct witnesses occurring in p. We call p a <u>condition</u> iff $(B,J) \models Qx\exists\bar{y}\phi$. We write N for the set of all conditions.

LEMMA 7.1.2. *Suppose* (B,J) *satisfies Axioms 0-4. Then:*

(a) N *is not empty.*

(b) If p *is a condition then* p *has a weak model.*

(c) If p *is a condition and* ψ *is a sentence of* L(Q)' *such that* $p \vdash \psi$ *(first-order implication), then* $p \cup \{\psi\}$ *is a condition.*

(d) If p *is a condition and* t *is a closed term of* L(Q)', *then for any witness* c *which doesn't occur in* p *or* t,

p ∪ {t=c} *is a condition.*

(e) *If* p *is a condition and* $\exists y\psi(y) \in p$, *then for any witness* c *which doesn't occur in* p *or* ψ, $p \cup \{\psi(c)\}$ *is a condition.*

(f) *If* p *is a condition and* ψ *is a sentence of* L(Q)', *then either* $p \cup \{\psi\}$ *or* $p \cup \{\neg\psi\}$ *is a condition.*

(g) N *is a notion of forcing (except that the formulas involved are not all first-order), and it satisfies (6)-(8) of section 2.3.*

Proof. Exactly as Lemma 6.1.1 and Theorem 6.1.2, except for clause (e) which is left to the reader. □

A construction sequence is an increasing chain $\bar{p} = (p_i)_{i<\omega}$ of conditions. We define the weak L(Q)-structure $(A^+(\bar{p}), I(\bar{p}))$ as follows. $A^+(\bar{p})$ is the canonical model of the least =-closed set containing all the atomic sentences in $\bigcup\bar{p}$, as before. If $\phi(x)$ is any formula of L(Q)', write $\phi(\bar{p})$ for the set $\{c^{A^+(\bar{p})} : c$ is a witness and $\phi(c) \in \bigcup\bar{p}\}$. Then define $I(\bar{p})$ to be $\{\phi(\bar{p}) : \neg Qx\phi \in \bigcup\bar{p}\}$. We shall sometimes abbreviate $(A^+(\bar{p}), I(\bar{p}))$ to (A^+, I).

Now it makes sense to talk about forcing a property of $\bigcup\bar{p}$ or of (A^+, I).

THEOREM 7.1.3. *Suppose the weak* L(Q)*-structure* (B,J) *satisfies Axioms 0-4. Then the following are enforceable (separately and hence also together):*

(a) $\bigcup\bar{p}$ *is =-closed and contains every sentence which is true in* (B,J).

(b) $\bigcup\bar{p}$ *is a maximal* $\vdash_{0-4}$*-consistent set of sentences of* L(Q)', *containing every sentence of* L(Q)' *or its negation, and if a sentence* $\exists y\phi(y)$ *is in* $\bigcup\bar{p}$ *then* $\phi(c)$ *is in* $\bigcup\bar{p}$ *for some witness* c.

(c) *Every element of* $A^+(\bar{p})$ *is named by infinitely many witnesses, and the element* ∞^{A^+} *is not equal to* b^{A^+} *for any element* B *of* B.

(d) (A^+, I) *is a weak model of* $\bigcup\bar{p}$.

(e) *If* (B,J) *also satisfies Axiom 5 and* $\phi(x)$ *is a formula of* L(Q,B) *such that* $(B,J) \models \neg Qx\phi$, *and* $(A^+, I) \models \phi(a)$, *then*

$a = b^{A^+}$ *for some element* b *of* B.

Proof. (a) and (c) are enforced just as in Theorem 6.1.3 (full first-order version). When (a) holds, all the axioms 0-4 are in $\mathbf{U}\bar{p}$; so to enforce (b) we can use Lemma 7.1.2(b,e,f).

To show that (d) is enforceable, assume that player ∃ plays to make (a)-(c) true. Then (d) will be true too. To show this, we prove by induction on the complexity of ϕ that for every sentence ϕ of L(Q)',

$$(4) \qquad \phi \in \mathbf{U}\bar{p} \Rightarrow (A^+,I) \models \phi, \quad \text{and} \quad \neg\phi \in \mathbf{U}\bar{p} \Rightarrow (A^+,I) \models \neg\phi.$$

The proof is the same as that of (16) in section 2.1, except that we now have to add a clause for the case where ϕ is $Qx\psi(x)$, as follows.

For the first half, suppose $(A^+,I) \not\models Qx\phi$. Then by (2), $\{a : (A^+,I) \models \psi(a)\} \in I$. By induction hypothesis, (b) and (c) we infer that $\psi(\bar{p}) \in I$. By the definition of I this implies that for some formula $\chi(x)$, $\neg Qx\chi \in \mathbf{U}\bar{p}$ and $\psi(\bar{p}) = \chi(\bar{p})$. We claim that $\forall x(\psi \to \chi) \in \mathbf{U}\bar{p}$. For otherwise by (b) there is a witness c such that $\psi(c), \neg\chi(c) \in \mathbf{U}\bar{p}$. Now by induction hypothesis this implies $(A^+,I) \models \psi(c)$, so $c^{A^+} \in \psi(\bar{p}) = \chi(\bar{p})$ and hence $c^{A^+} = d^{A^+}$ for some witness d with $\chi(d) \in \mathbf{U}\bar{p}$. Then $\mathbf{U}\bar{p}$ contains $\chi(d)$, c=d and $\neg\chi(c)$, contradicting (b). So $\forall x(\psi \to \chi) \in \mathbf{U}\bar{p}$ as claimed. Hence by (b) and Axiom 1, if $Qx\psi \in \mathbf{U}\bar{p}$ then $Qx\chi \in \mathbf{U}\bar{p}$. But $Qx\chi \notin \mathbf{U}\bar{p}$ since $\neg Qx\chi \in \mathbf{U}\bar{p}$. So $Qx\psi \notin \mathbf{U}\bar{p}$.

For the second half, suppose $\neg Qx\psi \in \mathbf{U}\bar{p}$. Then $\psi(\bar{p}) \in I$ by the definition of I. Using (b), (c), induction hypothesis and (2) it follows that $(A^+,I) \models \neg Qx\psi$.

Finally the argument for (e) is essentially the same as Theorem 6.1.5, using Axiom 5. □

In the light of (a) and (d) of the theorem, we can assume in future that the L(Q)-reduct of the compiled structure (A^+,I) is an elementary extension of (B,J). Also in any game of the form G(-;odds), if $\phi(x,\bar{y})$ is a formula of L(Q) such that $(B,J) \models Qx\exists\bar{y}\phi$, and $\bar{c}$ is a tuple of distinct witnesses, then player ∀ can make his first move $\{\phi(\infty,\bar{c})\}$. Hence we have shown:

COROLLARY 7.1.4. *Let* (B,J) *be any countable weak* L(Q)-*structure satisfying Axioms 0-5. Let* $\phi(x)$ *be any formula of* L(Q,B) *such that* $(B,J) \models Qx\phi$. *Then there is a countable elementary extension* (A,I) *of* (B,J) *such that*

(a) $(A,I) \models \phi(a)$ *for some element* a *not in* B;

(b) *if* $\psi(x)$ *is any formula of* $L(Q,B)$ *such that* $(B,J) \models \neg Qx\psi$, *then for every element* a *of* A, $(A,I) \models \psi(a)$ *implies* a *is already in* B. □

Keisler's completeness theorem is only a stone's throw away. We have all the machinery needed to extend weak models into a chain, but so far nobody has given us a weak model to start the chain.

LEMMA 7.1.5. *Let* T *be any* $\vdash_{0-4}$*-consistent theory in* L(Q). *Then there exists a countable weak* L(Q)*-structure* (A,I) *which is a model of* T.

Proof. We revert to a simpler kind of forcing. Let L' be L(Q) with countably many added witnesses and an overflow constant ∞. A condition shall be a finite set p of sentences of L' such that if $\bigwedge p$ is $\phi(\infty,\bar{c})$ for some formula $\phi(x,\bar{y})$ of L(Q) and some tuple $\bar{c}$ of distinct witnesses, then $T \cup \{Qx\exists\bar{y}\phi(x,\bar{y})\}$ is $\vdash_{0-4}$-consistent. Let N_T be the set of all conditions. Formally N_T behaves very much like the notion of forcing N in Lemma 7.1.2. We can define construction sequences $\bar{p}$ and weak L(Q)-structures $(A^+(\bar{p}),I(\bar{p}))$ just as above. By the immediate analogues of Theorem 7.1.3(a,d), it is enforceable that the compiled structure is a weak model of T. The model is countably infinite by Axioms 2-4. □

THEOREM 7.1.6. *Let* L *be a countable first-order language.*

(a) (Completeness) *Every* $\vdash_{0-5}$*-consistent theory in* L(Q) *has an uncountable standard model.*

(b) (Compactness) *Let* T *be a theory in* L(Q) *such that every finite subset of* T *has a standard model. Then* T *has a standard model.*

Proof. (a) Let T be a $\vdash_{0-5}$-consistent theory in L(Q). By Lemma 7.1.5, T has a countable weak model (A_0,I_0). We can assume that (A_0,I_0) satisfies Axioms 0-5, since these axioms can be added to T without damaging $\vdash_{0-5}$-consistency. Now construct an elementary chain $(A_\alpha,I_\alpha)_{\alpha<\omega_1}$ as follows. At limit ordinals take unions (cf. Lemma 7.1.1). For each $\alpha < \omega_1$, choose $(A_{\alpha+1},I_{\alpha+1})$ so that it is related to (A_α,I_α) as (A,I) is related to (B,J) in Corollary 7.1.4, for some formula $\phi = \phi_\alpha$. Since there are ω_1 steps in the chain, we can arrange that for each $\beta < \omega_1$, if $(A_\beta, I_\beta) \models Qx\psi(x)$, then ψ is ϕ_α for uncountably many $\alpha > \beta$.

Consider the union $(A^*, I^*) = (A_{\omega_1}, I_{\omega_1})$ of this chain. If $\phi(x,\bar{y})$ is any formula of L(Q) and $\bar{a}$ is a tuple of elements of A^*, then the construction ensures that $(A^*, I^*) \models Qx\phi(x,\bar{a})$ if and only if uncountably many elements of A^* satisfy $\phi(x,\bar{a})$ in (A^*, I^*). It follows that for any sentence ψ of L(Q) with parameters from A^*, $(A^*, I^*) \models \psi$ in the weak interpretation if and only if $A^* \models \psi$ in the ω_1-interpretation. Finally $A^* \models T$ since $(A_0, I_0) \preccurlyeq (A^*, I^*)$.

(b) Suppose first that every finite subset of T has an uncountable standard model. It follows that every finite subset of T is $\vdash_{0-5}$-consistent. But then T itself is $\vdash_{0-5}$-consistent, by the definition of $\vdash_{0-5}$-consistency and the compactness theorem for first-order logic. So T has a standard model by (a).

On the other hand suppose some finite subset T_0 of T has no uncountable standard model. Then by the hypothesis of (b), every finite subset of T containing T_0, and hence every finite subset, must have an (at most) countable standard model. It follows that every finite subset of T' has a countable model, where T' comes from T by replacing each maximal subformula of form $Qx\phi$ by a contradiction. So by the compactness theorem for first-order logic and the downward Löwenheim-Skolem theorem (Lemma 1.2.6), T' has a countable model, and this model will be a standard model of T. □

One of the chief joys of forcing constructions is that you can add new pieces without having to buy a complete new outfit. In section 5.1 we saw how to make the compiled structure satisfy some infinitary rules (cf. the Henkin-Orey theorem, Theorem 5.1.8), and Exercise 10 of that section used the same idea to generalise forcing from a first-order language L to a countable fragment of $L_{\omega_1\omega}$.

We can extend the completeness theorem, Theorem 7.1.6(a), in just the same way. If L is a countable first-order language, then it is clear how to form the language $L_{\omega_1\omega}(Q)$. A countable fragment of $L_{\omega_1\omega}(Q)$ is the same as for $L_{\omega_1\omega}$ (cf. section 1.2), except that the set of formulas is also closed under quantifying by Q. As axioms and rules of proof for $L_{\omega_1\omega}(Q)$ we need all those of $L_{\omega_1\omega}$ and all those of L(Q). Besides these, just one more axiom is needed, to relate Q to the infinitary symbols:

$$\forall \bar{z}(Qx \bigvee_{i<\omega}\phi_i(x,\bar{z}) \rightarrow \bigvee_{i<\omega} Qx\phi_i(x,\bar{z})).$$

We can write $T \vdash_\Omega \psi$ to mean that ψ is deducible from T by these rules and axioms. Then it is not hard to adapt the proof of Theorem 7.1.6(a) to show: If ϕ is a sentence of $L_{\omega_1\omega}(Q)$, then ϕ is $\vdash_\Omega$-consistent if and only if ϕ has an uncountable standard model.

It follows that if ϕ is a sentence of $L_{\omega_1\omega}(Q)$ and M is a transitive model of ZFC containing ϕ, then the question whether ϕ has a model in M is independent of the choice of M. For by the analogue of Lemma 7.1.5, either $\neg\phi$ has a proof P in M, or ϕ has a countable weak model (A,I) in M. But if P is a proof of $\neg\phi$ in M, then it remains one in the set-theoretic universe V; and if (A,I) is a countable weak model of ϕ in M, then it remains one in V, so that the proof of the completeness theorem finds a standard model of ϕ in V.

There are sorrows as well as joys. The compactness theorem for L(Q), Theorem 7.1.6(b), only holds for countable languages. This makes it almost useless for model theory. If the structures are countable, the quantifier Qx has nothing to say. But if they are uncountable, then so are their diagrams, and this prevents us from using the compactness theorem to deduce even an upward Löwenheim-Skolem theorem; cf. Exercise 3(b) below. The compactness theorem holds only because the quantifier Qx cannot see what flowers are at its feet.

Exercises for 7.1.

Throughout these exercises, L *is a countable first-order language.*

1. Show that in the ω_1-interpretation, $QxQy\phi$ need not be equivalent to $QyQx\phi$. (If this leads you to look for a symmetrical quantifier Qxy, compare your candidate with section 7.3 below.)

2. (Downward Löwenheim-Skolem theorem for L(Q)) Let B be an uncountable L-structure and X a set $\subseteq$ dom(B) of cardinality $\leqslant \omega_1$. Show that there is an L-structure $A \subseteq B$ such that $|A| = \omega_1$, $X \subseteq$ dom(A) and for all formulas $\phi(\bar{x})$ of L(Q) and tuples $\bar{a}$ from A, $A \models \phi(\bar{a})$ iff $B \models \phi(\bar{a})$.

3. (a) Show that if L^* is a first-order language with uncountably many constants, then both the completeness theorem and the compactness theorem (Theorem 7.1.6(a,b)) fail for $L^*(Q)$. (b) Describe an L-structure A of cardinality ω_1, for a suitable countable first-order language L, such that if $A \subseteq B$ and A satisfies the same sentences of L(Q) as B, then A = B.

4. Show that the compactness theorem for L(Q), Theorem 7.1.6(b), is false if Q is given the ω-interpretation.

5. Choose some standard proof calculus for first-order logic, and assume that $\vdash_{0-5}$ is defined in terms of that calculus, using Axioms 0-5. Prove the following without assuming the completeness theorem. (a) If T is a theory in L(Q) and ϕ, ψ are sentences of L(Q), then $T \cup \{\phi\} \vdash_{0-5} \psi$ iff $T \vdash_{0-5} \phi \rightarrow \psi$. (b) If $\phi(\bar{z})$ and $\psi(x,\bar{z})$ are formulas of L(Q), then $\vdash_{0-5} \forall\bar{z}(Qx(\phi \wedge \psi) \leftrightarrow \phi \wedge Qx\psi)$. (c) If $\phi(x,\bar{z})$ and $\psi(x,\bar{z})$ are formulas of L(Q), then $\vdash_{0-5} \forall\bar{z}(Qx\phi \wedge \neg Qx\psi \rightarrow Qx(\phi \wedge \neg\psi))$. (d) If $\phi(w,x_1,\ldots,x_n)$ is a formula of L(Q) and R is a quantifier string made up of quantifiers of the forms Qx_i or $\exists x_j$, then $\vdash_{0-5} R\exists w\phi \rightarrow \exists wR\phi \vee Qw\exists x_1\ldots x_n\phi$.

As Exercises 6-10 indicate, there is plenty of scope for a model theory of weak L(Q)*-structures. A lot more is known than I indicate here.*

6. Complete the proof of Lemma 7.1.5.

7. Let T be a $\vdash_{0-5}$-consistent theory in L(Q), and suppose that for each $n < \omega$, $\Phi_n(\bar{x}_n)$ is a set of formulas of L(Q) such that for each formula $\psi(\bar{x}_n)$ of L(Q), if $T \cup \{\exists\bar{x}_n\psi\}$ is $\vdash_{0-5}$-consistent then $T \cup \{\exists\bar{x}_n(\psi \wedge \neg\phi)\}$ is $\vdash_{0-5}$-consistent for some $\phi \in \Phi_n$. Show that T has a weak model which omits each Φ_n $(n < \omega)$.

8. An <u>ideal</u> L(Q)-<u>structure</u> is a countable weak L(Q)-structure (B,J) in which J is a proper ideal in the algebra of subsets of dom(B), and all finite subsets of dom(B) are in J. If T is a theory in L(Q), an <u>ideal model</u> of T is an ideal L(Q)-structure which is a weak model of T. Show: (a) For every theory T in L(Q), T has an ideal model iff T is $\vdash_{0-4}$-consistent. (b) For every theory T in L(Q), T has an ideal model iff every finite subset of T has an ideal model.

9. Let Axiom $\frac{1}{2}$ be the axiom schema $\forall\bar{z}(\forall x(\phi \leftrightarrow \psi) \rightarrow (Qx\phi \leftrightarrow Qx\psi))$, and define $\vdash_{0,\frac{1}{2}}$ accordingly. Show that if T is a theory in L(Q), then T has a weak model iff T is $\vdash_{0,\frac{1}{2}}$-consistent.

10. Let ϕ be a sentence of L(Q), and suppose that for all countable L-structures A, if I, J are ideals on dom(A) such that $\bigcup I = \bigcup J = \mathrm{dom}(A)$, then $(A,I) \models \phi$ iff $(A,J) \models \phi$. Show that ϕ is equivalent to a sentence of L.

11. Let $\vdash_{0-3,5}$ be defined like $\vdash_{0-5}$ but leaving out Axiom 4. Show that a theory in L(Q) has a standard model iff it is $\vdash_{0-3,5}$-consistent.

12. Where is (5) used in the proof of the completeness theorem for countable fragments of $L_{\omega_1\omega}(Q)$?

13. Let L be a countable first-order language, A an L-structure of cardinality ω_1, and $\phi(s)$ a sentence of L with an extra 1-ary relation symbol s. Identifying the elements of A with the ordinals $< \omega_1$, we define $A \models Qs\phi(s)$ to hold iff $\{\alpha : (A,\alpha) \models \phi(s)\}$ is stationary in ω_1 (interpreting s by α). Show that the truth of $A \models Qs\phi(s)$ is independent of the way we identify dom(A) with ω_1. [Use Lemma 5.2.2(b).]

Given a countable first-order language L, *we write* L(*aa*) *for the language got from* L *by allowing extra 1-ary relation symbols* s, r, s_1, s_2, ... *etc. and quantifiers* Qs, Qr *etc. The* *sentences* *of* L(*aa*) *are the formulas of* L(*aa*) *in which every first-order variable and every new relation symbol is bound by a quantifier* ∀x, ∃x *or* Qs. *In* L-*structures of cardinality* ω_1 *we interpret formulas of* L(*aa*) *by using Exercise 13 above to interpret* Q. Qs *is usually written stats, and the dual* ¬Qs¬ *is written aas (for "almost all").* L(*aa*) *with this interpretation is called* *stationary logic*. *Part (c) of Exercise 14 below is proved like Theorem 7.1.6, but the adjustments are not routine. Cf. Barwise et al. (1978) and Bruce (1980).*

14. (a) Write a formula of stationary logic L(*aa*) which is equivalent to the formula $Qx\phi$ of L(Q) in the ω_1-interpretation. [Try $\neg Qs\forall x(\phi \to s(x))$.]
(b) Verify that the following are true in every L-structure of cardinality ω_1 (where $aas_1 \ldots s_n$ means $aas_1 \ldots aas_n$, and ϕ, ψ are any formulas of L(*aa*) with the variables indicated):

(Axiom 0') $aa\bar{r}\forall\bar{z}(stats\phi(s,\bar{r},\bar{z}) \leftrightarrow statt\phi(t,\bar{r},\bar{z}))$.

(Axiom 1') $aa\bar{r}\forall\bar{z}(aas(\phi \to \psi) \to (stats\phi \to stats\psi))$.

(Axiom 2') $aa\bar{r}\forall\bar{z}(stats(\phi \vee \psi) \to stats\phi \vee stats\psi)$.

(Axiom 3') $\forall x\, aas\ s(x)$, $aar\,aas\forall x(r(x) \to s(x))$.

(Axiom 4') $aa\bar{r}\forall\bar{z}(stats\phi(\bar{r},\bar{z}) \leftrightarrow \phi(\bar{r},\bar{z}))$.

(Axiom 5') $aa\bar{r}\forall\bar{z}(stats\exists x(s(x) \wedge \phi) \to \exists x\, stats\phi)$.

(c) Let $\vdash_{0'-5'}$ be defined by analogy with $\vdash_{0-5}$ of the text, using Axioms 0'-5' above. Show that a theory in L(*aa*) is $\vdash_{0'-5'}$-consistent if and only if it has a model. Deduce that if T is a theory in L(*aa*) and every finite subset of T has a model then T has a model.

7.2 OMITTING TYPES IN L(Q)

The previous section showed how to construct models of consistent theories in L(Q). To analyse that construction, we turn it into a game G. The length of G is ω_1, and it is played as a succession of games G_α ($\alpha < \omega_1$), each of length ω. Each game G_α constructs a countable weak L(Q)-structure (A_α, I_α). For every positive $\alpha < \omega_1$ we write $(A_{<\alpha}, I_{<\alpha})$ for the union $\bigcup_{\beta<\alpha}(A_\beta, I_\beta)$, assuming this is defined. Then G_α constructs (A_α, I_α) as an elementary extension of $(A_{<\alpha}, I_{<\alpha})$ in exactly the same way as the games of section 7.1 built (A, I) as an elementary extension of (B, J). At the end of the game, the <u>compiled structure</u> is the union $(A^*, I^*) = \bigcup_{\alpha<\omega_1}(A_\alpha, I_\alpha)$. The game G is called a <u>hypergame</u>, to distinguish it from the countable games G_α inside it.

To be a little more formalistic: we always assume that player $\forall$ wins if some (A_α, I_α) fails to be an elementary extension of $(A_{<\alpha}, I_{<\alpha})$. Since it is enforceable that $(A_{<\alpha}, I_{<\alpha}) \preccurlyeq (A_\alpha, I_\alpha)$, player $\exists$ can always make sure that at least player $\forall$ doesn't win on <u>these</u> grounds. So without loss we can assume henceforth that an elementary chain $((A_\alpha, I_\alpha) : \alpha < \omega_1)$ always does get constructed in a play of G.

What happens at the first game G_0? Here we allow two possibilities. <u>Either</u> the players are given a countable weak L(Q)-structure (B, J) which is a model of Axioms 0-5, and they play to construct (A_0, I_0) as a proper elementary extension of (B, J) just as in section 7.1; <u>or</u> they are given a $\vdash_{0-5}$-consistent theory T and they play as in Lemma 7.1.5 to construct a countable weak model (A_0, I_0) of T which satisfies Axioms 0-5. In the first case we call (B, J) the <u>starting structure</u> and we put $(A_{<0}, I_{<0}) = (B, J)$. In the second case we call T the <u>starting theory</u>.

Compare this for a moment with the games of section 5.3, which could be of length ω_1. In those games the conditions were allowed to be infinite, and in each play they were built up as a chain $(p_\alpha : \alpha < \omega_1)$. But in the hypergames that we have just introduced, the conditions are all finite and the players start a fresh chain at each limit ordinal. In short the structure (A^*, I^*) is built up two-dimensionally. Picture (7) of section 1.1 applies.

To specify a hypergame G, we have to state (i) the starting structure (B, J) or starting theory T, (ii) the set X of ordinals $\alpha < \omega_1$

such that player ∃ chooses at the α-th step, and (iii) the property P which must hold at the end in order for player ∃ to win. We write G as G(P;X,(B,J)) or G(P;X,T) accordingly. Usually (B,J) or T is fixed and we abbreviate to G(P;X).

Sometimes in this chapter and the next, the two players will be allowed to specify P gradually as the hypergame proceeds, by adding new tasks that player ∃ must carry out. This complicates the notion of enforceability, so for simplicity I assume in the next few definitions that P is stated before the hypergame begins.

The first ω steps of the hypergame G belong to the game G_0, the next ω steps belong to G_1, and so on. Recall that the games G_α are said to be standard iff player ∀ makes the opening move and each player has infinitely many moves. We call the hypergame G standard iff each subgame G_α is standard. Consider for example the hypergame G(P;odds) in which the players move at alternate steps and player ∀ moves at step δ whenever δ is 0 or a limit ordinal. This hypergame is standard.

Let the starting structure (B,J) or starting theory T be fixed, and let P be a property. We say that P is hyperenforceable iff player ∃ has a winning strategy for every standard hypergame G(P;X). Just as in section 2.3, this is equivalent to: player ∃ has a winning strategy for G(P;odds).

If we look back at the details of the proof of Theorem 7.1.6, we find a slightly disconcerting fact. In order to make sure that if $(A^*,I^*) \models Qx\phi(x)$ then uncountably many elements satisfy ϕ, we used Corollary 7.1.4 to make the overflow element satisfy ϕ in uncountably many of the games G_α. This can only be done by the player who has the opening move of each game G_α. For this reason the wanted properties of A^* will not always be hyperenforceable, and we have to introduce a weaker notion. We say that a property P is weakly hyperenforceable iff player ∃ has a winning strategy for G(P;evens). (An ordinal is even iff it is of form $\delta+2n$ for some $n < \omega$, with δ equal to 0 or a limit.)

Let us call a weak L(Q)-structure (A,I) candid iff for every formula $\phi(\bar{x})$ of L(Q) and every tuple $\bar{a}$ from A, $(A,I) \models \phi(\bar{a})$ iff $A \models \phi(\bar{a})$ (so that the weak interpretation agrees with the ω_1-interpretation). Let us call (A,I) a precise extension of (B,J) iff $(B,J) \preccurlyeq (A,I)$ and for every formula $\phi(x)$ of L(Q,B), $(B,J) \models Qx\phi$ iff there is an element a of A such that $(A,I) \models \phi(a)$ and a is not in B.

LEMMA 7.2.1. *Suppose we play hypergames starting with a countable weak* L(Q)*-structure* (B,J) *which is a weak model of Axioms 0-5 (resp. with a* $\vdash_{0-5}$*-consistent theory* T *in* L(Q)*).*

(a) *The conjunction of countably many hyperenforceable properties is hyperenforceable; the conjunction of a hyperenforceable property and a weakly hyperenforceable property is weakly hyperenforceable.*

(b) *It is hyperenforceable that* $(B,J) \leqslant (A^*,I^*)$ *(resp. that* (A^*,I^*) *is a weak model of* T*). It is weakly hyperenforceable that* (A^*,I^*) *is a precise extension of* (B,J).

(c) *It is weakly hyperenforceable that* (A^*,I^*) *is candid.*

<u>Proof</u>. The first half of (a) is by the conjunction lemma (Lemma 2.3.3(e)), and the second half is direct from the definitions. (b) and (c) summarise the proof of the completeness theorem in section 7.1. □

What other properties are hyperenforceable?

Let us ruminate a little. If player ∃ wants some property P to hold, say, at (A_{53},I_{53}), then she would like "P is enforceable" to hold at (A_{52},I_{52}). Then in turn she would like ""P *is enforceable*" is enforceable" to hold at (A_{51},I_{51}). And so on. Looking at it from player ∀'s point of view, in order to frustrate player ∃, what he wants is a condition, say over (A_4,I_4), which forces there to be a condition over (A_5,I_5) which forces there to be a condition over (A_6,I_6) which ... which forces P to fail in (A_{53},I_{53}). Formally speaking, what he wants is that there should be some "hypercondition" $p(\infty_4,\bar{c}_4,\infty_5,\bar{c}_5,\ldots,\bar{c}_{52})$ which satisfies $(A_4,I_4) \models Qx_4\exists\bar{y}_4Qx_5\ldots\exists\bar{y}_{52}\bigwedge p(x_4,\bar{y}_4,x_5,\ldots,\bar{y}_{52})$ and prevents P. This is the underlying idea from now to the end of this chapter - and indeed to the end of the book.

A quantifier prefix consisting entirely of Qx_i's and $\exists y_j$'s is known as a <u>quexistential prefix</u>, or as a <u>quexistential quantifier</u>.

Let $\Phi(\bar{w})$ be a set of formulas of L(Q), where $\bar{w}$ is a tuple of variables. When is it hyperenforceable that (A^*,I^*) omits Φ? One would naturally expect that Φ can be omitted unless it has a "hypersupport", which should be the same thing as a support (cf. section 5.1) except that a quexistential prefix slips in somewhere. This expectation is correct.

The definitions are a little tiresome, because we need to define "hypersupport" three times: once for theories, once for weak L(Q)-structures and once for L-structures.

First let T be a $\vdash_{0-5}$-consistent theory in L(Q). We define a hypersupport of $\Phi(\bar{w})$ in T to be a formula $\psi(x_0,\bar{y}_0,\ldots,x_n,\bar{y}_n,\bar{w})$ of L(Q) such that

(1) $T \cup \{Qx_0\exists\bar{y}_0\ldots Qx_n\exists\bar{y}_n\exists\bar{w}\psi\}$ is $\vdash_{0-5}$-consistent, but

$T \vdash_{0-5} \neg Qx_0\exists\bar{y}_0\ldots Qx_n\exists\bar{y}_n\exists\bar{w}(\psi \wedge \neg\phi)$ for every $\phi(\bar{w}) \in \Phi$.

Secondly, let (B,J) be a weak L(Q)-structure. We define a hypersupport of $\Phi(\bar{w})$ in (B,J) to be a formula $\psi(x_0,\bar{y}_0,\ldots,x_n,\bar{y}_n,\bar{w})$ of L(Q,B) such that

(3) $(B,J) \models Qx_0\exists\bar{y}_0\ldots Qn_x\exists\bar{y}_n\exists\bar{w}\psi$, but

(4) $(B,J) \models \neg Qx_0\exists\bar{y}_0\ldots Qx_n\exists\bar{y}_n\exists\bar{w}(\psi \wedge \neg\phi)$ for every $\phi(\bar{w}) \in \Phi$.

And thirdly, if B is an L-structure, then we define a hypersupport of $\Phi(\bar{w})$ in B to be a formula ψ as in (3), (4) but with B in place of (B,J).

We say that T strongly omits $\Phi(\bar{w})$ iff there is no hypersupport of Φ in T. This is equivalent to the simpler statement:

(5) If $\psi(\bar{x},\bar{w})$ is a formula of L(Q) and $S\bar{x}$ is a quexistential prefix such that $T \cup \{S\bar{x}\exists\bar{w}\psi\}$ is $\vdash_{0-5}$-consistent, then for some $\phi(\bar{w}) \in \Phi$, $T \cup \{S\bar{x}\exists\bar{w}(\psi \wedge \neg\phi)\}$ is $\vdash_{0-5}$-consistent.

I gave the more complicated definition (1), (2) above because the pattern of blocks $Qx_i\exists\bar{y}_i$ puts one in the right frame of mind for the proofs to come.

Likewise we say that an L-structure or weak L(Q)-structure strongly omits a set $\Phi(\bar{w})$ iff it contains no hypersupport of Φ. Observe that if $(B,J) \preccurlyeq (A,I)$ and (A,I) strongly omits Φ then (B,J) strongly omits Φ.

The central fact about strongly omitting is the next theorem. It comes in two forms, according as we start with a theory or a structure.

THEOREM 7.2.2. *Let* L *be a countable first-order language,* $\bar{w}$ *a tuple of variables and* $\Phi(\bar{w})$ *a set of formulas of* L(Q).

(a) *If* T *is a* $\vdash_{0-5}$*-consistent theory in* L(Q) *which strongly omits* Φ*, then with* T *as starting theory it is hyperenforceable that* (A^*,I^*) *omits* Φ.

(b) *If* (B,J) *is a countable weak* L(Q)*-structure obeying Axioms 0-5, and* (B,J) *strongly omits* Φ*, then with* (B,J) *as starting structure it is hyperenforceable that* (A^*,I^*) *omits* Φ.

Proof. Let P be the property "(A^*,I^*) omits Φ". Let players $\forall$ and $\exists$ play G(P;odds), with starting theory T or starting structure (B,J) as assumed in (a) or (b). We have to show that player $\exists$ can always win this hypergame. We do it by showing

(6) Player $\exists$ can play so that each of the weak L(Q)-structures (A_α,I_α) $(\alpha < \omega_1)$ strongly omits Φ,

and

(7) If (A^*,I^*) strongly omits Φ, then it omits Φ.

This will suffice for the theorem. Since the structures (A_α,I_α) $(\alpha < \omega_1)$ form an elementary chain with union (A^*,I^*), and hypersupports are finite, any hypersupport of Φ in (A^*,I^*) must already be a hypersupport of Φ in some (A_α,I_α). So by (6) it is hyperenforceable that (A^*,I^*) strongly omits Φ, and hence by (7) that (A^*,I^*) omits Φ.

We prove (6) by induction on α. If α is positive then we can assume that $(A_{<\alpha},I_{<\alpha})$ obeys Axioms 0-5 and (by induction hypothesis) that $(A_{<\alpha},I_{<\alpha})$ strongly omits Φ. I claim that if we play the game G_α to construct (A_α,I_α) as a proper elementary extension of $(A_{<\alpha},I_{<\alpha})$, then it is enforceable in G_α that (A_α,I_α) strongly omits Φ too. By the conjunction lemma this claim follows if we show:

(8) If $\psi(x_0,\ldots,\bar{y}_n,\bar{z},\bar{w})$ is a formula of L(Q) and $\bar{d}$ is a tuple of distinct witnesses, then it is enforceable that $\psi(x_0,\ldots,\bar{y}_n,\bar{d},\bar{w})$ is not a hypersupport of Φ in (A_α,I_α).

We prove (8) as follows.

Suppose player $\forall$ opens G_α by choosing a condition p_0, and

$\bigwedge p_0$ is $\chi(\infty,\bar{c},\bar{d})$ where $\chi(x,\bar{y},\bar{z})$ is a formula of $L(Q,A_{<\alpha})$ and $\bar{c}$ is a tuple listing the distinct witnesses which are in p_0 but not in $\bar{d}$. We have $(A_{<\alpha},I_{<\alpha}) \models Qx\exists\bar{y}\bar{z}\chi$. There are two cases.

<u>Case 1</u> is that

$$(9) \qquad (A_{<\alpha},I_{<\alpha}) \models Qx\exists\bar{y}\bar{z}(\chi \wedge \neg Qx_0\ldots\exists\bar{y}_n\exists\bar{w}\psi).$$

In this case player $\exists$ can choose p_1 to contain $\neg Qx_0\ldots\exists\bar{y}_n\exists\bar{w}\psi(x_0,\ldots,\bar{y}_n,\bar{d},\bar{w})$ and then play to make this sentence true in (A_α,I_α).

<u>Case 2</u> is that (9) fails. Then by Axiom 2 and the assumption on χ,

$$(10) \qquad (A_{<\alpha},I_{<\alpha}) \models Qx\exists\bar{y}\bar{z}(\chi \wedge Qx_0\ldots\exists\bar{y}_n\exists\bar{w}\psi).$$

As in Exercise 5(b) of section 7.1, this implies

$$(11) \qquad (A_{<\alpha},I_{<\alpha}) \models Qx\exists\bar{y}\bar{z}Qx_0\ldots\ \bar{y}_n\exists\bar{w}(\chi \wedge \psi).$$

Since $(A_{<\alpha},I_{<\alpha})$ strongly omits Φ, we infer that for some $\phi(\bar{w}) \in \Phi$,

$$(12) \qquad (A_{<\alpha},I_{<\alpha}) \models Qx\exists\bar{y}\bar{z}Qx_0\ldots\exists\bar{y}_n\exists\bar{w}(\chi \wedge \psi \wedge \neg\phi).$$

Rearranging again as in Exercise 5(b) of section 7.1,

$$(13) \qquad (A_{<\alpha},I_{<\alpha}) \models Qx\exists\bar{y}\bar{z}(\chi \wedge Qx_0\ldots\exists\bar{y}_n\exists\bar{w}(\psi \wedge \neg\phi)).$$

Then player $\exists$ can choose p_1 to contain $Qx_0\ldots\exists\bar{y}_n\exists\bar{w}(\psi(x_0,\ldots,\bar{y}_n,\bar{d},\bar{w}) \wedge \neg\phi)$, and again she can play to make this sentence true in (A_α,I_α). In either case she prevents $\psi(x_0,\ldots,\bar{y}_n,\bar{d},\bar{w})$ from becoming a hypersupport of Φ.

This proves (8) when α is positive. In case (b) of the theorem the argument for $\alpha = 0$ is just the same. For (a) the argument at $\alpha = 0$ is not much different: Case 1 is that $T \cup \{Qx\exists\bar{y}\bar{z}(\chi\wedge\neg Qx_0\ldots\exists\bar{y}_n\exists\bar{w}\psi)\}$ is $\vdash_{0-5}$-consistent, Case 2 is that it is not. Thus (6) is proved.

It remains to prove (7). For this, suppose that some tuple $\bar{a}$ realises Φ in (A^*,I^*). Let $\psi(\bar{w},\bar{z})$ be $\bar{w}=\bar{z}$ (a conjunction of equations). Then $(A^*,I^*) \models \exists\bar{w}\psi(\bar{w},\bar{a})$, but for all $\phi(\bar{w})$ in Φ,

$(A^*,I^*) \models \neg\exists\bar{w}(\psi(\bar{w},\bar{a}) \wedge \neg\phi)$. Hence $\psi(\bar{w},\bar{a})$ is a hypersupport of Φ in (A^*,I^*). □

With the aid of Lemma 7.2.1 we deduce at once:

COROLLARY 7.2.3. *Let* L *be a countable first-order language and for each* $i < \omega$ *let* $\Phi_i(\bar{w})$ *be a set of formulas of* L(Q). *(The tuple* $\bar{w}$ *may vary with* i.*)*

(a) *If* T *is a* $\vdash_{0-5}$*-consistent theory in* L(Q) *and for each* $i < \omega$, T *strongly omits* Φ_i, *then* T *has a (standard) model which omits all the sets* Φ_i $(i < \omega)$.

(b) *If* (B,J) *is a countable weak* L(Q)*-structure satisfying Axioms 0-5 and strongly omitting each* Φ_i $(i < \omega)$, *then there is a candid weak* L(Q)*-structure* (A,I) *such that* $(B,J) \preccurlyeq (A,I)$ *and* A *omits every* Φ_i $(i < \omega)$. □

Corollary 7.2.3 can be extended in three ways. First, in part (b) we can assume that the sets Φ_i have parameters from B. This follows from the corollary as stated, just by adding constants to the language L.

Second, we can complicate the hypergame which builds (A^*,I^*). At each ordinal $\alpha < \omega_1$, before (A_α,I_α) is constructed, let the players choose a countable family $\{\Phi_i^\alpha : i < \omega\}$ of sets $\Phi(\bar{w})$ of formulas of $L(Q,A_{<\omega})$, each of which is strongly omitted by $(A_{<\alpha},I_{<\alpha})$. Then player $\exists$ can still play so that all the sets Φ_i^α $(i < \omega;\ \alpha < \omega_1)$ are omitted in (A^*,I^*). This is because at each step during the hypergame she still has only countably many sets Φ to take care of.

Third, we can replace L(Q) by a countable fragment of $L_{\omega_1\omega}(Q)$ (cf. Exercise 10 of section 5.1 and the remarks at the end of section 7.1). This is quite routine and I leave the details to the reader.

The following beautiful theorem of Keisler (1970) combines all three of the extensions just described. If $\Phi(\bar{x})$ is a set of formulas and $\bar{a}$ a tuple of elements of A, the Φ-type of $\bar{a}$ in A is the set of formulas $\{\phi : \phi \in \Phi \text{ and } A \models \phi(\bar{a})\} \cup \{\neg\phi : \phi \in \Phi \text{ and } A \models \neg\phi(\bar{a})\}$. A Φ-type is realised in A iff it is the Φ-type of some tuple in A.

THEOREM 7.2.4. *Let* L *be a countable first-order language,* ψ *a sentence of* $L_{\omega_1\omega}(Q)$ *and* $\Phi(\bar{x})$ *a countable set of formulas of* $L_{\omega_1\omega}(Q)$. *Suppose* ψ *has a model in which uncountably many different* Φ*-types are realised. Then there is a family* $(A_i : i < 2^{\omega_1})$ *of models of* ψ, *each of cardinality* ω_1, *such that if* $i \neq j$ *then some* Φ*-type is realised in* A_i *but not in* A_j.

Proof. Let A be a model of ψ in which uncountably many different Φ-types are realised. Without loss we can assume that for some $m < \omega$, all these Φ-types are types of m-tuples. We begin with some plastic surgery. Add to L an m-ary relation symbol R and (if $m > 1$) a symbol for a pairing function. Expand A so that

(14) R^A is an uncountable set of m-tuples, each of which has a different Φ-type.

Then add to ψ a conjunct which expresses that (14) holds. Since we are in $L_{\omega_1\omega}(Q)$, this is straightforward. Henceforth we can assume without loss that every (standard) model of ψ realises uncountably many different Φ-types.

We choose a countable fragment L^+ of $L_{\omega_1\omega}(Q)$ containing ψ and all the formulas of Φ. Write ${}^{<\omega_1}2$ for the set of all sequences σ of 0's and 1's, of length $< \omega_1$; $\sigma \subset \tau$ means that σ is a proper initial segment of τ.

For each $\sigma \in {}^{<\omega_1}2$ we shall construct a countable weak L(Q)-structure (B_σ, J_σ) and a finite or countable set Π_σ of sets of formulas of Φ, so that:

(15) (B_σ, J_σ) satisfies the same sentences of L^+ as A;

(16) If $\sigma \subset \tau$ then (B_τ, J_τ) is a precise extension of (B_σ, J_σ) and $\Pi_\sigma \subseteq \Pi_\tau$;

(17) $(B_{\sigma^\frown 0}, J_{\sigma^\frown 0})$ strongly omits every set in $\Pi_{\sigma^\frown 0}$, but some set in $\Pi_{\sigma^\frown 0}$ is the Φ-type of some m-tuple in $(B_{\sigma^\frown 1}, J_{\sigma^\frown 1})$;

(18) As (17) but with 0 and 1 transposed.

Assuming (15)-(18) hold, let σ be any sequence $\sigma: \omega_1 \to \{0,1\}$. Define (A_σ, I_σ) to be the union of the chain $((B_\tau, J_\tau) : \tau \subset \sigma)$. By (16), (A_σ, I_σ)

is candid (in the sense of L^+). By (15), A_σ is a model of ψ. By (16)-(18), if σ': $\omega_1 \to \{0,1\}$ and $\sigma \neq \sigma'$, then some Φ-type is realised in A_σ but strongly omitted (and hence omitted, cf. (7) above) by $A_{\sigma'}$. Thus the structures A_σ verify the theorem.

The structures (B_σ, J_σ) and sets Π_σ are chosen by induction on the length of σ. To begin the construction, let I be the set of all finite or countable subsets of dom(A). Then (A,I) is candid (in the sense of L^+). Writing $\langle\rangle$ for the empty sequence, let $(B_{\langle\rangle}, J_{\langle\rangle})$ be any countable elementary substructure of (A,I) (in the sense of L^+ - the reader should verify the appropriate downward Löwenheim-Skolem theorem). Let $\Pi_{\langle\rangle}$ be empty.

If the length of σ is a limit ordinal, take (B_σ, J_σ) to be the union of the (B_τ, J_τ) with $\tau \subset \sigma$ (the reader should verify that (B_σ, J_σ) is an elementary extension of each (B_τ, J_τ) in the sense of L^+). Let Π_σ be the union of the corresponding sets Π_τ.

Finally suppose $\sigma \in {}^{<\omega_1}2$, and (B_σ, J_σ) and Π_σ have just been chosen. By (15) and Corollary 7.2.3(b) there is a candid weak L(Q)-structure (C_0, K_0) such that $(B_\sigma, J_\sigma) \preccurlyeq (C_0, K_0)$ (all this in the sense of L^+) and (C_0, K_0) omits all the sets in Π_σ. Then $C_0 \models \psi$, and so uncountably many Φ-types are realised in C_0. Since B_σ is countable, at most countably many of these Φ-types can have hypersupports in (B_σ, J_σ). (This should be checked from the definition of a hypersupport.) Hence some m-tuple $\bar{c}$ in C_0 realises a Φ-type Φ_1 which has no hypersupport in (B_σ, J_σ). Let $(B_{\sigma^\frown 0}, J_{\sigma^\frown 0})$ be a countable precise extension of (B_σ, J_σ) which is $\preccurlyeq (C_0, K_0)$ (all in the sense of L^+) and contains $\bar{c}$, and let $\Pi_{\sigma^\frown 1}$ be $\Pi_\sigma \cup \{\Phi_1\}$. Since $(B_{\sigma^\frown 0}, J_{\sigma^\frown 0}) \preccurlyeq (C_0, K_0)$ (in the sense of L^+), $(B_{\sigma^\frown 0}, J_{\sigma^\frown 0})$ strongly omits all the sets in Π_σ, by Exercise 1(c) (or by the proof of Corollary 7.2.3).

Since (B_σ, J_σ) strongly omits all the sets in $\Pi_{\sigma^\frown 1}$, by Corollary 7.2.3(b) again there is a candid weak L(Q)-structure (C_1, K_1) such that $(B_\sigma, J_\sigma) \preccurlyeq (C_1, K_1)$ (all in the sense of L^+) and C_1 omits all the sets in $\Pi_{\sigma^\frown 1}$. By the same argument as for (C_0, K_0), some m-tuple $\bar{d}$ from C_1 realises a Φ-type Φ_0 which has no hypersupport in $(B_{\sigma^\frown 0}, J_{\sigma^\frown 0})$. Let $(B_{\sigma^\frown 1}, J_{\sigma^\frown 1})$ be a countable precise extension of (B_σ, J_σ) which is $\preccurlyeq (C_1, K_1)$ (all in the sense of L^+) and contains $\bar{d}$, and let $\Pi_{\sigma^\frown 0}$ be $\Pi_\sigma \cup \{\Phi_0\}$. By Exercise 1(c) again, $(B_{\sigma^\frown 1}, J_{\sigma^\frown 1})$ strongly omits all the sets in $\Pi_{\sigma^\frown 1}$. This completes the construction. □

I mention two corollaries, one model-theoretic and the other algebraic. Cf. section 5.1 for the notion of a complete type, and section 4.1 for ∃-types.

COROLLARY 7.2.5. *Let* L *be a countable first-order language and suppose* A *is an* L*-structure in which uncountably many complete types are realised. Then there is a family* $(A_i : i < 2^{\omega_1})$ *of* L*-structures, each of cardinality* ω_1 *and elementarily equivalent to* A*, such that if* $i \neq j$ *then some complete type is realised in* A_i *but not in* A_j.

Proof. In the theorem, take ψ to be $\bigwedge Th(A)$ and Φ to be the set of formulas of L. □

COROLLARY 7.2.6. *For each* $k \geq 2$*, there is a family* $(G_i : i < 2^{\omega_1})$ *of e.c. nilpotent groups of class* k*, each of cardinality* ω_1*, such that if* $i \neq j$ *then some* ∃*-type is realised in* G_i *but not in* G_j.

Proof. For simplicity I take k to be 2. Let L be the language of groups. By Exercise 6 of section 3.2, the class of e.c. nilpotent groups of class 2 (= e.c. N_2 groups for brevity) is axiomatised by a sentence ψ of $L_{\omega_1\omega}$. Let Φ be the set of all formulas of L of the form "x is a p-th power" or the form "There is y such that $y^p = 1$ and $[y,x] \neq 1$" for primes p. In the terminology of the proof of Theorem 4.4.5, if g and g' are elements of e.c. N_2 groups which represent different sets of primes, then g and g' have different Φ-types. Furthermore elements with different Φ-types have different ∃-types too.

So by the theorem it suffices to produce an e.c. N_2 group in which uncountably many different sets of primes are represented. One way to do this is to take the direct sum of the groups G(X) in the proof of Theorem 4.4.5, as X ranges over all sets of primes, and then to extend this direct sum to an e.c. N_2 group by Theorem 3.2.1.

For the case $k > 2$, cf. Exercise 6(d) of section 4.4. □

Let me draw out two implications of this corollary. First, if $i \neq j$ then G_i is not embeddable in G_j. This follows from the fact that in an e.c. model all ∃-types of elements are maximal (cf. Lemma 4.1.2).

And second, if $i \neq j$ then G_i is not equivalent to G_j in the language $L_{\omega_1\omega}$. This is because in $L_{\omega_1\omega}$ one can say what first-order types are realised. Note that by Exercise 12(b) of section 4.4, all periodic e.c. N_2 groups are $L_{\infty\omega}$-equivalent. Note also that Lemma 4.1.10, which is the best-known method of getting many non-isomorphic models, is no good for producing models that are not $L_{\infty\omega}$-equivalent. One needs different mass-production methods for different purposes.

Exercises for 7.2.

L *is a countable first-order language throughout these exercises.*

1. Let A be an uncountable L-structure and X a finite or countable set of elements of A. An element a of A is said to be small over X iff there are a formula $\phi(x,\bar{w})$ of L(Q) and a tuple $\bar{d}$ from X such that $A \models \phi(a,\bar{d}) \wedge \neg Qx\phi(x,\bar{d})$. If a is not small over X we say it is large over X. Show: (a) At most countably many elements of A are small over X. (b) For all formulas $\phi(x,\bar{w})$ of L(Q) and tuples $\bar{d}$ from A, $A \models Qx\phi(x,\bar{d})$ iff $A \models \phi(a,\bar{d})$ for some element a which is large over $\{\bar{d}\}$. (c) If $\Phi(\bar{w})$ is a set of formulas of L(Q) and A omits Φ then A strongly omits Φ. [Given a hypersupport $\psi(x_0,\bar{y}_0,\ldots,x_n,\bar{y}_n,\bar{d},\bar{w})$ of Φ, use (b) inductively to find a_0, $\bar{b}_0$, ..., a_n, $\bar{b}_n$, $\bar{c}$ so that $A \models \psi(a_0,\ldots,\bar{b}_n,\bar{d},\bar{c})$. If A omits Φ then for some $\phi(\bar{w})$ in Φ, $A \models \psi(a_0,\ldots,\bar{b}_n,\bar{d},\bar{c}) \wedge \neg\phi(\bar{c})$. Using (b) again, quantify out the elements to get a contradiction.]

2. Let $((B_n,J_n) : n < \omega)$ be an elementary chain of countable weak models of Axioms 0-5, such that every step in the chain is precise. Suppose $\Phi(\bar{w})$ is a set of formulas of L(Q) which is omitted by every (B_n,J_n) $(n < \omega)$. Show that in hypergames with starting structure (B_0,J_0) it is hyperenforceable that (A^*,I^*) omits Φ.

3. Let T be a $\vdash_{0-5}$-consistent theory in L(Q), and suppose a hypergame is played with T as starting theory. Let $\Phi(w)$ be a set of formulas of L(Q) such that for some formula $\theta \in \Phi$, $T \vdash_{0-5} \neg Qw\theta$. Suppose also that for each formula $\psi(w)$ of L(Q), if $T \cup \{\exists w\psi\}$ is $\vdash_{0-5}$-consistent then so is $T \cup \{\exists w(\psi \wedge \neg\phi)\}$ for some formula $\phi \in \Phi$. Show that it is hyperenforceable that (A^*,I^*) omits Φ.

4. In a two-sorted first-order language L, let B be a countable model of second-order arithmetic, including the full schemas of comprehension and choice. Suppose B is an ω-model, i.e. the numbers of B have order-type ω in the ordering $<^B$. Show that B has an elementary extension of cardinality ω_1 which is also an ω-model. [Define $QX\phi(X)$ to mean $\forall Y\exists X(\phi(X) \wedge \forall n\, X \neq Y_n)$, where Y_n means $\{m : 2^m3^n \in Y\}$. Omit the set

$\{x \neq n : n < \omega\} \cup \{x \text{ is a number}\}$.]

5. Suppose we redefine hypersupport as in (1)-(4) except that the variables $y_0, \ldots, y_{n-1}$ are missing. (Equivalently, we require that in the quexistential prefix $S\bar{x}$ in (5), all Q quantifiers precede all quantifiers.) Show that Theorem 7.2.2 and its corollary remain true. [In the proof of (8), given $\chi(x,\bar{y},\bar{z})$, suppose that $(A_{<\alpha}, I_{<\alpha}) \models Q\bar{x}'\exists\bar{y}'\bar{z}'\chi$ where $Q\bar{x}'$ is a quantifier string $QxQy_i\ldots Qz_j$ absorbing as many of the variables in $\bar{y}$, $\bar{z}$ as possible, and $\bar{y}'$, $\bar{z}'$ are the remaining variables in $\bar{y}$, $\bar{z}$. Case 1 is now that $(A_{<\alpha}, I_{<\alpha}) \models Q\bar{x}'\exists\bar{y}'\bar{z}'(\chi \wedge \neg Qx_0 \ldots \exists\bar{y}_n \exists\bar{w}\psi)$, and Case 2 is that $(A_{<\alpha}, I_{<\alpha}) \models Q\bar{x}'Qx_0 \ldots Qx_n \exists\bar{y}_n \exists\bar{w}\exists\bar{y}'\bar{z}'(\chi \wedge \psi)$. To move the quantifiers $Qx_0, \ldots, Qx_n$ back over $\exists\bar{y}'\bar{z}'$ after adding $\neg\phi$, use Axiom 5 and the choice of $Q\bar{x}'$ as maximal.]

6. Let T be a $\vdash_{0-5}$-consistent theory in L(Q) and $\Phi(w)$ a set of formulas of L(Q). Suppose that for all formulas $\psi(\bar{y},w,\bar{z})$ of L(Q) and all quexistential prefixes $R\bar{y}$ and $S\bar{z}$, if $T \cup \{R\bar{y}QwS\bar{z}\psi\}$ is $\vdash_{0-5}$-consistent then so is $T \cup \{R\bar{y}QwS\bar{z}(\psi \wedge \neg\phi)\}$ for some $\phi \in \Phi$. Show that with T as starting theory it is hyperenforceable that at most countably many elements realise Φ.

7. Let L^+ be a countable fragment of $L_{\omega_1\omega}(Q)$, and write $(B,J) \leqslant^+ (A,I)$ to mean that $B \subseteq A$ and for all formulas $\phi(\bar{x})$ of L and tuples $\bar{b}$ from B, $(B,J) \models \phi(\bar{b})$ iff $(A,I) \models \phi(\bar{b})$. Show: (a) If $((B_i, J_i) : i < \gamma)$ form a $\leqslant^+$-chain of weak L(Q)-structures with union (A,I), then $(B_i, J_i) \leqslant^+ (A,I)$ for all $i < \gamma$. (b) If (A,I) is an infinite weak L(Q)-structure and X is a countable set of elements of A, then there is a countable weak L(Q)-structure (B,J) such that $(B,J) \leqslant^+ (A,I)$ and $X \subseteq \text{dom}(B)$.

8. Let T be a theory in L, and suppose T has a model A with a countable set X of elements such that uncountably many complete types over X are realised in A. Show: (a) If $2^\omega < 2^{\omega_1}$ then T has a family of 2^{ω_1} pairwise non-isomorphic models, each of cardinality ω_1. (b) If $2^\omega = \omega_1$ then T has a family $(B_i : i < 2^{\omega_1})$ of models of cardinality ω_1, such that if $i \neq j$ then B_i is not elementarily embeddable in B_j.

9. Let ψ be a sentence of $L_{\omega_1\omega}$ which has at least one but fewer than 2^{ω_1} isomorphism types of models of cardinality ω_1. Show that there are a countable first-order language $L^* \supseteq L$ and a theory T in L^* such that (a) T is complete and model-complete, (b) if A is an atomic model of T

then $A \models \psi$, and (c) T has an atomic model of cardinality ω_1. [Use Corollary 7.2.4 to show that some model of ψ of cardinality ω_1 satisfies a Scott sentence in some countable fragment of $L_{\omega_1\omega}$ (cf. Theorem 4.2.7); form L^* by adding relation symbols for the formulas in this fragment.]

10. Show that there is a family $(G_i : i < 2^{\omega_1})$ of e.c. groups, each of cardinality ω_1, such that if $i \neq j$ then some finitely generated group is embeddable in G_i but not in G_j.

7.3 MAGIDOR-MALITZ QUANTIFIERS

The upward and downward Löwenheim-Skolem theorems say that first-order logic can't distinguish between one infinite cardinal and another. This means that many important things can't be expressed in first-order logic, which is bad. On the other hand it also means that first-order theories usually have lots of models, which is good.

If we add new logical symbols to first-order logic, we can increase its expressive power, but at the cost of losing some ways of making models. In the subject known as Generalised Model Theory one studies different ways of striking the balance. Among those generalisations of first-order logic which still allow some form of Henkin construction, one of the strongest is the Magidor-Malitz logic.

Menachem Magidor and Jerome Malitz introduced a sequence of quantifiers Q^1x_1, $Q^2x_1x_2$, $Q^3x_1x_2x_3$, ..., so that Q^1x_1 is equivalent to the quantifier Qx_1 which we discussed in the two previous sections.

These quantifiers are understood as follows. Let A be an L-structure and $\phi(x_1,\ldots,x_m)$ a formula of L. A ϕ-cube is a set X of elements of A, such that

(1) $A \models \phi(a_1,\ldots,a_m)$ for all $a_1, \ldots, a_m \in X$.

Let λ be an infinite cardinal. Then in the λ-interpretation, $Q^m x_1\ldots x_m\phi$ is understood to say: there exists a ϕ-cube of cardinality λ.

EXAMPLE 1. Let A be a group. Then A has an uncountable abelian subgroup if and only if

(2) $A \models Q^2xy(xy = yx)$ (in the ω_1-interpretation).

Literally, (2) says that there is an uncountable set of pairwise commuting elements. The subgroup generated by such a set is an uncountable abelian subgroup of A.

EXAMPLE 2. Write $[X]^m$ for the set of all strictly increasing m-tuples of elements of X, where X is any linearly ordered set. Suppose $[\omega]^m = R \cup S$. Then

(3) $$Q^m x_1 \ldots x_m (x_1 < \ldots < x_m \rightarrow Rx_1 \ldots x_m) \quad \text{(in the } \omega\text{-interpretation)}$$

expresses that there is an infinite $X \subseteq \omega$ such that $[X]^m \subseteq R$. (The reader can supply the appropriate L-structure in (3).) In this setting, the conclusion of Ramsey's theorem is that

(4) $$Q^m \bar{x} (x_1 < \ldots < x_m \rightarrow R\bar{x}) \vee Q^m \bar{x} (x_1 < \ldots < x_m \rightarrow S\bar{x}).$$

Because of this connection with Ramsey's theorem, the quantifiers $Q^m \bar{x}$ in the ω-interpretation are often known as the Ramsey quantifiers.

Some writers use Q^m_λ to mean Q^m in the λ-interpretation. Also some writers require $a_1, \ldots, a_m$ to be pairwise distinct in (1) - this defines slightly different quantifiers which we shall write $Q^m_{\neq} \bar{x}$. The quantifiers $Q^m \bar{x}$ are definable in terms of the $Q^m_{\neq} \bar{x}$, and *vice versa*.

We call a formula $\phi(x_1, \ldots, x_m)$ strongly symmetrical iff $\phi(x_1, \ldots, x_m)$ implies $\phi(x_{f(1)}, \ldots, x_{f(m)})$ for every map $f: \{1, \ldots, m\} \rightarrow \{1, \ldots, m\}$. Let ϕ^{sts} be the conjunction of all these formulas $\phi(x_{f(1)}, \ldots, x_{f(m)})$. Then clearly ϕ^{sts} is strongly symmetrical. Also it is obvious from the definitions that $Q^m \bar{x} \phi$ is logically equivalent to $Q^m \bar{x} (\phi^{sts})$. This will allow us to assume sometimes that the quantifiers $Q^m \bar{x}$ are only applied to strongly symmetrical formulas.

Let L be a first-order language. Then L^n is defined to be the language got from L by adding $Q^1, \ldots, Q^n$ as quantifier symbols. The Magidor-Malitz language $L^{<\omega}$ is defined to be $\bigcup_{n<\omega} L^n$. Magidor and Malitz proved:

THEOREM 7.3.1. *Assume* $\Diamond$ *(cf. section 5.2). Let* L *be a countable first-order language, and give* $L^{<\omega}$ *the* ω_1*-interpretation. Then:*

(a) (Compactness) *If* T *is a theory in* $L^{<\omega}$ *and every finite subset of* T *has a model then* T *has a model.*

(b) (Completeness) *The set of all sentences of* $L^{<\omega}$ *which are true in every uncountable* L*-structure is recursively enumerable.*

Let me first explain why this is harder to prove than the corresponding result for L(Q). The problem is to construct a structure A so that for every formula $\phi(\bar{x})$, $A \models Q^m\bar{x}\phi$ if and only if there is an uncountable ϕ-cube in A. In the case where m = 1, if A does contain such a ϕ-cube, this means that there is an uncountable set X of elements of A, all of which satisfy ϕ; it follows at once that there is a definable such set X, namely $\phi(A)$. For this reason, when we proved Theorem 7.1.6 we only had to control the sizes of definable sets of elements.

However, when m > 1 it is quite possible for A to contain uncountable ϕ-cubes but no definable ones. (Cf. Exercise 4.) Hence we have to keep track of the sizes of many more sets than before. The basic idea will be that whenever $Q^m\bar{x}\phi$ is supposed to be false, we keep catching maximal ϕ-cubes Z during the course of the construction, and whenever we find one, we arrange that from that point onwards we omit a set Ψ_Z of formulas which express "$x \notin Z$ but $Z \cup \{x\}$ is a ϕ-cube". Thus maximal ϕ-cubes never grow, provided that we catch them in time. The principle $\diamondsuit$ is used to make sure that we do catch them.

For most of the proof, all the structures are models of Zermelo-Fraenkel set theory, ZFC. It turns out that there is no loss of generality, because all model theory can be formalised within ZFC.

Until further notice, L is the language of ZFC. We shall regard $Q^m x_1 \ldots x_m \phi(x_1,\ldots,x_m,\bar{y})$ as shorthand for a set-theoretic formula which expresses: "There is an uncountable set z such that for all $x_1, \ldots, x_m \in z$, $\phi(x_1,\ldots,x_m,\bar{y})$ holds". We identify Q^1 with Q. Since ZFC implies "ω_1 is regular", every model of ZFC satisfies all the axioms 0-5 for L(Q) under this interpretation of Q. (Cf. Exercise 1(c) of section 6.1.) Also our interpretation makes every formula of $L^{<\omega}$ equivalent to a first-order formula, so that we can identify $L^{<\omega}$ with L.

Let B be a countable model of ZFC. We shall construct a countable elementary extension A of B by forcing. We write L' for the language L with the following constants added: (i) the elements of B, (ii) an overflow constant ∞, (iii) a countable set of constants called witnesses. We write L(B) for L with just the constants (i) added. A condition is a finite set p of sentences of L' such that if $\bigwedge p$ is $\theta(\infty,\bar{c})$ for some $\theta(x,\bar{y})$ in L(B) and some tuple $\bar{c}$ of distinct witnesses, then $B \models Qx\exists\bar{y}\theta$. This puts us into the setting of section 6.1, and we can read off the definitions of "forcing" etc., together with:

LEMMA 7.3.2. *It is enforceable that:* $\bigcup\bar{p}$ *contains the elementary diagram of* A^+, *every element of* A^+ *is named by infinitely many witnesses,* A *is an elementary extension of* B *and* ∞ *is an element of* A *which is not in* B. □

When can a set $\Psi(\bar{z})$ of formulas of L(B) be omitted in A? Since ordered n-tuples are definable in ZFC, there will be no loss of generality if we assume that $\bar{z}$ is a single variable z. The answer to the question will use the following definitions.

We write $(aax\in y)\phi$ as a shorthand for a formula which expresses: "y is an uncountable set, and for all but at most countably many elements x of y, ϕ holds". (*aa* = almost all.) If $\Psi(z)$ is any set of formulas of L(B), we define its derivative $\partial\Psi(w)$ to be the set of formulas

$$\{(aaz\in w)\psi : \psi(z) \in \Psi\}. \tag{5}$$

We define inductively: $\partial^0\Psi = \Psi$, $\partial^{n+1}\Psi = \partial\partial^n\Psi$.

LEMMA 7.3.3. *Let* $\Psi(z)$ *be a set of formulas of* L(B). *Then the following are equivalent:*

(a) It is enforceable that A *omits* Ψ.

(b) B *omits both* Ψ *and* $\partial\Psi$.

Proof. (a) $\Rightarrow$ (b): Clearly if Ψ is already realised in B, then it is not enforceable that A omits Ψ. Suppose some element b realises $\partial\Psi$ in B. Then (ignoring the trivial case where Ψ is empty) b is uncountable in the sense of B. Hence player $\forall$ can play "$\infty \in. b$" as his first move. Let him continue the game so that for each formula $\phi(z)$ of L(B), if $B \models \neg Qz\phi$ then every element of A which satisfies ϕ is already in B. (Cf. Theorem 6.1.4.) Then ∞ will not lie in any set which B thinks is countable, and hence ∞ will realise Ψ. Thus (a) fails.

(b) $\Rightarrow$ (a): Assume (b) and let Ψ^* be the set $\Psi \cup \{z \neq a : a \in \mathrm{dom}(B)\}$. It suffices to show that it is enforceable that A omits Ψ^*. For then no new element of A realises Ψ, and (b) already says that no element of B realises Ψ. By Lemma 7.3.2 and the conjunction lemma, it suffices to show for each witness d that the statement "d^{A^+} doesn't realise Ψ^*" is enforceable.

Let player $\forall$ begin the game with p_0, where $\bigwedge p_0$ is $\theta(\infty,\bar{c},d)$ for some formula $\theta(x,\bar{y},z)$ of L(B). We have

(6) $$B \models Qx\exists\bar{y}z\theta.$$

We claim that for some formula $\psi(z) \in \Psi^*$, $B \models Qx\exists\bar{y}z(\theta \wedge \neg\psi)$. Suppose not. Then in particular $B \models \neg Qx\exists\bar{y}z(\theta \wedge z=a)$ for every element a of B. We infer

(7) $$B \models \forall z\neg Qx\exists\bar{y}\theta.$$

By (6) and (7), since B is a model of ZFC,

(8) $B \models$ "There are an uncountable set e and a 1-1 function f with domain e such that for every $x \in e$, $\exists\bar{y}\theta(x,\bar{y},fx)$".

Consider the image b of f in B. Since f is 1-1, b is uncountable in the sense of B. Since by assumption b fails to realise $\partial\Psi$, there must be some $\psi(z) \in \Psi$ such that $\neg\psi(z)$ holds for uncountably many elements z of b (counting inside B). We infer

(9) $$B \models Qx\exists\bar{y}z(\theta \wedge \neg\psi(z)),$$

proving the claim. By (9), player $\exists$ can ensure that d^{A^+} fails to realise Ψ, by putting $p_1 = p_0 \cup \{\neg\psi(d)\}$. □

We say that $\Psi(z)$ is <u>totally unsupported</u> (in B) iff none of Ψ, $\partial\Psi$, $\partial^2\Psi$, ... is realised in B. By Lemma 7.3.3 and the conjunction lemma, if Ψ is totally unsupported in B, then it is enforceable that Ψ is totally unsupported in A.

Let Z be a set of elements of B, not necessarily definable in

B. We write Ψ_Z for the following set of formulas:

(10) $\{\psi(z) : \psi \in L(B) \text{ and for all but finitely many } a \in Z,\ B \models \psi(a)\}$.

The next lemma is the technical heart of the proof of Theorem 7.3.1. It tells us how we can prevent maximal ϕ-cubes from growing.

LEMMA 7.3.4. *Let* $\phi(x_1,\ldots,x_m)$ *be a strongly symmetrical formula of* L(B) *such that* $B \models \neg Q^m x_1 \ldots x_m \phi$, *and let* $Z \subseteq \mathrm{dom}(B)$ *be a maximal* ϕ*-cube in* B. *Then the set* Ψ_Z *is totally unsupported in* B.

Proof. Suppose to the contrary that some derivative of Ψ_Z, say $\partial^n \Psi_Z$, is realised in B by an element b. There are two cases to consider. Suppose first that n = 0. For each element a of Z the formula "$z \neq a$" is in Ψ_Z, so b is not in Z. But also for all $a_1, \ldots, a_i$ in Z the formula $\phi(a_1,\ldots,a_i,z,\ldots,z)$ is in Ψ_Z, so that $Z \cup \{b\}$ is a ϕ-cube. This contradicts the maximality of Z.

Suppose on the other hand that n > 0. Then for each $\psi \in \Psi_Z$ we have

(11) $$B \models (aax_1 \in b)(aax_2 \in x_1)\ldots(aax_n \in x_{n-1})\psi(x_n).$$

In particular, since Z is a ϕ-cube, if $a_1, \ldots, a_{m-1} \in Z$ then $B \models \phi(a_1,\ldots,a_{m-1},a)$ for all $a \in Z$, and hence by the definition of Ψ_Z,

(12) $$B \models (aax_1 \in b)\ldots(aax_n \in x_{n-1})\phi(a_1,\ldots,a_{m-1},x_n).$$

Next, if $a_1, \ldots, a_{m-2} \in Z$ then (12) holds for all elements a_{m-1} of Z, and so again by the definition of Ψ_Z, relabelling some variables,

(13) $$B \models (aax_1^{m-1} \in b)\ldots(aax_n^{m-1} \in x_{n-1}^{m-1})$$
$$(aax_1^m \in b)\ldots(aax_n^m \in x_{n-1}^m)\phi(a_1,\ldots,a_{m-2},x_n^{m-1},x_n^m).$$

Applied m times, this argument gives:

(14) $$B \models (aax^1_1 \in b) \ldots (aax^1_n \in x^1_{n-1}) \quad \cdot \quad \cdot \quad \cdot \quad \cdot \quad \cdot \quad (aax^m_1 \in b) \ldots (aax^m_n \in x^m_{n-1}) \phi(x^1_n, \ldots, x^m_n).$$

Now we shall forget about Z, and use (14) together with some set-theoretical reasoning inside B, to deduce that in B "there is an uncountable ϕ-cube".

Call a set e in B <u>solid</u> iff for all k ($0 \leq k \leq m$) and all $a_1, \ldots, a_k \in e$ the following holds:

(15) $$B \models (aax^{k+1}_1 \in b) \ldots (aax^{k+1}_n \in x^{k+1}_{n-1}) \quad \cdot \quad \cdot \quad \cdot \quad \cdot \quad \cdot \quad (aax^m_1 \in b) \ldots (aax^m_n \in x^m_{n-1}) \phi(a_1, \ldots, a_k, x^{k+1}_n, \ldots, x^m_n).$$

Now (14) asserts that (15) holds for any e when k = 0; so in particular the empty set is solid. The property of being solid is of finite character, and hence by Zorn's lemma in B, there is a "maximal solid set" e in B.

We claim that $B \models$ "e is uncountable". For suppose otherwise. Then inside B, e is countable. Since e is solid, for each k ($0 \leq k < m$) and all $a_1, \ldots, a_k \in e$ there are at most countably many elements x^{k+1}_1 in b such that

(16) $$B \models \neg (aax^{k+1}_2 \in x^{k+1}_1) \ldots (aax^{k+1}_n \in x^{k+1}_{n-1}) \quad \cdot \quad \cdot \quad \cdot \quad \cdot \quad \cdot \quad (aax^m_1 \in b) \ldots (aax^m_n \in x^m_{n-1}) \phi(a_1, \ldots, a_k, x^{k+1}_n, \ldots, x^m_n).$$

Let b' be b with all such elements x^{k+1}_1 removed (for all choices of k, $a_1, \ldots, a_k$ at once). Since e is countable and b is uncountable, b' is not empty. Choose $b_1 \in b'$. Then (for all choices of k, $a_1, \ldots, a_k$)

(17) $$B \models (aax^{k+1}_2 \in b_1) \ldots (aax^{k+1}_n \in x^{k+1}_{n-1}) \quad \cdot \quad \cdot \quad \cdot \quad \cdot \quad \cdot \quad (aax^m_1 \in b) \ldots (aax^m_n \in x^m_{n-1}) \phi(a_1, \ldots, a_k, x^{k+1}_n, \ldots, x^m_n).$$

Repeat this argument with b_1 in place of b, then with b_2 in place of b_1, and so on up to b_{n-1}. Then for all k ($0 \leqslant k < m$) and all $a_1, \ldots, a_k \in e$,

(18) $$B \models (aax_n^{k+1} \in b_{n-1}) \quad \cdot \quad \cdot \quad \cdot \quad \cdot \quad \cdot \quad (aax_1^m \in b) \ldots (aax_n^m \in x_{n-1}^m)\phi(a_1,\ldots,a_k,x_n^{k+1},\ldots,x_n^m).$$

Hence, since b_{n-1} is uncountable and e is countable, there is some $b_n \in b_{n-1}$ distinct from all elements of e, such that for all k ($0 \leqslant k < m$) and all $a_1, \ldots, a_k \in e$,

(19) $$B \models (aax_1^{k+2} \in b) \ldots (aax_n^{k+2} \in x_{n-1}^{k+2}) \quad \cdot \quad \cdot \quad \cdot \quad \cdot \quad \cdot \quad (aax_1^m \in b) \ldots (aax_n^m \in x_{n-1}^m)\phi(a_1,\ldots,a_k,b_n,x_n^{k+2},\ldots,x_n^m).$$

Comparing (15) with (14) and (19) together, and recalling that ϕ is strongly symmetrical, we see that $e \cup \{b_n\}$ is solid, contradicting the maximality of e. This proves the claim.

Since e is solid, putting $k = m$ in (15) we have $B \models \phi(a_1,\ldots,a_m)$ for all $a_1, \ldots, a_m \in e$. Since e is uncountable in B by the claim, we have shown that $B \models Q^m x_1 \ldots x_m \phi(x_1,\ldots,x_m)$, which proves the lemma. □

Putting together the lemmas proved so far, we can show:

THEOREM 7.3.5. *Assume* $\diamondsuit$. *Let* B *be a countable model of* ZFC. *Then there is an uncountable elementary extension* A *of* B *such that for every formula* $\phi(x_1,\ldots,x_m,\bar{y})$ *and all tuples* $\bar{a}$ *from* A,

(20) $A \models Q^m\bar{x}\phi(\bar{x},\bar{a})$ *iff there is an uncountable* $\phi(\bar{x},\bar{a})$*-cube in* A.

Proof. We build up A as the union of a continuous elementary chain $(A_\alpha : \alpha < \omega_1)$ of countable structures. We put $A_0 = B$. When A_α has been constructed, we build $A_{\alpha+1}$ by games, and in particular $A_{\alpha+1}$ is related to A_α as A to B in Lemma 7.3.2. The same argument as in section

7.1 will ensure that if $A \models Qx\phi(x,\bar{a})$ then uncountably many elements of A satisfy $\phi(x,\bar{a})$. Bearing in mind our set-theoretic interpretation of $Q^m\bar{x}$, this also proves left to right in (20).

Right to left needs more work. While the chain $(A_\alpha : \alpha < \omega_1)$ is being built, as soon as A_α is constructed we define a countable set Γ_α of sets of formulas which are totally unsupported in A_α, and we build $A_{\alpha+1}$ so that these sets are still totally unsupported in $A_{\alpha+1}$. Since Γ_α will always contain $\bigcup_{\beta<\alpha}\Gamma_\beta$, and the chain $(A_\alpha : \alpha < \omega_1)$ is continuous, this will guarantee that all sets in any Γ_α remain totally unsupported in A, and hence A will omit all these sets.

We use $\Diamond$ to guide the choice of the Γ_α. By $\Diamond$ there is a family of sets $S_\alpha \subseteq \alpha$ $(\alpha < \omega_1)$ such that for each $X \subseteq \omega_1$ the set $\{\alpha : S_\alpha = X\cap\alpha\}$ is stationary in ω_1. We identify the elements of A_0 with the ordinals $< \omega$, the elements of A_1 with the ordinals $< \omega+\omega$, and so on through the construction, until finally $\mathrm{dom}(A) = \omega_1$. As soon as A_α has been constructed, it is determined whether or not:

(21) There is a formula $\phi(x_1,\ldots,x_m)$ of $L(A_\alpha)$ such that $A_\alpha \models \neg Q^m\bar{x}\phi$ and S_α is a maximal ϕ-cube in A_α.

If (21) holds, we choose Γ_α to be $\bigcup_{\beta<\alpha}\Gamma_\beta \cup \{\Psi_{S_\alpha}\}$, invoking Lemma 7.3.4 for ϕ^{sts}. If (21) fails, we take Γ_α to be $\bigcup_{\beta<\alpha}\Gamma_\beta$. This ensures:

(22) If (21) holds then S_α is a maximal ϕ-cube in A.

We deduce right to left in (20) as follows.

Suppose that at the end of the construction, $A \models \neg Q^m\bar{x}\phi(\bar{x},\bar{a})$ and X is a $\phi(\bar{x},\bar{a})$-cube in A. We have to show that X is at most countable. Without loss we can assume (using Zorn's lemma in the real world, not in A!) that X is a maximal $\phi(\bar{x},\bar{a})$-cube in A. There is a club C in ω_1 such that for each $\alpha \in C$, (i) $\alpha = \mathrm{dom}(A_\alpha)$, (ii) all the parameters $\bar{a}$ lie in α, and (iii) the expanded structure $(A_\alpha, X\cap\alpha)$ (with $X\cap\alpha$ as an added 1-ary relation) is an elementary substructure of (A,X). By (iii), $X\cap\alpha$ is a maximal $\phi(\bar{x},\bar{a})$-cube in A_α and $A_\alpha \models \neg Q^m\bar{x}\phi(\bar{x},\bar{a})$. Since C is a club, there is some $\alpha \in C$ such that $S_\alpha = X\cap\alpha$, and (21) holds for this α. Hence by (22), $X\cap\alpha$ is a maximal $\phi(\bar{x},\bar{a})$-cube in A, so that $X = X\cap\alpha$ and thus X is at most countable. □

Now we drop the assumption that L is the language of set theory. Let L be any countable first-order language. Without loss we can assume that L has countably many constants and countably many relation and function symbols of each arity, so that L is easy to define set-theoretically. Let T be a theory in $L^{<\omega}$. We shall say that T <u>has an uncountable set model</u> iff there exists a model B of ZFC with an element b such that B ⊨ "b is an uncountable L-structure" and for every sentence ϕ in T, B ⊨ "b ⊨ ϕ". This definition is incomplete until we have explained what set-theoretical formulas are "b is an uncountable L-structure" and "b ⊨ ϕ". Here the reader should follow his nose, using any standard set-theoretical definition of L-structures and ⊨, for example as in section 1.3 of Chang & Keisler (1973). The definition can be made so that for sentences ϕ of $L^{<\omega}$, the map $\phi \mapsto$ "x ⊨ ϕ" is recursive and commutes with ¬ and ∧. Write ϕ^x for "x ⊨ ϕ"; this is a formula of the language of ZFC with one free variable.

LEMMA 7.3.6. *Assume* ◊. *If* T *has an uncountable set model then* T *has an uncountable model in the* ω_1*-interpretation.*

<u>Proof</u>. Suppose T has an uncountable set model. Then by the downward Löwenheim-Skolem theorem there exists a countable model B of ZFC with an element b satisfying "b is an uncountable L-structure" and "b ⊨ ϕ" for every sentence ϕ in T. By Theorem 7.3.5, B has an elementary extension A in which the set-theoretic interpretation of each quantifier $Q^m\bar{x}$ agrees with the ω_1-interpretation. In A the element b encodes an L-structure D whose elements are the "elements of dom(B)" in the sense of A. Since A ⊨ "b is uncountable", D is uncountable.

We claim that for every formula $\phi(\bar{y})$ of $L^{<\omega}$ and every tuple $\bar{d}$ of elements of D,

(23) A ⊨ "b ⊨ $\phi(\bar{d})$" iff D ⊨ $\phi(\bar{d})$ in the ω_1-interpretation.

The proof is by induction on the complexity of ϕ. Suppose for example that ϕ is $Q^m\bar{x}\psi(\bar{x},\bar{y})$, and D ⊨ $\phi(\bar{d})$. Then D contains an uncountable $\psi(\bar{x},\bar{d})$-cube X. Now X is an uncountable θ-cube in A, where $\theta(\bar{x})$ is the formula "$x_1, \ldots, x_m$ are elements of the structure b and b ⊨ $\psi(\bar{x},\bar{d})$". So by choice of A, A ⊨ $Q^m\bar{x}\theta$ in the set-theoretic interpretation of Q^m. But this says precisely that A ⊨ "b ⊨ $\phi(\bar{d})$".

For each sentence ϕ in T we have $B \models$ "$b \models \phi$" by choice of B, and so $A \models$ "$b \models \phi$" since $B \preccurlyeq A$, and finally $D \models \phi$ by the claim. □

<u>Proof of Theorem 7.3.1</u>. Introducing a new constant s, let Σ be the set of all sentences ϕ of $L^{<\omega}$ such that

(24) $\quad$ ZFC $\cup$ {s is an uncountable L-structure} $\vdash \phi^s$.

Since ZFC is a recursive set and the map $\phi \mapsto \phi^s$ is recursive, Σ is recursively enumerable.

We claim that for every theory T in $L^{<\omega}$,

(25) $\quad$ T has an uncountable model in the ω_1-interpretation if and only if $T \cup \Sigma$ is consistent in first-order logic.

For left to right, suppose D is an uncountable model of T in the ω_1-interpretation. Consider the set-theoretic universe V as a model of ZFC, with the constant s naming D. By (24), $V \models \phi^s$ for all $\phi \in \Sigma$, and hence $D \models \Sigma$. We deduce that $T \cup \Sigma$ is first-order consistent.

Conversely suppose T has no uncountable model in the ω_1-interpretation. Then by Lemma 7.3.6 (and here we use $\diamondsuit$), T has no uncountable set model. In other words, ZFC $\cup$ {s is an uncountable L-structure} $\cup$ $\{\phi^s : \phi \in T\}$ has no model, and so by the compactness theorem for first-order logic there are $\phi_1, \ldots, \phi_n \in T$ such that $\neg(\phi_1 \wedge \ldots \wedge \phi_n) \in \Sigma$ (using (24) and the choice of the map $\phi \mapsto \phi^s$). This gives a first-order inconsistency in $T \cup \Sigma$. (25) is proved.

Claim (25) gives both parts of the theorem. Compactness is immediate from (25) and the compactness theorem for first-order logic, adding a short special argument as in the proof of Theorem 7.1.6(b) for the case where some finite subset of T has no uncountable model.

To prove completeness, let ϕ be a sentence of $L^{<\omega}$ which is true in every uncountable L-structure under the ω_1-interpretation. By (25), $\Sigma \vdash \phi$ by a first-order implication. So the set of all such sentences ϕ is the closure of Σ under first-order implication, and clearly this set is recursively enumerable. □

The use of the set-theoretic universe in the proof of (25) may seem a bit alarming, since there is no hope of formalising this argument in ZFC. However, much less set theory than ZFC is needed to make the proof go through, so that it would suffice to consider D as an element of some transitive model of a finite fragment of ZFC.

The reader may wonder whether this translation into models of set theory has prevented us from finding some simpler set of axioms for $L^{<\omega}$, like Keisler's Axioms 0-5 for L(Q). It seems not. Shelah & Steinhorn (198-) show that it is impossible to axiomatise even L^2 in the ω_1-interpretation by a finite set of axiom schemas.

Exercises for 7.3.

1. Let m be a positive integer. (a) Show that $Q^m_{\neq}$ is definable in terms of Q^m. (b) Show that Q^m is definable in terms of $Q^1_{\neq}, \ldots, Q^m_{\neq}$. In both cases the definition can be chosen uniformly for all λ-interpretations ($\lambda \geqslant \omega$).

2. Show that $Q^2xy(Rxy \vee Ryx)$ doesn't imply $Q^2xyRxy \vee Q^2xyRyx$ in any λ-interpretation ($\lambda \geqslant \omega$).

3. For suitable languages L, find the following: (a) A sentence of L^2 which has a model in the ω_1-interpretation iff there exists a Suslin tree. *(A Suslin tree is an Aronszajn tree in which there are no uncountable sets of pairwise incomparable elements; cf. section 8.1 below. The existence of Suslin trees follows from* $\lozenge$ *but is independent of* ZFC*, so that this exercise finds a sentence of* L^2 *which has models in some models of* ZFC *but not in others.)* (b) A sentence of L^2 which is true in every L-structure in the λ-interpretation iff every linear ordering of cardinality λ contains a subset which is ordered in order-type either λ or the reverse of λ. [The condition holds iff: for any two orderings of a set of cardinality λ, there is a subset of cardinality λ on which they agree everywhere or disagree everywhere.] *(Cardinals* λ *with this property are said to be* <u>*weakly compact*</u>. ω *is weakly compact but* ω_1 *is not.)*

4. Find a structure A and a first-order formula $\phi(x,y)$ such that A contains an uncountable ϕ-cube but no uncountable ϕ-cube is first-order definable in A with parameters. [Let A be the group of all permutations of ω_1 which move at most finitely many elements, and let $\phi(x,y)$ be xy=yx.]

5. (Downward Löwenheim-Skolem theorem) Let L be a countable first-order

language, A an uncountable L-structure and X a set of elements of A of cardinality $\leqslant \omega_1$. Show that there is a substructure B of A such that B has cardinality ω_1, $X \subseteq \mathrm{dom}(B)$, and for every formula $\phi(\bar{x})$ of $L^{<\omega}$ and tuple $\bar{b}$ of elements of B, $B \models \phi(\bar{b})$ iff $A \models \phi(\bar{b})$ (both in the ω_1-interpretation).

6. Let λ be an infinite regular cardinal, L a countable first-order language and T a theory in $L^{<\omega}$. Assuming $\Diamond$, show that if T has a model in the λ-interpretation then T has a model in the ω_1-interpretation. [Let ZFC^{κ} be ZFC together with an extra constant κ and an axiom "κ is an infinite regular cardinal". Interpret $Q^m\bar{x}$ in ZFC^{κ} using κ in place of ω_1, and use the resulting version of Lemma 7.3.6.]

7. Prove Theorem 7.3.1(b) with the word "uncountable" deleted.

There are a number of other quantifiers that look a bit like the Magidor-Malitz quantifiers. The next two exercises introduce some.

8. Let L be a countable first-order language and m a positive integer. Define $Q^{m,1}x_1 \ldots x_m y\phi(x_1,\ldots,x_m,y)$ to be true in A (in the λ-interpretation) iff: There is a proper subset X of dom(A) such that $|X| \geqslant \lambda$ and for all $a_1, \ldots, a_m \in X$ and all $b \notin X$, $A \models \phi(a_1,\ldots,a_m,b)$. Write $L^{m,1}$ for L with the quantifier $Q^{m,1}$ added. (a) Show that there is a set T of sentences of a suitable $L^{1,1}$ such that for every infinite cardinal λ, each finite subset of T has a model in the λ-interpretation but T itself has none. [Start from the fact that $\neg Q^{1,1}xy(F(x)\neq y \wedge x\neq 0)$ expresses that every large proper subset of the domain which is closed under F contains 0.] (b) Show that for every sentence ψ of $L^{1,1}$ there is a sentence θ of $L_{\omega_1\omega}(Q)$ such that if λ is any infinite cardinal then ψ and θ have exactly the same models in the λ-interpretation. [$Q^{1,1}xy\phi(x,y)$ says $\exists yQx(x\neq y \wedge \bigwedge_{n<\omega}$(there is no chain $\neg\phi(x,x_1)$, $\neg\phi(x_1,x_2)$, ..., $\neg\phi(x_n,y)$)).] (c) Show that if T is a theory in $L^{1,1}$, λ is a regular uncountable cardinal and T has a model in the λ-interpretation, then T has a model in the ω_1-interpretation. (d) Show that there is a sentence ϕ of a suitable $L^{2,1}$ such that for every infinite cardinal λ, $A \models \phi$ in the λ-interpretation iff A is a group of cardinality λ which has no proper subgroup of cardinality λ. *(Such a group is said to be λ-Jónsson.)*

9. Let L be the language of first-order arithmetic. In models A of arithmetic, interpret $Q^m x_1 \ldots x_m \phi(x_1,\ldots,x_m)$ to mean that there is a set X which is unbounded in A, such that $A \models \phi(a_1,\ldots,a_m)$ for all $a_1, \ldots, a_m$ in

X. *(This is the Macintyre interpretation of* Q^m.*)* Let $PA(Q^2)$ (resp. $PA(Q^{<\omega})$) be the theory in L^2 (resp. $L^{<\omega}$) consisting of Peano arithmetic with induction for all formulas of L^2 (resp. $L^{<\omega}$). (a) Show that if A is a model of $PA(Q^2)$ then there is no map from a proper initial segment of A onto an unbounded subset of A; hence in particular $PA(Q^2)$ has no models of singular cardinality, and in regular cardinality λ all models of $PA(Q^2)$ are λ-like. (b) Show that for every infinite cardinal λ the Macintyre interpretation of each Q^m coincides with the λ-interpretation in all models of $PA(Q^{<\omega})$ of cardinality λ. (c) Show that for every formula $\phi(\bar{x})$ of $L^{<\omega}$ there is a formula $\psi(\bar{x})$ of L^2 which is equivalent to ϕ in all models of $PA(Q^{<\omega})$. (d) Show (using just ZFC) that if λ is a regular uncountable cardinal then $PA(Q^{<\omega})$ has a model of cardinality λ. (e) Show that the consequences of $PA(Q^2)$ in L are exactly the consequences in L of second-order arithmetic with Π^1_1 comprehension and restricted induction.

REFERENCES FOR CHAPTER 7

7.1: Mostowski (1957) introduced the quantifier Qx together with the λ-interpretation for any infinite cardinal λ, and proved Exercises 2 and 4. The compactness theorem for the ω_1-interpretation (Theorem 7.1.6(b)) is due to Fuhrken (1964), and Vaught (1964) anticipated Theorem 7.1.6(a) by showing that the set of everywhere-true sentences in this interpretation is recursively enumerable. All the remaining results in the text, including Theorem 7.1.6(a), are due to Keisler (1970). This pathfinding paper also included Exercises 3(a), 5, 7, 9, 11. See Bruce (1978 a) and Bruce & Keisler (1979) for infinitary generalisations; Bruce (1978 b) and Kaufmann (1981 b; 1984) for a general model theory of weak L(Q)-structures; and Kaufmann (198-) for a survey.

Exercise 3(b): Malitz & Reinhardt (1972). Exercise 10: Kaufmann (1984). Exercise 14: Barwise et al. (1978); cf. Schmerl (1976) for related material and Bruce (1980) for a treatment by forcing.

7.2: The central results, Corollary 7.2.3 and Theorem 7.2.4, are from Keisler (1970). Shelah (1978 a) Theorem VIII.1.5(4) contains an alternative proof of Corollary 7.2.5 by way of the Halpern-Läuchli-Laver-Pincus theorem. In fact Shelah finds a family of continuum many types such that we can choose arbitrarily what subset of the family shall be realised, and the models can be arbitrarily large. I don't know if Theorem 7.2.4 can be improved along the same lines.

Exercise 1: Keisler (1970), but this proof is from Shelah (1980 a). Exercises 2-4 are from Keisler (1970). Exercise 5: Bruce (1978 a). Exercise 6 is a special case of a generalisation of Corollary 7.2.3 by Kaufmann (1979). Exercise 8(b): Shelah in Hodges (198- b). Exercise 9 is the first step of Shelah's argument showing that a sentence of $L_{\omega_1\omega}$ with few models in cardinality ω_{n+1} must have at least one model in cardinality ω_{n+2}, cf. Shelah (1975 a, 1983 b). Exercise 10: Ziegler (1980).

7.3: Theorem 7.3.1 is due to Magidor & Malitz (1977 a). The proof given above uses ideas in Malitz & Rubin (1980); Kaufmann (1979) suggests another proof by way of properties of Q. Shelah has announced in Rubin & Shelah (1983) that there are models of ZFC $\cup \{2^{\omega} = \omega_1\}$ in which (a) of Theorem 7.3.1 (compactness) fails; though Magidor & Malitz (1977 a) showed that (b) holds in some models of ZFC in which $\Diamond$ is not true. There are interesting connections between Ramsey quantifiers and stability theory; cf. Baldwin & Kueker (1980). Presumably Barwise & Feferman (198-) will discuss $L^{<\omega}$ and its applications.

Exercises 3, 5-7: Magidor & Malitz (1977 a). Exercise 8: Malitz (1983). Exercise 9: (a-c) Macintyre (1980), cf. Morgenstern (1982); (d,e) Schmerl & Simpson (1982). Strictly this paper proves (d) for $PA(Q^2)$, but the same proof extends; cf. Schmerl (198-) for a stronger result of the same kind.

8 L(Q) IN HIGHER CARDINALITIES

That will never be:
Who can impress the forest; bid the tree
Unfix his earth-bound root?
Macbeth IV.1.94ff.

Building models by games is easy when the models are countable. Models of cardinality ω_1 are not much harder - they can be reached as the limits of chains of countable models. The technology for building models of cardinality ω_2 as limits of chains of limits of chains is still relatively undeveloped, and there are signs that it may only work smoothly under certain set-theoretic assumptions (cf. Shelah (1983 b,c; 198- d)).

For larger structures the going gets very rough. To build them by games we usually need to invoke strong set-theoretic assumptions and the combinatorics involved are quite challenging. Naturally everyone hopes that the next generation of model theorists will weaken the assumptions and simplify the proofs. But there are some constraints on possible improvements. One can find models of set theory in which certain types of combinatorial structure (for example some kinds of tree) exist in cardinality ω_1 but not in various other cardinalities. In mathematics, if a structure just doesn't exist then no amount of model theorists will succeed in constructing it.

Section 8.1 explains why there is a problem about transporting the completeness theorem for L(Q) from the ω_1-interpretation to higher cardinalities. Section 8.2 solves the problem in two different ways. The first way uses the assumption $\lambda^{<\lambda} = \lambda$. The second uses the stronger assumption $\Diamond_\lambda$, but it also yields an omitting types theorem.

8.1 *THERE IS A PROBLEM*

There are sentences of L(Q) which have models in the λ-interpretation for some but not all uncountable cardinals λ. It is only reasonable to expect this. Axiom 5 in section 7.1 more or less said "ω_1 is regular"; in fact it is easy to write down a sentence of L(Q) which has a model in the λ-interpretation if and only if λ is singular.

But if we look closer we find phenomena that are not at all so reasonable. One such phenomenon, discovered by James Schmerl, is that L(Q) distinguishes between cardinals quite far up the Mahlo hierarchy. These are cardinals which are too big to be familiar and too small to be exciting, so that one is surprised to meet them in a theorem about something else. Jack Silver and William Mitchell discovered something even more distressing: the question whether L(Q) distinguishes between ω_1 and ω_2 depends on which model of set theory we work in.

To understand the results of Silver and Mitchell, one should know what a tree is. A <u>tree</u> is a structure consisting of a set X and a partial ordering $<^X$ of X such that for every element x of X, the set $\hat{x}$ of $<^X$-predecessors of x is well-ordered by $<^X$. The <u>height</u> of the element x is the order-type of $\hat{x}$. The <u>height</u> of the tree is the least ordinal greater than the heights of all the elements of the tree. We say that y is an <u>immediate successor</u> of x iff $x <^X y$ and height(y) = height(x) + 1. A <u>branch</u> of the tree is a maximal linearly ordered subset of the tree.

A well-known combinatorial lemma due to D. König says that if T is a tree of infinite height with only finitely many elements of height 0, and every element of T has at most finitely many immediate successors, then T has an infinite branch. (Cf. Exercise 7 of section 2.1.)

The natural analogue of König's lemma, replacing "infinite" by "of cardinality at least ω_1", turns out to be false:

THEOREM 8.1.1. *There is a tree* $(X,<^X)$ *such that*

(a) the height of the tree is ω_1*;*

(b) for every ordinal $\alpha < \omega_1$*, there are at most countably many elements of height* α*;*

(c) there are no uncountable branches, and in fact:

(c^+) there is a map from X *to the rationals which is 1-1 on each branch of the tree.*

Proof. Let Q be the set of rational numbers. A map f from an ordinal α to Q is said to be strictly increasing iff $i < j < \alpha$ implies $f(i) < f(j)$ in the usual ordering of the rationals.

We shall define, for each ordinal $\alpha < \omega_1$, a set $X(\alpha)$ of strictly increasing maps $f: \alpha \to Q$, so that

(1) each set $X(\alpha)$ is non-empty and at most countable;

(2) for each pair of ordinals $\alpha < \beta < \omega_1$, each $f \in X(\alpha)$ and each rational $\varepsilon > 0$, there is an element $g \in X(\beta)$ such that $f = g|\alpha$ and $\sup(\text{im } g) \leqslant \sup(\text{im } f) + \varepsilon$;

(3) if $\alpha < \beta < \omega_1$ and $g \in X(\beta)$ then $g|\alpha \in X(\alpha)$.

The definition of $X(\alpha)$ is by induction on α. $X(0)$ contains just the empty map. When a set $X(\alpha)$ has been chosen, we define $X(\alpha+1)$ to be the set of all strictly increasing maps $g: \alpha+1 \to Q$ such that $g|\alpha \in X(\alpha)$. For a limit ordinal δ we first express δ as the supremum of a strictly increasing sequence $(\alpha_n : n < \omega)$ of length ω. Then for each $\alpha < \delta$, each $f \in X(\alpha)$ and each rational $\varepsilon > 0$, we choose a strictly increasing sequence $(\varepsilon_n : n < \omega)$ of positive rationals converging to ε. If m is the least integer such that $\alpha_m > \alpha$, we extend f successively to $f_0 \subseteq f_1 \subseteq \ldots$ so that each f_n is in $X(\alpha_{m+n})$ and $\sup(\text{im } f_n) \leqslant \sup(\text{im } f) + \varepsilon_n$; we write $g_{f,\varepsilon} = \bigcup_{n<\omega} f_n$. Finally we take $X(\delta)$ to be the set of all $g_{f,\varepsilon}$ as f ranges through $\bigcup_{\alpha<\delta} X(\alpha)$ and ε ranges through the positive rationals.

This defines the sets $X(\alpha)$ so that (1)-(3) are satisfied. We take X to be $\bigcup\{X(\alpha+1) : \alpha < \omega_1\}$, and $<^X$ to be the extension relation $\subset$. Then the $<^X$-predecessors of an element are simply its initial segments with last element, so that $X(\alpha+1)$ is exactly the set of elements of height α. Clauses (a) and (b) then follow from (1). For each map $f: \alpha+1 \to Q$, write $\max(f)$ for $f(\alpha)$; then max is a map from X to the rationals which is 1-1 on each branch of the tree. This proves (c^+). Since ω_1 is uncountable, (c) follows. □

A tree with the properties (a)-(c) of the theorem is called an Aronszajn tree; if it also has the stronger property (c^+) it is called a special Aronszajn tree. For any infinite cardinal λ a λ-Aronszajn tree is a tree of height λ with no branches of length λ and with fewer than λ elements of height α for each $\alpha < \lambda$; it is called special iff there are a set S of cardinality $< \lambda$ and a map from the tree to S which is 1-1 on each branch.

By König's lemma there is no ω-Aronszajn tree. By Theorem 8.1.1 there is a special ω_1-Aronszajn tree. About ω_2 the set theorists tell us the following:

FACT 8.1.2. *(a) If $2^\omega = \omega_1$ then there is a special ω_2-Aronszajn tree.*

(b) If there exists a Mahlo cardinal (cf. Exercise 9 of section 5.2), then it is consistent with ZFC to suppose that there is no special ω_2-Aronszajn tree.

(c) If there exists an uncountable weakly compact cardinal then it is consistent with ZFC to suppose that there is no ω_2-Aronszajn tree.

Now it is not hard to write down a sentence ψ of L(Q) such that for every infinite cardinal λ, ψ has model in the λ^+-interpretation if and only if there exists a special λ^+-Aronszajn tree. In fact in section 8.2 we shall use such a sentence ψ to prove Fact 8.1.2(a). The distressing discoveries of Mitchell and Silver were parts (b) and (c) of Fact 8.1.2, and they show that there is no hope of generalising the completeness theorem for L(Q) from the ω_1-interpretation to the ω_2-interpretation without using some special set-theoretic hypothesis.

It turns out that a sufficient condition for extending the completeness theorem for L(Q) to the λ^+-interpretation is that $\lambda^{<\lambda} = \lambda$. A sufficient condition for extending Keisler's omitting types theorem (Corollary 7.2.3) to the λ^+-interpretation is that $\diamondsuit_\lambda$ holds. A sufficient condition for extending the Magidor-Malitz completeness theorem to the λ^+-interpretation is that both $\diamondsuit_\lambda$ and $\diamondsuit_{\lambda^+}$ hold. Very little is known about how far these sufficient conditions are necessary. One complicating fact is that even when these sufficient conditions hold, the previous proofs don't work. For the rest of this section I shall generalise the methods of Chapters 6 and 7, up to the point where they break down and we have to think of something new.

For the rest of the section, λ is an uncountable cardinal such that $\lambda^{<\lambda} = \lambda$, and L is a first-order language of cardinality at most λ.

If the reader feels tempted to see how far he can get with weaker assumptions on λ, I wish him luck. Recall from Lemma 5.2.1 that if $\lambda^{<\lambda} = \lambda$ then λ is regular.

The first task is to adapt section 6.1. Let B be an L-structure of cardinality at most λ. We write L(B) for L with the elements of B added as new constants, and L' for L(B) with an added overflow constant ∞ together with λ new constants called witnesses. Let Q be a largeness property over B (this means exactly what it meant in section 6.1).

Some new notation will be needed. Let Φ be a set of formulas. We write $\bigwedge\!\!\!\uparrow\Phi$ for the set of all finite conjunctions of formulas in Φ. If S is a string of quantifiers (maybe infinitely long) and ϕ is a formula, then we identify $S\phi$ with the formula $S'\phi$, where S' is the quantifier string got from S by striking out those quantifiers whose variables are not free in ϕ. We write $S\Phi$ for $\{S\phi : \phi \in \Phi\}$. Thus for example $\exists\bar{y}\bigwedge\!\!\!\uparrow\Phi$ is the set of all formulas $\exists\bar{y}(\phi_1 \wedge \ldots \wedge \phi_n)$ with $\phi_1, \ldots, \phi_n \in \Phi$.

In this setting we define a condition to be a set p of form $\Phi(\infty,\bar{c})$, where $\Phi(x,\bar{y})$ is a set of fewer than λ formulas of L(B) and $\bar{c}$ is a non-repeating sequence of witnesses, such that $B \models Qx\exists\bar{y}\bigwedge\!\!\!\uparrow\Phi(x,\bar{y})$. Note that $\bar{c}$ and $\bar{y}$ may be infinite. We write N for the set of conditions.

LEMMA 8.1.3. N *is a notion of forcing for* L', *and* N *has the properties (6)-(8) of section 2.3. The union of a chain of fewer than members of* N *is again a member of* N.

Proof. This is an analogue of Lemma 5.3.5. As a specimen I prove (7) of section 2.1, viz.

(4) If $\phi\vee\psi \in p \in N$ then either $p \cup \{\phi\} \in N$ or $p \cup \{\psi\} \in N$.

Suppose $p \in N$ and the conclusion of (4) fails. Write p as $\Phi(\infty,\bar{c})$ for a suitable set $\Phi(x,\bar{y})$ of formulas of L(B). Since $p \cup \{\phi\} \notin N$, there is a finite subset Ψ_1 of Φ such that $B \not\models Qx\exists\bar{y}\bigwedge(\Psi_1 \cup \{\phi\})$. Similarly there is a finite subset Ψ_2 of Φ such that $B \not\models Qx\exists\bar{y}\bigwedge(\Psi_2 \cup \{\psi\})$. By Axioms 1, 2 for largeness properties it follows that $B \not\models Qx\exists\bar{y}(\bigwedge(\Psi_1 \cup \Psi_2) \wedge (\phi \vee \psi))$ and hence $B \not\models Qx\exists\bar{y}\bigwedge\!\!\!\uparrow\Phi(x,\bar{y})$. This contradicts the assumption that $p \in N$.

□

By Lemma 8.1.3 we can play games using N, to build construction sequences $\bar{p}$ of length λ and compiled structures $A^+(\bar{p})$, just as in section 5.3. As before, $A(\bar{p})$ is the L-reduct of $A^+(\bar{p})$, and where the context allows we write these structures as A^+, A. Also as before, a property P is <u>enforceable</u> iff player $\exists$ has a winning strategy for G(P;odds). The conjunction lemma holds as in Lemma 5.3.6(c): the conjunction of λ enforceable properties is enforceable.

LEMMA 8.1.4. *The following is enforceable: "The compiled structure* $A(\bar{p})$ *is an elementary extension of* B; *every element of* $A^+(\bar{p})$ *is named by* λ *distinct witnesses;* $\bigcup\bar{p} = \text{eldiag}(A^+(\bar{p}))$; ∞^{A^+} *is not in* B; $A(\bar{p})$ *is* λ*-compact".*

<u>Proof</u>. All but the last clause follows from the arguments of Lemma 2.3.1, Lemma 5.3.7(a) and Theorem 6.1.3 (full first-order version).

For λ-compactness we argue as follows. It is sufficient to show that if $\Psi(\bar{z},\bar{w})$ is a set of fewer than λ formulas of L and $\bar{d}$ is a sequence of witnesses, then the following property P is enforceable:

(5) Either $A^+ \models \exists\bar{w}\bigwedge\Psi(\bar{d},\bar{w})$ or $A^+ \not\models \exists\bar{w}\bigwedge\Psi(\bar{d},\bar{w})$.

Since $\lambda^{<\lambda} = \lambda$, the number of such properties is λ, so that if each is enforceable, then so is the conjunction of them all.

We show that P is enforceable. Suppose players $\forall$ and $\exists$ play G(P;odds), and player $\forall$ begins with a condition p_0 which is $\Phi(\infty,\bar{c},\bar{d})$ for a suitable set $\Phi(x,\bar{y},\bar{z})$ of formulas of L(B). There are two cases.

<u>Case 1</u>: $B \models Qx\exists\bar{y}\bar{z}\bar{w}\bigwedge(\Phi \cup \Psi)$. Then player $\exists$ can choose new witnesses $\bar{e}$ and put $p_1 = p_0 \cup \Psi(\bar{d},\bar{e})$, to be sure of winning.

<u>Case 2</u>: $B \not\models Qx\exists\bar{y}\bar{z}\bar{w}\bigwedge(\Phi \cup \Psi)$. Then there is a finite subset Ψ' of Ψ such that $B \not\models Qx\exists\bar{y}\bar{z}\bigwedge(\Phi \cup \{\exists\bar{w}\bigwedge\Psi'\})$. Since p_0 is a condition, $B \models Qx\exists\bar{y}\bar{z}\bigwedge\Phi$, and hence by Axioms 1, 2 for largeness properties we infer that $B \models Qx\exists\bar{y}\bar{z}\bigwedge(\Phi \cup \{\neg\exists\bar{w}\bigwedge\Psi'\})$. In this case player $\exists$ should choose p_1 to be $p_0 \cup \{\neg\exists\bar{w}\bigwedge\Psi'(\bar{d},\bar{w})\}$, and again she can arrange to win. □

We turn next to the language L(Q). Let (B,J) be a weak L(Q)-structure of cardinality at most λ, satisfying Axioms 0-5 of section 7.1. We write L(Q,B) for L(Q) with the elements of B added as new constants, and L(Q)' for L(Q,B) with an added overflow constant ∞ together with λ new constants called witnesses. Sentences of $L(Q,B)$ are interpreted in (B,J) by the weak interpretation, as in section 7.1.

A condition is now a set p of form $\Phi(\infty,\bar{c})$, where $\Phi(x,\bar{y})$ is a set of fewer than λ formulas of $L(Q,B)$ and $\bar{c}$ is a non-repeating sequence of witnesses, such that $(B,J) \models Qx\exists\bar{y}\bigwedge\Phi(x,\bar{y})$. Exactly as in Lemma 8.1.3, the set of all conditions is a notion of forcing, so that we can construct a weak $L(Q)$-structure (A,I) by playing games of length λ over (B,J).

By analogy with L-structures, we say that a weak $L(Q)$-structure (A,I) is λ-compact iff:

(6) For every set $\Psi(\bar{x},\bar{y})$ of formulas of $L(Q)$ with $|\Psi| < \lambda$, and every sequence $\bar{a}$ of elements of A, if $(A,I) \models \exists\bar{y}\bigwedge\Psi(\bar{a},\bar{y})$ then $(A,I) \models \exists\bar{y}\bigwedge\Psi(\bar{a},\bar{y})$.

Making allowances for the difference between L-structures and weak $L(Q)$-structures, the next lemma is proved exactly like Lemma 8.1.4. (eldiag (A^+,I) is the set of all sentences of $L(Q)'$ which are true in (A^+,I).)

LEMMA 8.1.5. *Suppose a weak* $L(Q)'$*-structure* (A^+,I) *is compiled by playing games of length* λ *over* (B,J). *The following is enforceable: "*$(B,J) \preccurlyeq (A,I)$*; every element of* A^+ *is named by* λ *distinct witnesses;* $\bigcup\bar{p} = \text{eldiag}(A^+,I)$*;* ∞^{A^+} *is not in* B*;* (A,I) *is* λ*-compact".*

□

Moving on now to section 7.2, we introduce the notion of a hypergame of length λ^+ with starting structure (B,J). The players $\forall$ and $\exists$ play to construct an elementary chain of weak $L(Q)$-structures, $((A_\alpha,I_\alpha) : \alpha < \lambda^+)$, as follows. (A_0,I_0) is constructed over (B,J) just as (A,I) in Lemma 8.1.5, by a game G_0 of length λ. For each non-zero ordinal $\alpha < \lambda^+$, when the structures (A_β,I_β) have been constructed for all $\beta < \alpha$, we put $(A_{<\alpha},I_{<\alpha}) = \bigcup_{\beta<\alpha}(A_\beta,I_\beta)$. The players then construct (A_α,I_α) over $(A_{<\alpha},I_{<\alpha})$ by a game G_α of length λ, again as in Lemma 8.1.5. Finally the weak $L(Q)$-structure (A^*,I^*) compiled by this hypergame is the union $\bigcup((A_\alpha,I_\alpha) : \alpha < \lambda^+)$.

As in section 7.2, a property P is hyperenforceable iff player $\exists$ has a winning strategy for the hypergame $G(P;\text{odds})$, and weakly hyperenforceable iff she has one for the hypergame $G(P;\text{evens})$. A conjunction of λ hyperenforceable properties is hyperenforceable, and a conjunction of a hyperenforceable property with a weakly hyperenforceable property is weakly hyperenforceable.

Just as in section 7.2, one can define hypergames that start with a theory T in L(Q) instead of a weak L(Q)-structure (B,J). This raises no new issues and I leave it to the reader.

We say that a weak L(Q)-structure (A,I) is λ-candid iff for every formula $\phi(\bar{x})$ of L(Q) and every tuple $\bar{a}$ from A, $(A,I) \models \phi(\bar{a})$ iff $A \models \phi(\bar{a})$ in the λ-interpretation. (Thus candid means ω_1-candid.)

If the world was a beautiful place, we would now be able to show that it is weakly hyperenforceable that (A^*,I^*) is a λ^+-candid elementary extension of (B,J). By a routine extension of the arguments of section 7.1, it would follow that Keisler's completeness theorem holds for L(Q) in the λ^+-interpretation. The reader will see that one crucial item is missing from our previous arguments: we have done nothing to ensure that small sets stay small as the chain of structures (A_α, I_α) is built up. The next lemma is a first attempt at filling that gap.

LEMMA 8.1.6. *Suppose* (B,J) *is* λ-*compact and* $\psi(z)$ *is a formula of* L(Q,B) *such that* $(B,J) \models \neg Qz\psi$. *Then in the game to compile* (A,I) *over* (B,J), *it is enforceable that for every element* a *of* A, *if* $(A,I) \models \psi(a)$ *then* $a \in \mathrm{dom}(B)$.

Proof. As in the proof of Theorem 6.1.4, it suffices to show that for each witness d the following property P is enforceable:

(7) either $\bigcup\bar{p} \vdash \neg\psi(d)$, or there is an element b of B such that $(d=b) \in \bigcup\bar{p}$.

We play G(P;odds). Suppose player $\forall$ offers p_0 and p_0 is $\Phi(\infty,\bar{c},d)$, so that $(B,J) \models Qx\exists\bar{y}z\bigwedge\!\!\!\uparrow\Phi(x,\bar{y},z)$. As always, there are two cases.

Case 1: $(B,J) \models Qx\exists\bar{y}z\bigwedge\!\!\!\uparrow(\Phi \cup \{\neg\psi(z)\})$. Then player $\exists$ can choose p_1 to be $p_0 \cup \{\neg\psi(d)\}$, and win.

Case 2: $(B,J) \not\models Qx\exists\bar{y}z\bigwedge\!\!\!\uparrow(\Phi \cup \{\neg\psi(z)\})$. Then for some finite set $\Delta \subseteq \Phi$, $(B,J) \not\models Qx\exists\bar{y}z(\bigwedge\Delta \wedge \neg\psi(z))$. By Axiom 2 it follows that for every finite subset Ψ of Φ containing Δ, $(B,J) \models Qx\exists z\exists\bar{y}(\bigwedge\Psi \wedge \psi(z))$; hence the same holds for every finite subset Ψ of Φ, by Axiom 1. But by assumption $(B,J) \models \neg Qz\psi$, and so by Axioms 1 and 5 we infer that for every finite subset Ψ of Φ, $(B,J) \models \exists z Q x \exists \bar{y}(\bigwedge\Psi \wedge \psi)$ and hence $(B,J) \models \exists w Q x \exists \bar{y} z(\bigwedge\Psi \wedge \psi \wedge z=w)$.

Let $\Theta(w)$ be the set $Qx\exists\bar{y}z\bigwedge(\Phi \cup \{\psi, z=w\})$. It is easily verified that $(B,J) \models \exists w\bigwedge\Theta$. Since p_0 has cardinality $< \lambda$, so does Θ. By the λ-compactness of (B,J) it follows that there is an element b of B such that $(B,J) \models \bigwedge\Theta(b)$. Hence $p_0 \cup \{\psi(d), d=b\}$ is a condition. If player $\exists$ chooses p_1 to be this condition, she can guarantee to win. □

The blemish in Lemma 8.1.6 is the assumption that (B,J) is λ-compact. True, by Lemma 8.1.5 player $\exists$ can play the hypergame so that each structure (A_α, I_α) is λ-compact. But what happens at limit ordinals δ? Even if (A_α, I_α) is λ-compact for every $\alpha < \delta$, there is no reason why $(A_{<\delta}, I_{<\delta})$ should be λ-compact. Hence new elements may creep into small sets at any limit ordinal. Since there are λ^+ limit ordinals below λ^+, this loophole could allow a small set to achieve cardinality λ^+, and then (A^*, I^*) would not be λ^+-candid. Our previous construction fails.

In the next section I describe two different ways around this problem.

Exercises for 8.1.

The first three exercises explore consequences of the existence of Aronszajn trees. Lest any reader should think that these trees are just a device dreamed up by moon-crazed set theorists, I give in Exercise 3 one of their earliest applications. It is in ring theory.

1. A linearly ordered set $(I,<)$ is called a Specker ordering iff (i) $|I| = \omega_1$, (ii) neither ω_1 nor ω_1^* (the reverse of ω_1) is order-embeddable in $(I,<)$, and (iii) no uncountable subordering of the reals is order-embeddable in $(I,<)$. Show that there exists a Specker ordering. [Take the Aronszajn tree constructed in the proof of Theorem 8.1.1, so that the elements are sequences of rationals; order the sequences lexicographically by first differences.]

2. A graph on a set X is a set of pairs $\{a,b\}$ with a, b in X and $a \neq b$; the pairs are called the edges of the graph. Show that there is a graph G on ω_1 so that $|G| = \omega_1$, and the edges in G can each be coloured with one of four colours, so that if Y is any uncountable subset of X and H the restriction of G to Y, then H contains edges of each of the four colours. (In symbols this says $\omega_1 \not\rightarrow [\omega_1]^2_4$. It is unknown whether the same can be

proved with 5 in place of 4.) [Let $(X,<_1)$ be a Specker ordering, $<_2$ an ordering of X in order-type ω_1 and $<_3$ an ordering of X in the order-type of some set of reals. Colour each edge according to which of the orderings $<_1$, $<_2$, $<_3$ agree on the edge.]

3. Let $(X,<^X)$ be an Aronszajn tree, and write $X(\alpha)$ for the set of all elements of height α. Let K be a countable or finite field. For each $\alpha < \omega_1$, let A_α be the polynomial ring $K[X(\alpha)]$, with the elements of $X(\alpha)$ as indeterminates. For all $\alpha \leqslant \beta < \omega_1$, let $\chi_{\alpha\beta}: A_\beta \to A_\alpha$ be the ring homomorphism which is the identity on K and such that for each $x \in X(\beta)$, $\chi_{\alpha\beta}(x)$ is the unique element z of height α such that $z \leqslant^X x$. For each α, let R_α be the subring of A_α consisting of the polynomials with zero constant term, and let the maps $\phi_{\alpha\beta}: R_\beta \to R_\alpha$ be the restrictions of the maps $\chi_{\alpha\beta}$. (a) Show that the rings R_α and the maps $\phi_{\alpha\beta}$ form a commutative diagram. (b) Show that the inverse (i.e. projective) limit of the diagram in (a) is the trivial ring.

4. Show that the following are equivalent, for any infinite cardinal λ: (a) There is a special λ^+-Aronszajn tree. (b) There is a λ^+-Aronszajn tree $(X,<^X)$ such that each element of X is a 1-1 map from an ordinal to λ, and $<^X$ is function extension. *(Some writers use the term special for Aronszajn trees of type (b) rather than (a).)*

5. Write down a sentence ψ of a language $L(Q)$, such that for every infinite cardinal λ, ψ has a model in the λ^+-interpretation iff there exists a special λ^+-Aronszajn tree. [Generalise the definition of tree to allow the levels to be linearly ordered rather than well-ordered. If there exists a special λ^+-Aronszajn tree in this extended sense, and the set of levels is λ^+-like, then there exists a special λ^+-Aronszajn tree in the original sense.]

6. Let μ and λ be infinite cardinals; show that the following are equivalent. (a) For every countable first-order language L whose symbols include a 2-ary relation symbol $<$, if T is any theory in L with a model A in which $<^A$ is a μ-like linear ordering, then T has a model B in which $<^B$ is a λ-like linear ordering. (b) For every countable first-order language L, if U is any theory in $L(Q)$ which has a model in the μ-interpretation then U has a model in the λ-interpretation.

7. Let μ and λ be infinite cardinals; show that the following are equivalent. (a) For every countable first-order language L whose symbols

include a 1-ary relation symbol P, if T is any theory in L with a model A of cardinality > μ and $|P^A| = \mu$, then T has a model B of cardinality > λ with $|P^B| = \lambda$. (b) For every countable first-order language L, if U is any theory in L(Q) which has a model in the μ^+-interpretation then U has a model in the λ^+-interpretation.

8. Show that if L is a countable first-order language and T is a theory in L(Q) which has a model in the ω-interpretation, then for every infinite cardinal λ, T has a model in the λ-interpretation. [Use Corollary 6.2.6.]

9. Write sentences ϕ_1, ϕ_2, ϕ_3 of languages L(Q) such that for every infinite cardinal λ: (a) ϕ_1 has a model in the λ-interpretation iff λ is singular; (b) ϕ_2 has a model in the λ-interpretation iff λ is a successor cardinal; (c) ϕ_3 has a model in the λ-interpretation iff λ is uncountable.

8.2 COMPLETENESS AND OMITTING TYPES

As in section 8.1, λ is an uncountable cardinal such that $\lambda^{<\lambda} = \lambda$, L is a first-order language of cardinality at most λ, and (B,J) is a weak L(Q)-structure of cardinality at most λ which satisfies Axioms 0-5 of section 7.1.

We pick up the thread from the end of section 8.1, where we had a problem about keeping small sets small. Two solutions to this problem are known. The first is essentially due to C. C. Chang. I shall only sketch it. It starts from the following definitions and observation:

Let (A,I) be a weak L(Q)-structure and $\psi(y)$ a formula of L(Q,A). We call ψ <u>small</u> iff $(A,I) \models \neg Qy\psi$. We say that (A,I) is λ-<u>compact over</u> ψ iff:

(1) For every set $\Psi(y)$ of formulas of L(Q,A) with $|\Psi| < \lambda$, if $\psi \in \Psi$ and $(A,I) \models \exists y \bigwedge\!\!\!\uparrow \Psi$ then $(A,I) \models \exists y \bigwedge \Psi$.

We say that (A,I) is λ-<u>compact over small sets</u> iff for every small formula ψ of L(Q,A), (A,I) is λ-compact over ψ.

OBSERVATION 8.2.1. *In Lemma 8.1.6, the hypothesis that* (B,J) *is* λ-*compact can be weakened to:* (B,J) *is* λ-*compact over small sets.*

<u>Proof</u>. In Case 2 of the proof, the set Θ contains ψ and ψ is small. □

It is still not clear that the union of a chain of weak L(Q)-structures which are λ-compact over small sets is itself λ-compact over small sets. But for our purposes this is unnecessary. What we need to know is:

THEOREM 8.2.2. *Let* $\psi(y)$ *be a small formula in* L(Q,B). *In hypergames of length* λ^+ *with starting structure* (B,J), *it is hyperenforceable that for every limit ordinal* $\delta < \lambda^+$, $(A_{<\delta}, I_{<\delta})$ *is* λ*-compact over* ψ.

<u>Proof</u>. The theorem is trivial when only finitely many elements of B satisfy ψ; so we assume that infinitely many elements satisfy ψ.

First we consider a special case. Suppose that the language L contains a 2-ary relation symbol ε, and that in (B,J) the set of elements satisfying ψ form a model of ZF^o, Zermelo-Fraenkel set theory with the axiom of infinity removed, with ε representing membership. This case is exactly covered by the coding device in Chang's proof of his two-cardinal theorem; cf. Chang & Keisler (1973) section 7.2 or Sacks (1972) Chapter 23. More precisely, Chang shows that if player $\exists$ can arrange that each (A_α, I_α) is λ-compact and no new elements satisfy ψ after (A_0, I_0), then each union $(A_{<\delta}, I_{<\delta})$ is automatically λ-compact over ψ.

We turn to the general case. The language L may have cardinality λ, but we can write it as the union of a chain $(L_i : i < \lambda)$ of languages of cardinality $< \lambda$, with ψ in L_0. Since λ is regular, any set of fewer than λ formulas of L(Q,B) lies already inside some $L_i(Q,B)$. So by the conjunction lemma, it suffices if we prove the theorem with $(A_{<\delta}|L_i, I_{<\delta})$ in place of $(A_{<\delta}, I_{<\delta})$, for an arbitrary ordinal $i < \lambda$. We write $L^\$$ for L with the new relation symbol ε added, and $L_i^\$$ for $L^\$$ L_i.

Player $\exists$ will find her strategy for the hypergame by playing another hypergame on the side, to construct an elementary chain of weak $L_i^\$(Q)$-structures $(A_\alpha^\$, I_\alpha^\$)$ $(\alpha < \lambda^+)$. In the subsidiary game, the structures will satisfy Axioms 0-5 for the language $L_i^\$$, and the elements satisfying ψ will form a model of ZF^o, so that by the special case already considered, player $\exists$ can ensure that each $(A_{<\delta}^\$, I_{<\delta}^\$)$ is λ-compact over ψ.

We have to show that player $\exists$ can keep the side game in step with the main one. Specifically, she must arrange that for each ordinal α, $A_\alpha^\$|L_i = A_\alpha|L_i$ and for every formula $\phi(\bar{x})$ of $L_i(Q)$ and every tuple $\bar{a}$

from A_α, $(A^\$_\alpha, I^\$_\alpha) \models \phi(\bar{a})$ iff $(A_\alpha, I_\alpha) \models \phi(\bar{a})$. Then it will follow at once that each structure $(A_{<\delta}|L_i, I_{<\delta})$ is λ-compact over ψ, as required.

At the outset, player $\exists$ must expand some elementary extension of $(B|L_i, J)$ to a weak $L^\$_i(Q)$-structure satisfying various axioms mentioned above. This can be done by first-order compactness and reduction to countable sublanguages. From that point on, player $\exists$'s tactics in the hypergame are the exact analogue of those in Exercise 23 of section 5.3. I leave the messy details to the reader. □

With Theorem 8.2.2 we have all the tools needed to infer the completeness theorem for L(Q) in the λ^+-interpretation:

COROLLARY 8.2.3. *Assuming* $\lambda^{<\lambda} = \lambda$, *let* L *be a first-order language of cardinality at most* λ, *and let* T *be a* $\vdash_{0-5}$*-consistent theory in* L(Q). *Then* T *has a model in the* λ^+*-interpretation.* □

The reader can deduce a compactness theorem. Meanwhile:

COROLLARY 8.2.4. *If* $\lambda^{<\lambda} = \lambda$ *then there exists a special* λ^+*-Aronszajn tree.*

Proof. Use the sentence ψ of Exercise 5 in section 8.1. □

Commenting on Chang's proof of the two-cardinal theorem, Shelah remarks "But sometimes we do not want to expand the language", and he goes on to give a new proof in which the language is not expanded (Shelah 1978 c). Theorem 8.2.2 shows that as far as constructions by hypergames are concerned, there is no point in worrying about the language being expanded. Player $\forall$ never need know that player $\exists$ is secretly adding an extra symbol. The crucial fact is that the property "Small sets stay small" is hyperenforceable, and in applications it is quite irrelevant how this fact was proved.

We turn to our second way of keeping small sets small. This is due to Shelah, and it rests on the observation that to stop new elements satisfying ψ is to omit a type. So we show how to omit types, and then the small sets will take care of themselves. This will need a stronger set-theoretic assumption. For the rest of this section we assume that $\Diamond_\lambda$ holds.

Let $\Phi(\bar{w})$ be a set of formulas of L(Q,B); $\bar{w}$ may be infinite but we assume it has length $< \lambda$. We define a hypersupport of Φ in (B,J) to be a set $\Psi(x_0,\bar{y}_0,x_1,\bar{y}_1,\ldots,\bar{w})$ of formulas of L(Q,B) such that $|\Psi| < \lambda$ and

(2) $(B,J) \models Qx_0\exists\bar{y}_0Qx_1\exists\bar{y}_1\ldots\exists\bar{w}\bigwedge\Psi$, but

(3) $(B,J) \not\models Qx_0\exists\bar{y}_0Qx_1\exists\bar{y}_1\ldots\exists\bar{w}\bigwedge(\Psi \cup \{\neg\phi\})$ for every $\phi(\bar{w}) \in \Phi$.

The quantifier prefix in (2) and (3) may be infinite; it is well-ordered and of length $< \lambda$. We say that (B,J) strongly omits Φ iff there is no hypersupport of Φ in (B,J).

THEOREM 8.2.5. *We assume that* $\Diamond_\lambda$ *holds,* L *is a first-order language of cardinality at most* λ, (B,J) *is a weak* L(Q)*-structure of cardinality at most* λ, *and* (B,J) *satisfies Axioms 0-5. For each* $i < \lambda$ *let* Φ_i *be a set of formulas which is strongly omitted by* (B,J). *Then in hypergames of length* λ^+ *with starting structure* (B,J), *it is weakly hyperenforceable that the compiled structure* (A^*,I^*) *is a* λ^+*-candid elementary extension of* (B,J) *which strongly omits all the sets* Φ_i $(i < \lambda)$.

The proof of this theorem will take us to the end of the section. In fact we shall prove something stronger. For each ordinal $\alpha < \lambda^+$, just before the players start the game G_α to construct (A_α,I_α), the players will lay on the table another λ sets which are strongly omitted by $(A_{<\alpha},I_{<\alpha})$. Player $\exists$ will manage to make (A^*,I^*) strongly omit all these sets too.

We check at once that this will stop small sets growing:

LEMMA 8.2.6. *Suppose* $\psi(y)$ *is a small formula* $\in$ L(Q,B). *Then* (B,J) *strongly omits the set* $\{\psi\} \cup \{y \neq b : b \in \mathrm{dom}(B)\}$.

Proof. Exercise (use Exercise 5(d) from section 7.1). □

So if player $\exists$ keeps putting onto the table the sets corresponding to small formulas as in Lemma 8.2.6, small sets will stay small.

As usual when diamonds are involved, we name the elements before we begin the construction. The overflow constant for the game G_α constructing (A_α, I_α) is ∞^α, and the witnesses for this game are $\bar{c}^\alpha = (c_i^\alpha : i < \lambda)$. We write the construction sequence of game G_α as $\bar{p}^\alpha = (p_i^\alpha : i < \lambda)$, and we write $p_{<i}^\alpha$ for $\bigcup_{j<i} p_j^\alpha$.

The idea of player ∃'s strategy will be as follows.

Consider a set $\Phi(\bar{w})$ which is strongly omitted by (B,J), a set $\Psi(x_0,\bar{y}_0,x_1,\ldots,x_0',\bar{y}_0',x_1',\ldots,\bar{w})$ of fewer than λ formulas of $L(Q,B)$, and an increasing sequence $(\beta_0,\beta_1,\ldots)$ of ordinals $< \lambda^+$. Player ∃ wants to prevent $\Psi(\infty^{\beta_0},\bar{c}^{\beta_0},\infty^{\beta_1},\ldots,x_0',\bar{y}_0',x_1',\ldots,\bar{w})$ from becoming a hypersupport of Φ. So she chooses a limit ordinal $\delta < \lambda$, and she takes the following steps.

<u>Step 0</u>: After $p_{<\delta}^{\beta_0}$ has been chosen, she checks whether or not

(4) $$p_{<\delta}^{\beta_0} \text{ is } Qx_1\exists\bar{y}_1Qx_2\ldots Qx_0'\exists\bar{y}_0'Qx_1'\ldots\exists\bar{w}\bigwedge\Psi(\infty^{\beta_0},\bar{c}^{\beta_0},x_1,\bar{y}_1,\ldots,\bar{w})$$

(up to choice of bound variables). If it isn't, she reckons nothing needs to be done, and she puts $p_\delta^{\beta_0} = p_{<\delta}^{\beta_0}$. But if it is, then by the definition of conditions,

(5) $$(A_{<\beta_0}, I_{<\beta_0}) \models Qx_0\exists\bar{y}_0Qx_1\ldots\exists\bar{w}\bigwedge\Psi.$$

Assuming inductively that Φ still has no hypersupport in $(A_{<\beta_0}, I_{<\beta_0})$, player ∃ considers (5) and chooses some formula $\phi(\bar{w}) \in \Phi$ such that

(6) $$(A_{<\beta_0}, I_{<\beta_0}) \models Qx_0\exists\bar{y}_0Qx_1\ldots\exists\bar{w}\bigwedge(\Psi \cup \{\neg\phi\}).$$

By (6) she can take $p_\delta^{\beta_0}$ to be

(7) $$Qx_1\exists\bar{y}_1Qx_2\ldots Qx_0'\exists\bar{y}_0'Qx_1'\ldots\exists\bar{w}\bigwedge(\Psi(\infty^{\beta_0},\bar{c}^{\beta_0},x_1,\ldots,\bar{w}) \cup \{\neg\phi\}),$$

and so $(A_{\beta_0}, I_{\beta_0}) \models (7)$.

<u>Step 1</u>: After $p_{<\delta}^{\beta_1}$ has been chosen, she checks whether or not

(8) $$p_{<\delta}^{\beta_1} \text{ is } Qx_2\exists\bar{y}_2Qx_3\ldots Qx_0'\exists\bar{y}_0'Qx_1'\ldots\exists\bar{w}\bigwedge\Psi(\infty^{\beta_0},\bar{c}^{\beta_0},\infty^{\beta_1},\bar{c}^{\beta_1},x_3,\ldots,\bar{w}).$$

Again if it isn't, she puts $p_\delta^{\beta_1} = p_{<\delta}^{\beta_1}$. But if it is, then she considers the set

(9) $$Qx_2\exists\bar{y}_2Qx_3\ldots Qx'_0\exists\bar{y}'_0Qx'_1\ldots\exists\bar{w}\bigwedge(\Psi(\infty^{\beta_0},\bar{c}^{\beta_0},\infty^{\beta_1},\bar{c}^{\beta_1},x_3,\ldots,\bar{w})\cup\{\neg\phi\}).$$

Since $(A_{<\beta_1}, I_{<\beta_1}) \models p_\delta^{\beta_0}$, the set (9) is certainly a condition over $(A_{<\beta_1}, I_{<\beta_1})$. By (8), this condition extends $p_{<\delta}^{\beta_1}$, and so player $\exists$ can choose $p_\delta^{\beta_1}$ to be (9). Then $(A_{\beta_1}, I_{\beta_1}) \models$ (9).

Step 2 etc.: Player $\exists$ repeats the action of step 2 after each of $p_{<\delta}^{\beta_2}$, $p_{<\delta}^{\beta_3}$ etc. has been chosen. If all of these sets have the right form, then eventually player $\exists$ will have played the whole of the set

(10) $$Qx'_0\exists\bar{y}'_0Qx'_1\ldots\exists\bar{w}\bigwedge(\Psi(\infty^{\beta_0},\bar{c}^{\beta_0},\infty^{\beta_1},\ldots,x'_0,\bar{y}'_0,x'_1,\ldots,\bar{w})\cup\{\neg\phi\}).$$

Hence $\Psi(\infty^{\beta_0},\bar{c}^{\beta_0},\infty^{\beta_1},\ldots,x'_0,\bar{y}'_0,x'_1,\ldots,\bar{w})$ will not be a hypersupport of Φ.

At first sight it seems very improbable that this approach will work, for two reasons.

(i) Even for a single choice of Φ and Ψ, there are at least λ^+ different choices of ordinal sequences $(\beta_0, \beta_1, \ldots)$, each of which might give rise to a different hypersupport for Φ. Since player $\exists$ only has λ moves during each game G_α, how can she cover all the possibilities?

Actually this problem is not serious on its own. During each game G_α player $\exists$ need not distinguish between two ordinal sequences which agree up as far as α. But some careful sorting will be needed.

(ii) More seriously, for a single choice of Φ, Ψ and ordinal sequence $(\beta_0, \beta_1, \ldots)$, how can player $\exists$ possibly hope to choose the right ordinal δ in advance? If player $\forall$ wants to slip in the unwelcome hypersupport for Φ, one might think, all he has to do is to wait until after the δ-th step in each game G_α.

This is where $\diamondsuit_\lambda$ is needed. During each game G_α, $\diamondsuit_\lambda$ will predict player $\forall$'s moves, and it will get them right a stationary number of times. So player $\exists$ can choose δ in the light of these predictions.

The details are as follows. First we write the language L as the union of a continuous chain $(L_i : i < \lambda)$ of languages of cardinality $< \lambda$. For simplicity we assume that every element of B already has a name in L, so that we can identify L(Q,B) with L(Q). For each ordinal $\alpha < \lambda^+$ we define L_i^α to be $L_i(Q)$ together with all the new constants ∞^β, c_j^β ($\beta \leq \alpha$; $j < i$).

For each $\alpha < \lambda^+$ we fix an enumeration $(\alpha_i : i < |\alpha+1|)$ of the ordinals $\leq \alpha$, so that $\alpha_0 = \alpha$ and no ordinal is repeated. For each $j < \lambda$ put $t(\alpha,j) = \{\alpha_i : i < j\}$. Let $L(\alpha,j)$ be the largest sublanguage of L_j^α such that if a constant ∞^β or c_k^β occurs in $L(\alpha,j)$ then $\beta \in t(\alpha,j)$. We say that the pair (α,j) is <u>full</u> iff for each $\beta \in t(\alpha,j)$,

(11) $\quad t(\beta,j) = t(\alpha,j) \cap (\beta+1)$, and

(12) $\quad p_{<j}^\beta$ is a complete theory in the language $L(\beta,j)$.

Clearly we know whether (α,j) is full as soon as $p_{<j}^\alpha$ has been chosen. Also when $\bar{p}^\alpha$ has been chosen, we can define D_α to be the set of all $j < \lambda$ such that $t(\alpha,j)$ is full.

LEMMA 8.2.7. *For each* $\alpha < \lambda^+$, D_α *is a club in* λ.
<u>Proof</u>. Exercise. □

In a moment we shall use $\diamondsuit_\lambda$ to assign to each limit ordinal $\delta < \lambda$ a pair of ordinals h_δ, $k_\delta < \lambda$, and a set

(13) $\quad \Psi_\delta(x_0,\bar{y}_0,x_1,\ldots,x_0',\bar{y}_0',x_1',\ldots,\bar{w})$

of fewer than λ formulas of L(Q); there is an ordinal $g_\delta < \lambda$ such that the unprimed variables in (13) are $x_i\bar{y}_i$ for all $i < g_\delta$. These assignments are made before the hypergame begins. Assume for the present that they have been made. Player ∃ works out her strategy from them as follows.

For any ordinal $\alpha < \lambda^+$ and limit ordinal $\delta < \lambda$, let $(\beta_i : i \leq e(\alpha,\delta))$ enumerate $t(\alpha,\delta)$ in increasing order. Writing e for $e(\alpha,\delta)$ when α and δ are fixed, we have $\beta_e = \alpha$. We write R^m for the quantifier prefix

(14) $\quad Qx_m\exists\bar{y}_m\ldots Qx_i\exists\bar{y}_i\ldots \qquad (m \leq i < g_\delta)$.

We write Ψ_δ^m for the set of formulas

(15) $\Psi_\delta(\infty^{\beta_0},\bar{c}^{\beta_0},\ldots,\infty^{\beta_i},\bar{c}^{\beta_i},\ldots,x_m,\bar{y}_m,x_{m+1},\ldots,\bar{w})$ $(i < m)$.

We say that the pair (α,δ) is <u>veridical</u> iff (α,δ) is full and

(16) $p^\alpha_{<\delta}$ is $R^{e+1}Qx'_0\exists\bar{y}'_0\ldots\exists\bar{w}\bigwedge\Psi^{e+1}_\delta$

(up to choice of bound variables).

LEMMA 8.2.8. *If* (α,δ) *is veridical, then for every* $\beta \in t(\alpha,\delta)$, (β,δ) *is veridical.*

<u>Proof</u>. If $\beta \in t(\alpha,\delta)$ then certainly (β,δ) is full. It remains to show, for example, that if $\psi(x_0,\bar{y}_0,x_e,\bar{y}_e,\ldots)$ is in Ψ_δ then $Qx_e\exists\bar{y}_eR^{e+1}Qx'_0\ldots\exists\bar{w}\psi(\infty^{\beta_0},\bar{c}^{\beta_0},x_e,\ldots)$ is in $p^\beta_{<\delta}$. If it is not, then because (β,δ) is full, its negation must be in $p^\beta_{<\delta}$. Hence

(17) $(A_{<\alpha},I_{<\alpha}) \models \neg Qx\exists\bar{y}R^{e+1}Qx'_0\ldots\exists\bar{w}\psi(\infty^{\beta_0},\bar{c}^{\beta_0},x,\bar{y},x_{e+1},\ldots,\bar{w})$.

But then no condition over $(A_{<\alpha},I_{<\alpha})$ can contain $R^{e+1}Qx'_0\ldots\exists\bar{w}\psi(\infty^{\beta_0},\bar{c}^{\beta_0},\infty^\alpha,\bar{c}^\alpha,x_{e+1},\ldots,\bar{w})$, contradicting (16). In this way we show that if β is β_i, then

(18) $p^{\beta_i}_{<\delta} \supseteq R^{i+1}Qx'_0\exists y'_0\ldots\exists\bar{w}\bigwedge\Psi^{i+1}_\delta$.

Since $p^\alpha_{<\delta}$ is complete, (16) implies that the inclusion in (18) is equality, proving the lemma. □

Now let the family of sets put on the table just before the construction of (A_α,I_α) be $(\Phi^\alpha_i : i < \lambda)$. Recall that $(A_{<\alpha},I_{<\alpha})$ strongly omits each set Φ^α_i.

Here is player $\exists$'s strategy when $p^\alpha_{<\delta}$ has just been chosen. In the notation above, if (α,δ) is not veridical, or $h_\delta > e$, then she chooses p^α_δ to be $p^\alpha_{<\delta}$. On the other hand if (α,δ) is veridical and $h_\delta \leqslant e$, then she considers she is arranging that $\Psi^{g_\delta}_\delta$ will not become a hypersupport for $\Phi^{\beta_{h_\delta}}_{k_\delta}$, where the sequence $(\beta_0,\beta_1,\ldots)$ is as above. We have already described how she plays to arrange this. Note that Step 0 occurs when $h_\delta = e$, and $h_\delta < e$ for the remaining steps.

This description of the choice of p^α_δ is incomplete until we have named the ordinals h_δ, k_δ and the set of formulas Ψ_δ for each limit ordinal δ. For this we assume $\diamondsuit_\lambda$ in the form which says that there is a family $((S_i: i \to i) : i < \lambda)$ of maps such that for any map $F: \lambda \to \lambda$ the set $\{i < \lambda : F|i = S_i\}$ is stationary. (Cf. Exercise 7 of section 5.2.)

Let P be the set of all sets of formulas of L(Q) of form

(19) $$\Psi(x_0,\bar{y}_0,x_1,\ldots,x_0',\bar{y}_0',x_1',\ldots,\bar{w})$$

with $|\Psi| < \lambda$. We write $\ell(\Psi)$ for the least ordinal j such that all of the unprimed variables occurring free in Ψ lie in $x_i\bar{y}_i$ for some $i < j$. Since $\lambda^{<\lambda} = \lambda$, we can list as $X^i = (\Psi^i,h^i,k^i,<^i)$ $(i < \lambda)$ all quadruples such that $\Psi^i \in P$, $h^i < \ell(\Psi^i)$, $k^i < \lambda$ and $<^i$ is a map which assigns to each $j < \ell(\Psi^i)$ a well-ordering $<^i_j$ of j+1. We make the listing so that each X^i occurs also as $X^{i'}$ for λ different $i' < \lambda$.

For any i, $i' < \lambda$, we say that $X^i \leq X^{i'}$ iff there is an order-preserving map $f: \ell(\Psi^i) \to \ell(\Psi^{i'})$ such that

(20) $$\Psi^i(x_{f(0)},\bar{y}_{f(0)},x_{f(1)},\bar{y}_{f(1)},\ldots,x_0',\bar{y}_0',x_1',\ldots) \subseteq \Psi^{i'}$$
(up to choice of bound variables),

(21) $f(h^i) = h^{i'}$ and $k^i = k^{i'}$,

(22) each $<^i_j$ is taken isomorphically to an initial segment of $<^{i'}_{f(j)}$ by f.

By (22), such a map f is unique if it exists.

Now let δ be a limit ordinal $< \lambda$. If there is an X^i such that for a final segment θ_0, θ_1, ... of δ,

(23) $$X^{S_\delta(\theta_0)} \leq X^{S_\delta(\theta_1)} \leq \ldots$$

converging to X^i, then clearly X^i is uniquely determined (up to choice of bound variables), and we put

(24) $$\Psi_\delta = \Psi^i, \quad h_\delta = h^i, \quad k_\delta = k^i.$$

If there is no such X^i, we choose Ψ_δ, h_δ, k_δ arbitrarily.

Finally we prove the theorem. We have said how player $\exists$ is to move at limit ordinals, and for her other moves we assume she does what is needed to make sure that $((A_\alpha, I_\alpha) : \alpha < \lambda^+)$ is an elementary chain of λ-compact elementary extensions of (B,J), and that sets which are not small keep growing. (We adduce Lemma 8.1.5 and the analogue of Corollary 7.1.4(a).) It remains only to show that if $\beta < \lambda^+$ and $k < \lambda$ then Φ_k^β will have no hypersupport in the union (A^*, I^*) of the chain.

Suppose to the contrary that Φ_k^β does have a hypersupport in (A^*, I^*). Then since hypersupports have cardinality $< \lambda$, there must be some first ordinal $\alpha < \lambda^+$ such that Φ_k^β has a hypersupport in (A_α, I_α). Calling this hypersupport H, let E be the set of limit ordinals $\delta \in D_\alpha$ such that H lies inside $L(\alpha,\delta)$ and $\beta \in t(\alpha,\delta)$. By Lemma 8.2.7, E is a club in λ.

For any $\delta \in E$, list $t(\alpha,\delta)$ in increasing order as $(\beta_i : i \leqslant e)$. In view of (12), $p^\alpha_{<\delta}$ lies in $L(\alpha,\delta)$, so we can write H, $p^\alpha_{<\delta}$ as

(25) $$\Psi(\infty^{\beta_0}, \bar{c}^{\beta_0}, \infty^{\beta_1}, \ldots, x_0', \bar{y}_0', x_1', \ldots, \bar{w}), \quad \Theta(\infty^{\beta_0}, \bar{c}^{\beta_0}, \infty^{\beta_1}, \ldots)$$

for some sets $\Psi(x_0, \bar{y}_0, x_1, \ldots, x_0', \ldots)$ and $\Theta(x_0, \bar{y}_0, x_1, \ldots)$ respectively. Define $F(\delta)$ to be the first ordinal $i \geqslant \sup\{F(\xi) : \xi < \delta\}$ such that

(26) Ψ^i is $\Psi \cup \Theta$, $\beta_{h^i} = \beta$, $k^i = k$, and $<^i_j$ (for each $i \leqslant e$) orders $t(\beta_j, \delta)$ as in the enumeration of $\beta_j + 1$ chosen earlier.

For other ordinals $\xi < \lambda$, let $F(\xi)$ be $\sup\{F(\delta) : \delta \in E \cap \xi\}$. Then F is monotone, and (since E is a club) F takes new values only at elements of E.

LEMMA 8.2.9. *If* δ, $\delta' \in E$ *and* $\delta < \delta'$ *then* $X^{F(\delta)} \preceq X^{F(\delta')}$.

<u>Proof</u>. Exercise. □

Now by $\diamondsuit_\lambda$ and Exercise 2 of section 5.2, there is a limit point δ of E such that $F|\delta = S_\delta$. Since δ is a limit point of E, Lemma 8.2.9 implies that we are in the situation of (23). By (24) it follows that Ψ_δ is the limit (in the sense given by (23)) of the sets $\Psi_{\delta'}$, with $\delta' \in E \cap \delta$. By the choice (26) of these sets $\Psi_{\delta'}$, we see that Ψ_δ is simply

$H \cup p^{\alpha}_{<\delta}$ with the overflow constants and witnesses replaced by variables; in fact Ψ_δ is $\Psi^{F(\delta)}$ (up to choice of bound variables). Since by assumption H is a hypersupport in (A_α, I_α) and lies in $L(\alpha,\delta)$, and (α,δ) is full, it follows that $p^{\alpha}_{<\delta}$ is $Qx'_0\exists\bar{y}'_0\ldots\exists\bar{w}\bigwedge\Psi^{e(\alpha,\delta)+1}$ (up to choice of bound variables), and hence that (α,δ) is veridical. We can also check that $\beta_{h_\delta} = \beta$, $k_\delta = k$ and $g_\delta = e(\alpha,\delta)+1 \geqslant h_\delta+1$.

In the light of Lemma 8.2.8, player $\exists$'s strategy will thus have ensured that for some $\phi(\bar{w}) \in \Phi^{\beta}_{k}$,

(27) $$(A_\alpha, I_\alpha) \models Qx'_0\exists\bar{y}'_0Qx'_1\ldots\exists\bar{w}\bigwedge(\Psi^{e+1}_{\delta} \cup \{\neg\phi\})$$

where $e = e(\alpha,\delta)$. (Recall (10).) But Ψ^{e+1}_{δ} includes H, and so (27) implies that H is not a hypersupport of Φ^{β}_{k} in (A_α, I_α) after all.

□ Theorem

Let T be a theory in L(Q) and $\Phi(\bar{w})$ a set of formulas of L(Q). We define a hypersupport of Φ in T to be a set $\Psi(x_0,\bar{y}_0,x_1,\bar{y}_1,\ldots,\bar{w})$ of formulas of L(Q) such that $|\Psi| < \lambda$ and

(28) $T \cup Qx_0\exists\bar{y}_0Qx_1\ldots\exists\bar{w}\bigwedge\Psi$ is $\vdash_{0-5}$-consistent, but

(29) for every $\phi(\bar{w}) \in \Phi$, $T \cup Qx_0\exists\bar{y}_0Qx_1\ldots\exists\bar{w}\bigwedge(\Psi \cup \{\neg\phi\})$ is not $\vdash_{0-5}$-consistent.

We say that T strongly omits Φ iff there is no hypersupport of Φ in T. Putting Theorem 8.2.5 together with the results of section 8.1, we infer the following analogue of Corollary 7.2.3(a):

COROLLARY 8.2.10. *Let λ be an uncountable cardinal such that $\diamondsuit_\lambda$ holds, let* L *be a first-order language of cardinality at most λ, let* T *be a $\vdash_{0-5}$-consistent theory in* L(Q), *and for each* $i < \lambda$ *let Φ_i be a set of fewer than λ formulas of* L(Q) *which is strongly omitted by* T. *Then* T *has a model of cardinality λ^+ (in the λ^+-interpretation) which omits all the sets Φ_i.* □

Before Corollary 8.2.10 was proved, people thought that the main difficulty in proving it would be to prevent hypersupports creeping in at unions $(A_{<\delta}, I_{<\delta})$ where δ is a limit ordinal of cofinality $< \lambda$. It seemed rather miraculous that Shelah's argument proved the result without even mentioning this particular difficulty.

Shelah has announced several further results along the lines of Corollary 8.2.10. For example:

THEOREM 8.2.11. *Let* λ *be an uncountable cardinal and suppose that* $\diamondsuit_\lambda$ *and* $\diamondsuit_{\lambda^+}$ *hold. Then there is a real-closed field of cardinality* λ^+ *which is rigid (i.e. has no non-trivial automorphism).*

For this one shows that there is a real-closed field A of cardinality λ^+, all of whose automorphisms are first-order definable with parameters. (Cf. Exercise 8.) Since A is elementarily equivalent to the reals, and every first-order definable automorphism of the reals is trivial, it follows that A is rigid. (We have already seen in section 6.2 how game constructions are good for making everything definable.)

The main new idea in the proof of Theorem 8.2.11 is a notion "A_β is generic over $A_{<\alpha}$", where $\alpha \leqslant \beta < \lambda^+$ and $(A_\alpha : \alpha < \lambda^+)$ is the chain of structures built in the hypergame. Shelah shows that the second player (the Generic Player in his terminology) can play so that for all $\alpha \leqslant \beta$, A_β is generic over $A_{<\alpha}$. In fact she can still do this even if the first player (the Random Player) keeps expanding the language to name new relations during the course of the construction.

Shelah has also extracted a combinatorial principle from the proofs of these theorems, and used it directly to generalise Theorem 6.3.1 (on boolean algebras with no large chains or pies) to higher cardinalities. The principle takes about a page to state, and its form is that certain partial orderings contain "generic" subsets. He comments that it is related to Jensen's morass.

Exercises for 8.2:

Throughout these exercises, λ *is an uncountable cardinal such that* $\lambda^{<\lambda} = \lambda$. L *is a first-order language, and where* λ *is mentioned,* L *is of cardinality at most* λ.

1. Let (B,J) be a weak L(Q)-structure of cardinality at most λ, satisfying Axioms 0-4. Show that in hypergames of length λ^+ with (B,J) as starting structure, it is hyperenforceable that the compiled structure (A^*,I^*) is λ-compact.

2. Verify the proof of Theorem 8.2.2.

3. Show that there are 2^{λ^+} pairwise non-isomorphic special λ^+-Aronszajn trees.

4. Show that if (A,I) is a λ^+-candid λ-compact weak $L(Q)$-structure in which Axioms 0-4 hold, then for every set $\Phi(\bar{w})$ of fewer than λ formulas of $L(Q,A)$, (A,I) omits Φ iff (A,I) strongly omits Φ (in the sense defined in this section). [Adapt Exercise 1 of section 7.2.]

5. Prove Lemmas 8.2.6, 8.2.7 and 8.2.9.

6. (a) Assuming $\Diamond_\lambda$ and $\Diamond_{\lambda^+}$, suppose T is a theory in the Magidor-Malitz language $L^{<\omega}$ such that every finite subset of T has a model in the ω_1-interpretation. Show that T has a model in the λ^+-interpretation.
(b) If E is a stationary subset of λ^+, we write $\Diamond_E$ for the statement: There is a family $\{S_\alpha \subseteq \alpha : \alpha \in E\}$ such that for every set $X \subseteq \lambda^+$, $\{\alpha : S_\alpha = X \cap \alpha\}$ is stationary. Show that in part (a) the assumption $\Diamond_{\lambda^+}$ can be weakened to $\Diamond_E$, where E is the set of all ordinals $< \lambda^+$ of cofinality λ.

7. Assuming $\Diamond_\lambda$ and $\Diamond_{\lambda^+}$, show that there exists a λ-saturated boolean algebra of cardinality λ^+ which has only λ^+ subalgebras.

8. Assuming $\Diamond_\lambda$ and $\Diamond_{\lambda^+}$, let T be a first-order theory of cardinality $< \lambda$ which has infinite models. Show that T has a λ-saturated model A of cardinality λ^+ such that for any ordered field B which is definable (first-order with parameters) in A, every automorphism of B is definable (first-order with parameters in A).

For constructions as in this section, the hypothesis $\Diamond_\lambda$ *can often be weakened as follows. Write* $(D\ell)_\mu$ *for the statement: There is a family* $\{S_i : i < \mu\}$ *of sets such that each* S_i *is a set of fewer than* μ *subsets of* i, *and for every set* $X \subseteq \mu$ *the set* $\{i < \mu : X \cap i \in S_i\}$ *is stationary in* μ.

9. Show: (a) If μ is strongly inaccessible (including ω) then $(D\ell)_\mu$ holds. (b) If μ is regular and $(D\ell)_\mu$ holds then $\mu^{<\mu} = \mu$. (c) The assumption $\Diamond_\lambda$ in Theorem 8.2.5 can be weakened to: λ is regular and $(D\ell)_\lambda$ holds.

10. Let μ be a regular cardinal such that $(D\ell)_\mu$ holds, and T a $\vdash_{0-5}$-consistent theory in $L(Q)$ of cardinality $\leqslant \mu$ such that for every model B of T, $<^B$ is a partial ordering in which the set of predecessors of any element is linearly ordered, and T contains $\forall x \neg Qy(y < x)$. Show that T has

a model A in the μ^+-interpretation such that every maximal linearly $<^A$-ordered subset of dom(A) is definable (in L(Q) with parameters).

11. Let μ be an infinite cardinal and consider the following statement (*): "For all first-order languages L of cardinality $\leqslant \mu$, every $\vdash_{0-5}$-consistent theory in L(Q) has a model in the μ^+-interpretation." Show: (a) If μ is a singular cardinal and $\Box_\mu$ and the GCH hold (cf. Devlin (1984); $\Box_\mu$ and the GCH both hold in the constructible universe), then (*) holds. (b) If μ is a strong limit cardinal (i.e. $\nu < \mu \Rightarrow 2^\nu < \mu$) and (*) holds, then for every theory T as in (*) and every set $\Phi(x)$ strongly omitted by T, there is a model of T in the μ^+-interpretation which omits Φ.

REFERENCES FOR CHAPTER 8

8.1: Aronszajn trees were discovered by N. Aronszajn (cf. Kurepa (1938)), and later independently by G. Higman & A. H. Stone (1954) who used them to prove Exercise 3. Mitchell showed that ZFC is consistent with "There is no special ω_2-Aronszajn tree" iff ZFC is consistent with "There exists a Mahlo cardinal". Silver showed that ZFC is consistent with "There is no ω_2-Aronszajn tree" iff ZFC is consistent with "There is an uncountable weakly compact cardinal". Both these results are in Mitchell (1972). There is a slicker proof of the consistency of "There is no ω_2-Aronszajn tree", due to Baumgartner, which uses proper forcing; cf. Devlin (1983) or p. 104ff of Shelah (1982). Chapter IX of Shelah (1982) discusses when all Aronszajn trees are special. Schmerl showed that if $k < \omega$ and λ is a k-inaccessible cardinal (cf. Exercise 9 of section 5.2), then the class of sentences of L(Q) which have a model in the λ-interpretation varies with the value of k; cf. Schmerl (1972, 1974) and Schmerl & Shelah (1972). (In connection with Schmerl (1972) see also Theorem 3.6 of Kaufmann (1983 b).)

Exercise 1: E. Specker (unpublished). Exercise 2: Galvin & Shelah (1973). On Exercise 3, see also the construction by Douady quoted on p. 312 of Cohn (1981). Exercise 5: J. Silver and F. Rowbottom (unpublished). Exercises 6, 7: Fuhrken (1964).

8.2: Chang's two-cardinal theorem is in Chang (1965). Corollary 8.2.3 stands in much the same relation to Chang's theorem as Keisler's completeness theorem for L(Q) does to the Vaught two-cardinal theorem; in fact Chang's proof adapts Vaught's. On Corollary 8.2.4 and

Exercise 3: Specker (1949) showed that if $\lambda^{<\lambda} = \lambda$ then λ^+-Aronszajn trees exist; Gaifman & Specker (1964) introduce a class of λ-Aronszajn trees which they call "normal", and under the assumption $\lambda^{<\lambda} = \lambda$ they show that there are 2^{λ^+} pairwise non-isomorphic normal λ^+-Aronszajn trees. Theorem 8.2.5 is from Shelah (1980 a), as are Corollary 8.2.10 and Exercises 5 and 7. Lemma 8.2.6 is essentially Lemma 4.4 of Keisler (1970). Theorem 8.2.11 is from Shelah (1983 d), which also states the combinatorial principle used in the proof. Proofs of this principle and Exercise 9 will appear in Shelah (198- c).

Exercise 6: Shelah (1979). See also the Concluding Remarks of Shelah (1981) for other results on boolean algebras which can be proved by the method of Theorem 8.2.5. Exercises 8, 10: Shelah (1983 d); on 10 cf. Keisler (1974). Exercise 11: (a) is essentially due to Jensen, cf. Devlin (1984) for the relevant two-cardinal theorem proved for countable languages; (b) Shelah and Grossberg (unpublished), cf. Grossberg (198-) for related results.

TWO LISTS

LIST OF TYPES OF FORCING

Set theorists find it convenient to baptise their various kinds of forcing, often by the names of their inventors. It was only a reluctance to introduce new terminology that stopped me doing the same in the body of the text. Here I am bolder.

1. Robinson (finite) forcing. Conditions are finite sets of basic sentences which are consistent with a given countable $\forall_2$ theory T. Used for constructing "thin" existentially closed models of T.
   Chapters 3 and 4.

2. Henkin forcing. Conditions are finite sets of first-order sentences consistent with a given countable first-order theory T. Used for constructing models of T which omit types. If "consistent" is defined syntactically or in various clever ways, this forcing can also be used to prove completeness and interpolation theorems.
   Section 5.1 including Exercise 9.

3. Henkin-Orey forcing. As Henkin forcing, but consistency is defined relative to some infinitary rules of proof. Used for constructing ω-models and for proving representability theorems.
   Section 5.1 including Exercise 13.

4. Makkai forcing. As Henkin forcing, but with finite sets of sentences from a fragment of $L_{\omega_1\omega}$. Used for proving completeness and interpolation theorems for $L_{\omega_1\omega}$.
   Exercise 10 of section 5.1.

5. λ-forcing. The analogue of Henkin forcing for first-order languages of cardinality λ. Gives λ-compact models.
   Section 5.3.

6. Robinson infinite forcing. Conditions are substructures of models of a $\forall_2$ theory T, with no restriction on cardinality. Gives λ-existentially compact models of T for any λ, and hence infinite-generic models of T.
Section 5.3 including Exercises 18, 19.

7. Q-forcing. Conditions are finite sets $p(\infty,\bar{c})$ where $B \models Qx\exists\bar{y}\bigwedge p(x,\bar{y})$, for some countable structure B and largeness property Q over B. Used for constructing proper extensions of B which reflect the properties of B.
Chapter 6.

8. Witness-free forcing (or Skolem forcing). As Q-forcing but without the parameters $\bar{c}$; equivalent to Skolem's definable ultrapowers. Generally weaker than Q-forcing. But can be internalised into models of arithmetic or set theory of any cardinality, to construct various kinds of end extension.
Sections 6.2, 6.3.

9. Keisler forcing. Q-forcing over a weak model of the axioms for L(Q); conditions are sentences of L(Q). Used for completeness and omitting types results for L(Q) and related languages.
Chapter 7.

10. Shelah forcing. Q-forcing iterated, using languages of any cardinality λ subject to some set-theoretic conditions. Used for building structures of cardinality λ^+ with various higher-order properties such as rigidity and chain conditions.
Not in this book, but sections 6.3 and 8.2 discuss forerunners.

LIST OF OPEN QUESTIONS

The questions below are a sample of questions I came across while writing the book and couldn't answer. There is no guarantee that the answers are really unknown, or that they are interesting.

The first two questions are special cases of a more general but vague question: How far do resultants exist for non-first-order theories? No prizes are offered for extending Theorem 3.1.3 to PC_Δ *classes (i.e. classes of form "All* L*-reducts of models of* T*" for a first-order theory* T *in a language extending* L*). The simplest interesting extension seems to* *be to classes of finite algebras.*

QUESTION 1. Is it true that for every positive primitive formula $\phi(\bar{x})$ in the language of groups, there is a recursive set $\Psi(\bar{x})$ of quantifier-free formulas such that for every finite group G and tuple $\bar{a}$ from G, $G \models \bigwedge\Psi(\bar{a})$ iff G can be extended to a finite group H such that $H \models \phi(\bar{a})$?
It is true with Π_1^0 *in place of recursive, by Hodges (198- a). A negative answer might use Slobodskoĭ (1981).*

QUESTION 2. Let p be a prime. Is it true that for every positive primitive formula $\phi(\bar{x})$ in the language of groups, there is a set $\Psi(\bar{x})$ of $\exists_1$ first-order formulas such that for every finite p-group G and tuple $\bar{a}$ from G, $G \models \bigwedge\Psi(\bar{a})$ iff G can be extended to a finite p-group H such that $H \models \phi(\bar{a})$?
The people in Freiburg have recently made good progress on e.c. p-groups from an algebraic point of view (e.g. Maier (1981), Leinen (1984)).

QUESTION 3. Does the statement "Every model of a countable $\forall_2$ first-order theory T with the amalgamation property can be extended to an existentially closed model of T" imply the axiom of choice?
The answer is Yes if we drop the amalgamation property. Actually the question makes better sense if "amalgamation property" is replaced by a syntactic criterion for this property, cf. Bacsich & Rowlands Hughes (1974); in the absence of choice, even the theory of fields lacks the amalgamation property.

QUESTION 4. Is there an e.c. group G whose elements are the natural numbers, such that if d is the Turing degree of eldiag(G) then $d < 0^{(\omega)}$?
The question is suggested by a similar result of Julia Knight (unpublished) about models of complete first-order arithmetic, together with Exercise 8 of section 3.3.

QUESTION 5 (Apps). Generalise Robinson forcing to allow some part of the compiled structure to be held fixed.
A test case is whether a forcing construction will build the groups of Exercise 14 in section 4.4, where the centre is fixed in advance.

QUESTION 6. Is it true that for every countable group G which is e.c. in some variety of groups, there are arbitrarily large groups which are back-and-forth equivalent to G?
Cf. Theorem 4.1.11. By Kueker (1975) p. 27, the answer Yes implies that the group G *has continuum many automorphisms.*

QUESTION 7. Is it true that for every variety V of groups, if V is non-abelian and not of bounded exponent then there are continuum many groups e.c. in V which are distinguished by their $\exists_3$ theories?
For quasivarieties of groups it is false, by Exercise 15(b) of section 4.4.

QUESTION 8. If T is the theory of commutative rings with 1, which side of the dichotomy of Theorem 4.2.6 does T lie on? In particular, is there an enforceable commutative ring of each finite characteristic?

QUESTION 9. How many non-elementarily-equivalent e.c. commutative rings with 1 are there?
By Exercise 6 of section 4.2 there are continuum many isomorphism types of countable e.c. commutative rings with 1.

QUESTION 10. Let V be a variety of groups and LFV the class of locally finite groups in V. Must LFV have a unique enforceable member? (This needs to be given a meaning. Let T be the set of all $\forall_1$ first-order sentences which are true in every group in LFV. In Robinson forcing with T, is there a unique enforceable model?)
The answer is Yes for groups (Exercise 15(c) of section 3.3), for abelian groups (Exercise 8 of section 4.2) and for N_2 groups (Theorems 4.4.6 and 4.4.7), but the reasons are quite different in the three cases.

QUESTION 11. Let G be the unique countable periodic e.c. nilpotent group of class 2. Is Th(G) decidable?

QUESTION 12. Let L be a countable first-order language and T a complete theory in L with ω_2 constants added. If principal types are dense for T, must T have an atomic model?
Cf. Knight (1978) and Exercise 5 of section 5.1. Maybe there is an answer

in Shelah (1983 b,c; 198- d).

QUESTION 13. Develop a classification theory for ω-e.s. models.
Shelah sketched out the first steps already in (1975 c). But first-order classification theory has moved on some way since then.

QUESTION 14. If $Qx\phi$ means "ϕ has Morley rank ∞", is Q a largeness property satisfying Axiom 5?
Cf. Exercise 1(d) of section 6.1. The question means: If Th(B) *is not totally transcendental, can we extend* B *keeping the totally transcendental part of* B *small?*

QUESTION 15. Let B be a countable locally finite group, and define a largeness property Q for $\exists_1$ formulas over B by putting $Qx\phi$ = "For some locally finite group $G \supseteq B$ there is an element $g \notin B$ such that $G \models \phi(g)$". Is it enforceable that the compiled extension of G is locally finite? (If it is, say more.)

QUESTION 16 (Shelah). Let T be a countable first-order theory containing the statement "< linearly orders all the elements and has no last element". Find necessary and sufficient conditions for every model of T of cofinality ω to have a proper elementary end extension.
Shelah (1978 b) answers the question with "of cofinality ω" deleted, and with it replaced by "of cardinality ω".

QUESTION 17. Prove the existence of rather classless recursively saturated models of Peano arithmetic in cardinality ω_1, without assuming $\Diamond$ at any stage in the proof.
Cf. the references of section 6.2. A number of people have asked for this.

QUESTION 18. Assuming only the continuum hypothesis, can we prove that there is an uncountable atomless boolean algebra with no uncountable chains or pies?
Rubin (1983) lists a number of other properties that we can impose on the algebra if we assume $\Diamond$.

QUESTION 19. Let ϕ be a sentence of $L_{\omega_1\omega}$ which has a model in which uncountably many Ψ-types are realised, for some set Ψ of first-order formulas. Is there a set S of countably many Ψ-types, such that for each $X \subseteq S$ there is an uncountable model of ϕ which realises the types in X but no other types in S?
Cf. the references of section 7.2.

QUESTION 20 (Malitz). Is it true (under some reasonable set-theoretic hypothesis) that if ϕ is a sentence of $L^{2,1}$ (cf. Exercise 8 of section 7.3) and ϕ has a model of cardinality $> \omega_1$ then ϕ has a model of cardinality ω_1?
This would imply for example that if there is a Jónsson group in some uncountable cardinality then there is one in cardinality ω_1. *In fact Shelah (1980 b) proved the existence of* ω_1*-Jónsson groups, using a two-dimensional construction much like those in Chapters 6 and 7, above, but with some very complicated amalgamations; also he constructed* λ^+*-Jónsson groups for every uncountable* λ *by a slightly different proof, using the GCH. I hear he has now built* ω_2*-Jónsson groups using ZFC alone.*

QUESTION 21 (Schmerl 198-). For the Macintyre interpretation of the quantifiers Q^m (cf. Exercise 9 of section 7.3), is $PA(Q^2)$ equivalent to $PA(Q^{<\omega})$?
The two theories have the same first-order consequences, and the arguments of Schmerl & Simpson (1982) apply equally well to either. In the other direction, Garavaglia (1978) and Shelah (unpublished, but cf. Rubin & Shelah (1983)) showed that for every n *there are structures which are equivalent in the Magidor-Malitz language* L^n *but not in* L^{n+1}.

QUESTION 22. Is there a good analogue of the dichotomy theorem, Theorem 4.2.6, for weakly hyperenforceable properties and hypergames?
I feel this is the chief gap in the general theory. The main reason for specifying <u>*weakly*</u> *hyperenforceable is that there are mass production methods which depend on doing the right thing at a stationary set of limit ordinals, and these are not what I have in mind. Possibly Grossberg & Shelah (1983) is relevant. An easier question to start on might be:*

QUESTION 23. Is there a good analogue of the dichotomy theorem for forcing proper extensions as in section 6.1?

QUESTION 24 (Shelah 1983 d). Can Corollary 8.2.10 be proved for $\lambda = \omega_1$ from the GCH alone?

A natural question, since by Lemma 5.2.7(b), if λ *is any other successor of a regular cardinal then the hypothesis* $\Diamond_\lambda$ *follows from the GCH. Strictly Shelah's question is whether his combinatorial principle (Theorem 2.17 in Shelah (1983 d)) can be proved from* $\lambda^{<\lambda} = \lambda$ *in place of* $(D\ell)_\lambda$.

BIBLIOGRAPHY

There is here no badd songe, but the best cann be hadd,
The chiefest from all men; yea there is not one badd,
And such sweet musicke as dothe much delite yeilde
Both unto men at home and birds abroade in fielde.
The autours for to name I maye not here forgett,
But will them now downe put and all in order sett.

John Baldwin (1591)

Numbers in square brackets are references to pages in the text above. To that extent, this is an author index as well as a bibliography. I have included several papers which are relevant to the main topics of the book but are not referred to in the text.

Allenby, R.B.J.T. (1966) Generalised regular products of groups. Ph.D. Thesis, University of Wales. [131]

Apps, A.B. (1983) On $\aleph_0$-categorical class two groups. J. Algebra 82, 516-538. [131]

Apps, A.B. (198-) Two counterexamples in $\aleph_0$-categorical groups. Algebra Universalis (to appear). [130]

Artin, E. & Schreier, O. (1926) Algebraische Konstruktion reeller Körper. Abh. Math. Sem. Univ. Hamburg 5, 83-115. [80]

Ash, C.J. (1971) Undecidable $\aleph_0$-categorical theories. Notices Amer. Math. Soc. 18, 423. [131]

Bacsich, P.D. (1972) Cofinal simplicity and algebraic closedness. Algebra Universalis 2, 354-360. MR 48:5962 A. Shafaat. [81]

Bacsich, P.D. (1975) The strong amalgamation property. Colloq. Math. 33, 13-23. MR 53:116 John T. Baldwin. [81]

Bacsich, P.D. & Fisher, E.R. (unpublished). [80]

Bacsich, P.D. & Rowlands Hughes, D. (1974) Syntactic characterisations of amalgamation, convexity and related properties. J. Symbolic Logic 39, 433-451. MR 52:71 John T. Baldwin. [80, 277]

Baker, K.A. (1969) Equational classes of modular lattices. Pacific J. Math. 28, 9-15. MR 39:5435 O. Frink. [82]

Baldwin, J. (1591) Manuscript collection of motets and instrumental music, Royal Library collection, British Museum. [282]

Baldwin, J.T. & Kueker, D.W. (1980) Ramsey quantifiers and the finite cover property. Pacific J. Math. 90, 11-19. MR 83e:03054 Alan Mekler. [249]

Baldwin, J.T. & Saxl, J. (1976) Logical stability in group theory. J. Austral. Math. Soc. Ser. A 21, 267-276. MR 53:10934 Paul C. Eklof. [130]

Barwise, K.J. (1975) Admissible Sets and Structures. Berlin: Springer-Verlag. MR 54:12519 Nigel J. Cutland. [94, 97, 102]

Barwise, K.J. (1981) The role of the omitting types theorem in infinitary logic. Arch. Math. Logik Grundlag. 21, 55-68. MR 83h:03052 Jouko Väänänen.

Barwise, K.J. & Feferman, S. (198-) Higher Model Theory: The Logic of Mathematical Concepts. Berlin: Springer-Verlag. [249]

Barwise, K.J., Kaufmann, M. & Makkai, M. (1978) Stationary logic. Ann. Math. Logic 13, 171-224. MR 82f:03031a. [222, 248]

Barwise, K.J. & Robinson, A. (1970) Completing theories by forcing. Ann. Math. Logic 2, 119-142. Reprinted in Robinson (1979), pp. 219-242. MR 42:7494 A. Máté. [34, 82, 131]

Barwise, K.J. & Schlipf, J. (1975) On recursively saturated models of arithmetic. In Model Theory and Algebra: A Memorial Tribute to Abraham Robinson, ed. D.H. Saracino & V.B. Weispfenning, Lecture Notes in Math. 498, pp. 42-55. Berlin: Springer-Verlag. MR 53:12934 S. Feferman. [210]

Baudisch, A., Seese, D.G., Tuschik, P. & Weese, M. (198-) Decidability and elimination for generalized quantifiers. In Barwise & Feferman (198-). [212]

Baumgartner, J.E. (1980) Chains and antichains in $\mathscr{P}(\omega)$. J. Symbolic Logic 45, 85-92. MR 81d:03054 Thomas J. Jech. [210]

Baumgartner, J.E. & Komjáth, P. (1981) Boolean algebras in which every chain and antichain is countable. Fund. Math. 111, 125-133. MR 82j:06023 Thomas J. Jech. [210]

Baumslag, B. & Levin, F. (1976) Algebraically closed torsion-free nilpotent groups of class 2. Comm. Algebra 4, 533-560. MR 53:5749 K.W. Weston. [131]

Baumslag, G. (1971) Lecture Notes on Nilpotent Groups. Regional Conference Series in Math. 2. Providence RI: Amer. Math. Soc. MR 44:315 M.F. Newman. [82]

Baur, W. (1976) Elimination of quantifiers for modules. Israel J. Math. 25, 64-70. MR 56:15409 O.V. Belegradek. [131]

Белеградек, О.В. (Belegradek, O.V.) (1974 a) Об алгебраически замкнутых группах. Алгебра и Логика 13 (3), 239-255. Tr. as On algebraically closed groups. Algebra and Logic 13, 135-143. MR 52:2859 Grigoriĭ V. Chudnovskiĭ.

Белеградек, О.В. (Belegradek, O.V.) (1974 b) Об определимости в алгебраически замкнутых группах. Мат. Заметки 16 (3), 375-380. Tr. as On definability in algebraically closed groups. Math. Notes 16, 813-816. MR 50:12710 Grigoriĭ V. Chudnovskiĭ. [81]

Белеградек, О.В. (Belegradek, O.V.) (1978) Элементарные свойства алгебраически замкнутых групп (Elementary properties of algebraically closed groups). Fund. Math. 98, 83-101. MR 57:9530 S.R. Kogalovskiĭ. [81, 130]

Белеградек, О.В. (Belegradek, O.V.) (1980) Разрешимие фрагменты универсальных теорий и экзистенциально замкнутые модели (Decidable fragments of universal theories and existentially closed models). Сибирск. Мат. Ж. (Sibirsk. Mat. Ž.) 21, 196-201. MR 82d:20033 A.T. Nurtazin.

Белеградек, О.В. (Belegradek, O.V.) (1981) Классы алгебр с внутренними отображениями (Classes of algebras with internal maps). In Исследования по теоретическому программированию (Investigations in Theoretical Programming), pp. 3-10. Alma-Ata: Kazakh State Univ. [81]

Bell, J.L. & Slomson, A.B. (1969) Models and Ultraproducts: An Introduction. Amsterdam: North-Holland. MR 42:4381 S. Garfunkel. [16]

Беляев, В.Я. (Belyaev, V.Ya.) (1977) Об алгебраически замкнутых полугруппах. Сибирск. Мат. Ж. 18 (1), 32-39. Tr. as Algebraically closed semigroups. Siberian Math. J. 18, 23-29. MR 58:939. [81]

Беляев, В.Я. (Belyaev, V.Ya.) (1978) Об алгебраической замкнутости в классах полугрупп и групп (On algebraic closure in classes of semigroups and groups). In Материалы Всесоюзной Научной Студенческой Конференции "Студент и Научно-Технический Прогресс" (Materials of the All-Union Scientific Student Conference "The Student and Scientific-Technical Progress"), pp. 95-105. Novosibirsk: Matematika. [81]

Belyaev, V.Y. (1982) On algebraic closure and amalgamation of semigroups. Semigroup Forum 25, 223-267. MR 84d:20059 Boris M. Schein. [81]

Беляев, В.Я. (Belyaev, V.Ya.) (1984) Несчетные расширения счетных алгебраически замкнутых полугрупп (Uncountable extensions of countable algebraically closed semigroups). Сибирск. Мат. Ж. (Sibirsk. Mat. Ž.) (to appear).

Беляев, В.Я. (Belyaev, V.Ya.) & Тайцлин, М.А. (Taĭtslin, M.A.) (1979) Об элементарных свойствах экзистенциально замкнутых систем. Успехи Мат. Наук 34 (2), 39-94. Tr. as On elementary properties of existentially closed systems. Russian Math. Surveys 34 (2), 43-107. MR 82a:03028 O.V. Belegradek. [80]

Berline, C. & Cherlin, G. (1981) QE rings in characteristic p. In Logic Year 1979-80, ed. M. Lerman et al., Lecture Notes in Math. 859, pp. 16-31. Berlin: Springer-Verlag. MR 84a:03032 U. Felgner. [131]

Blake, W. (c. 1790) Proverbs of Hell. In Sloss, D.J. & Wallis, J.P.R. (1926) The Prophetic Writings of William Blake, I. Oxford: Clarendon Press. [170]

Blass, A. (1979) Injectivity, projectivity, and the axiom of choice. Trans. Amer. Math. Soc. 255, 31-59. MR 81j:04004 U. Felgner. [81]

Bloomfield, L. (1933) Language. London: G. Allen & Unwin. [132]

Boffa, M. (1972 a) Modèles universels homogènes et modèles génériques. C.R. Acad. Sci. Paris Sér. A-B 274, A693-A694. MR 45:6606 U. Felgner. [169]

Boffa, M. (1972 b) Sur l'existence des corps universels-homogènes. C.R. Acad. Sci. Paris Sér. A-B 275, A1267-A1268. MR 47:1598 B. Węglorz.

Boffa, M. (1973) Structures extensionnelles génériques. Bull. Soc. Math. Belg. 25, 3-10. MR 50:101 L. Bukovský.

Boffa, M. (1975) A note on existentially complete division rings. In Model Theory and Algebra: A Memorial Tribute to Abraham Robinson, ed. D.H. Saracino & V.B. Weispfenning, Lecture Notes in Math. 498, pp. 56-59. Berlin: Springer-Verlag. MR 53:121 A.B. Slomson.

Boffa, M. & van Praag, P. (1972 a) Sur les corps génériques. C.R. Acad. Sci. Paris Sér. A-B 274, A1325-A1327. MR 45:6607 R.S. Pierce. [81]

Boffa, M. & van Praag, P. (1972 b) Sur les sous-champs maximaux des corps génériques dénombrables. C.R. Acad. Sci. Paris Sér. A-B 275, A945-A947. MR 48:101 R.S. Pierce.

Бокуть, Л.А. (Bokut', L.A.) (1963) Некоторые теоремы вложения для колец и полугрупп I (Some embedding theorems for rings and semigroups I). Сибирск. Мат. Ж. (Sibirsk. Mat. Ž.) 4 (3), 500-518. MR 27:2530 P.M. Cohn. [80, 81]

Bonnet, R. & Shelah, S. (198-) Narrow Boolean algebras. (Submitted) [210]

Boulez, P. (1952) Letter to John Cage. Quoted p. 82 of Peyser, J. (1976) Boulez, Composer, Conductor, Enigma. London: Cassell. [83]

Bouscaren, E. (1980) Existentially closed modules: types and prime models. In Model Theory of Algebra and Arithmetic, ed. L. Pacholski et al., Lecture Notes in Math. 834, pp. 31-43. Berlin: Springer-Verlag. MR 82j:03038 Bienvenido F. Nebres.

Bridge, J. (1977) Beginning Model Theory: The Completeness Theorem and Some Consequences. Oxford: Clarendon Press. MR 58:27171 James H. Schmerl. [16]

Broesterhuizen, G. (1975) Structures for a logic with additional generalized quantifier. Colloq. Math. 33, 1-12. MR 53:92 Lee Badger.

Bruce, K.B. (1978 a) Model-theoretic forcing in logic with a generalized quantifier. Ann. Math. Logic 13, 225-265. MR 80c:03033 M. Makkai. [248, 249]

Bruce, K.B. (1978 b) Ideal models and some not so ideal problems in the model theory of L(Q). J. Symbolic Logic 43, 304-321. MR 80a:03048 Lee Badger. [248]

Bruce, K.B. (1980) Model constructions in stationary logic, I, Forcing. J. Symbolic Logic 45, 439-454. MR 82f:03032 M. Makkai. [222, 248]

Bruce, K.B. & Keisler, H.J. (1979) $L_A(\text{Ⅎ})$. J. Symbolic Logic 44, 15-28. MR 80f:03037 Stephen G. Simpson. [248]

Cage, J. (1958) Lecture at Darmstadt. Quoted p. 167 of Peyser, J. (1976) Boulez, Composer, Conductor, Enigma. London: Cassell. [83]

Cantor, G. (1872) Ueber die Ausdehnung eines Satzes aus der Theorie der trigonometrischen Reihen. Math. Ann. 5, 123-132. [16]

Cardano, G. (1545) Ars Magna. Nürnberg. [80]

Carson, A.B. (1973) The model completion of the theory of commutative regular rings. J. Algebra 27, 136-146. MR 52:7893 G. Cherlin. [210]

Carson, A.B. (1974) Algebraically closed regular rings. Canad. J. Math. 26, 1036-1049. MR 50:13124 R.S. Pierce.

Chang, C.C. (1965) A note on the two cardinal problem. Proc. Amer. Math. Soc. 16, 1148-1155. MR 33:1238 G. Fodor. [273]

Chang, C.C. (1967) Omitting types of prenex formulas. J. Symbolic Logic 32, 61-74. MR 35:4094 R.S. Pierce.

Chang, C.C. & Keisler, H.J. (1973) Model Theory. Amsterdam: North-Holland. MR 53:12929 B. Węglorz. [11, 13, 14, 16, 18, 44, 136, 169, 186, 188, 210, 244, 261]

Chase, S.U. (1960) Direct products of modules. Trans. Amer. Math. Soc. 97 97, 457-473. MR 22:11017 D. Zelinsky. [45]

Cherlin, G.L. (1972) The model-companion of a class of structures. J. Symbolic Logic 37, 546-556. MR 52:73 Carol Wood. [169]

Cherlin, G.L. (1973) Algebraically closed commutative rings. J. Symbolic Logic 38, 493-499. MR 49:36 Paul C. Eklof. [80, 81]

Cherlin, G.L. (1975) Second order forcing, algebraically closed structures, and large cardinals. Fund. Math. 87, 141-160. MR 51:12515 J.E. Rubin.

Cherlin, G.L. (1976) Model Theoretic Algebra - Selected Topics. Lecture Notes in Math. 521. Berlin: Springer-Verlag. MR 58:27455. [80, 131, 169]

Cherlin, G.L. & Lachlan, A.H. (198-) Stable finitely homogeneous structures. (preprint) [130]

Cherlin, G.L. (see also under Berline, C.)

Чихачев, С.А. (Chikhachev, S.A.) (1975 a) О генерических моделях. Алгебра и Логика 14, 345-353. Tr. as Generic models. Algebra and Logic 14, 214-218. MR 58:27442. [131]

Чихачев, С.А. (Chikhachev, S.A.) (1975 b) Генерические модели счетных теорий. Алгебра и Логика 14, 704-721. Tr. as Generic models of countable theories. Algebra and Logic 14, 421-429. MR 56:11777 S.R. Kogalovskiĭ. [131]

Чихачев, С.А. (Chikhachev, S.A.) (1975 c) Об одном вопросе Симмонса. Сибирск. Мат. Ж. 16 (4) 876-880. Tr. as On a question of Simmons. Siberian Math. J. 16, 674-677. MR 52:5405 A. Sochor.

Church, A. & Rosser, J.B. (1936) Some properties of conversion. Trans. Amer. Math. Soc. 39, 472-482. [16]

Clapham, C.R.J. (1964) Finitely presented groups with word problems of arbitrary degrees of insolubility. Proc. London Math. Soc. 14, 633-676. MR 31:3486 M. Greendlinger. [91]

Cohen, P.J. (1963) The independence of the continuum hypothesis I. Proc. Nat. Acad. Sci. USA 50, 1143-1148. MR 28:1118 A. Mostowski. [34]

Cohn, P.M. (1962) Eine Bemerkung über die multiplikative Gruppe eines Körpers. Arch. Math. (Basel) 13, 344-348. MR 26:3774 K. Honda. [81]

Cohn, P.M. (1974 a) The class of rings embeddable in skew fields. Bull. London Math. Soc. 6, 147-148. MR 51:3216 L. Bokut'. [38]

Cohn, P.M. (1974 b) Algebra, I. London: John Wiley. MR 50:12496 M. Pearl. [36]

Cohn, P.M. (1977) Skew Field Constructions. Cambridge: Cambridge U.P. MR 57:3190 L.A. Koifman. [68, 81]

Cohn, P.M. (1981) Universal Algebra (2nd ed.) Dordrecht: Reidel. MR 82j:08001 W. Taylor. [6, 44, 273]

Craig, W. (1953) On axiomatizability within a system. J. Symbolic Logic 18, 30-32. MR 14 p. 1051 P. Lorenzen. [80]

Cusin, R. (1971 a) Sur le "forcing" en théorie des modèles. C.R. Acad. Sci. Paris Sér. A-B 272, A845-A848. MR 44:2589 E. Engeler.

Cusin, R. (1971 b) Structures préhomogènes et structures génériques. C.R. Acad. Sci. Paris Sér. A-B 273, A137-A140. MR 45:1742 V. Harnik.

Cusin, R. (1971 c) Recherche du "forcing-compagnon" et du "modèle-compagnon" d'une théorie, liée à l'existence de modèles $\aleph_\alpha$-universels. C.R. Acad. Sci. Paris Sér. A-B 273, A956-A959. MR 50:88 Gabriel Sabbagh.

Cusin, R. (1972) L'algèbre des formules d'une théorie complète et "forcing-compagnon". C.R. Acad. Sci. Paris Sér. A-B 275,

A1269-A1272. MR 47:3162 U. Felgner.
Cusin, R. (1973) Sur les structures génériques en théorie des modèles. Rev. Roumaine Math. Pures Appl. 18, 519-541. MR 53:7762 William H. Wheeler.
Cusin, R. (1974) The number of countable generic models for finite forcing. Fund. Math. 84, 265-270. MR 49:2349 Grigorii V. Čudnovskiĭ. [131]
Cusin, R. & Pabion, J.R. (1971) Structures génériques associées à une classe de théories. C.R. Acad. Sci. Paris Sér. A-B 272, A1620-A1623. MR 44:47 V. Harnik.
Devlin, K.J. (1983) The Yorkshireman's guide to proper forcing. In Surveys in Set Theory, ed. A.R.D. Mathias, pp. 60-115. Cambridge: Cambridge U.P. [273]
Devlin, K.J. (1984) Constructibility. Berlin: Springer-Verlag. [150, 168, 209, 273, 274]
Ehrenfeucht, A. (1958) Theories having at least continuum many non-isomorphic models in each infinite power. Notices Amer. Math. Soc. 5, 680. [130]
Ehrenfeucht, A. (1972) There are continuum ω_0-categorical theories. Bull. Acad. Polon. Sci. Sér. Math. Astron. Phys. 20, 425-427. MR 47:1596 S. Comer. [131]
Eklof, P.C. & Mez, H.-C. (198- a) The ideal structure of existentially closed algebras. (preprint) [80]
Eklof, P.C. & Mez, H.-C. (198- b) Additive groups of existentially closed rings. (preprint)
Eklof, P.C. & Sabbagh, G. (1970) Model-completions and modules. Ann. Math. Logic 2, 251-295. MR 43:3105 I.L. Gaal. [80, 81, 169]
Engeler, E. (1959) A characterization of theories with isomorphic denumerable models. Notices Amer. Math. Soc. 6, 161. [167]
Erdős, P. & Rado, R. (1960) Intersection theorems for systems of sets. J. London Math. Soc. 35, 85-90. MR 22:2554 Đ. Kurepa. [169]
Evans, T. (1969) Some connections between residual finiteness, finite embeddability and the word problem. J. London Math. Soc. 1, 399-403. MR 40:2589 F.B. Cannonito. [82]
Faber, V. (1978) Large abelian subgroups of some infinite groups, II. Rocky Mountain J. Math. 8, 481-490. MR 58:11128 R. Göbel. [148]
Feferman, S. (1968) Lectures on proof theory. In Proceedings of the Summer School in Logic, Leeds, 1967, ed. M.H. Löb, Lecture Notes in Math. 70, pp. 1-107. Berlin: Springer-Verlag. MR 38:4294 A.S. Troelstra. [81]
Feferman, S. (see also under Barwise, K.J.)
Felgner, U. (1975) On $\aleph_0$-categorical extra-special p-groups. Logique et Analyse 1975 volume, 407-428. Reprinted in Six Days of Model Theory, ed. P. Henrard, pp. 175-196. Albeuve, Switzerland: Editions Castella 1977. MR 57:16054 Paul C. Eklof. [131]
Fisher, E.R. (1970) Homogeneous universal models revisited. (preprint) [169]
Fisher, E.R. & Robinson, A. (1972) Inductive theories and their forcing companions. Israel J. Math. 12, 95-107. Reprinted in Robinson (1979) pp. 267-279. MR 47:3163 Constantin Milici.
Fisher, E.R., Simmons, H. & Wheeler, W. (1975) Elementary equivalence classes of generic structures and existentially complete structures. In Model Theory and Algebra: A Memorial Tribute to Abraham Robinson, ed. D.H. Saracino & V.B. Weispfenning, Lecture Notes in Math. 498, pp. 131-169. Berlin: Springer-

Verlag. MR 53:7763 B. Węglorz. [131]

Fisher, E.R. (see also under Bacsich, P.D.)

Fraissé, R. (1953) Sur certaines relations qui généralisent l'ordre des nombres rationnels. C.R. Acad. Sci. Paris 237, 540-542. MR 15 p. 192 B. Jónsson. [16, 130, 169]

Fraissé, R. (1954) Sur l'extension aux relations de quelques propriétés des ordres. Ann. Sci. École Norm. Sup. 71, 363-388. MR 16 p. 1006 G. Kurepa. [130]

Fuchs, L. (1970) Infinite Abelian Groups I. New York: Academic Press. MR 41: 333 J. Rotman. [112, 114, 115, 120, 129]

Fuhrken, G. (1964) Skolem-type normal forms for first-order languages with a generalized quantifier. Fund. Math. 54, 291-302. MR 29: 3363 A. Mostowski. [248, 273]

Fuhrken, G. & Taylor, W. (1971) Weakly atomic-compact relational structures. J. Symbolic Logic 36, 129-140. MR 44:2590 J.M. Plotkin. [168]

Gaifman, H. (1967) Uniform extension operators for models and their applications. In Sets, Models and Recursion Theory, ed. J.N. Crossley, pp. 122-155. Amsterdam: North-Holland. MR 36:3640 D. Monk. [210]

Gaifman, H. (1976) Models and types of Peano's arithmetic. Ann. Math. Logic 9, 223-306. MR 53:10577 S.R. Kogalovskiĭ. [209, 210]

Gaifman, H. & Specker, E.P. (1964) Isomorphism types of trees. Proc. Amer. Math. Soc. 15, 1-7. MR 29:5746 Đ. Kurepa. [274]

Galvin, F. & Shelah, S. (1973) Some counterexamples in the partition calculus. J. Combin. Theory Ser. A 15, 167-174. MR 48:8240 R.J. Wilson. [273]

Gandy, R.O., Kreisel, G. & Tait, W.W. (1960) Set existence. Bull. Acad. Polon. Sci. Sér. Sci. Math. Astronom. Phys. 8, 577-582. MR 28:2964a P. Axt. [168]

Garavaglia, S. (1978) Relative strength of Malitz quantifiers. Notre Dame J. Formal Logic 19, 495-503. MR 58:5018 Lee Badger. [280]

Giorgetta, D. & Shelah, S. (1984) Existentially closed structures in the power of the continuum. Ann. Pure Appl. Logic 26, 123-148.

Glass, A.M.W. (1981) Ordered Permutation Groups. London Math. Soc. Lecture Notes. Cambridge: Cambridge U.P. MR 83j:06004 J. Jakubik. [81]

Glass, A.M.W. & Pierce, K.R. (1980 a) Existentially complete abelian lattice-ordered groups. Trans. Amer. Math. Soc. 261, 255-270. MR 81k:03032 B.I. Zil'ber. [131]

Glass, A.M.W. & Pierce, K.R. (1980 b) Existentially complete lattice-ordered groups. Israel J. Math. 36, 257-272. MR 82b:06016 Carol Wood. [81]

Glass, A.M.W. (see also under van den Dries, L.P.D.)

Glassmire, W. Jr. (1971) There are $2^{\aleph_0}$ countably categorical theories. Bull. Acad. Polon. Sci. Sér. Sci. Math. Astronom. Phys. 19, 185-190. MR 45:8513 W. Schwabhäuser. [131]

Gödel, K. (1930) Die Vollständigkeit der Axiome des logischen Funktionenkalküls. Monatshefte für Math. Phys. 37, 349-360. Tr. in From Frege to Gödel, ed. J. v. Heijenoort, pp. 583-591. Cambridge Mass.: Harvard U.P. 1967. [34, 136]

Gödel, K. (1931) Über formal unentscheidbare Sätze der Principia Mathematica und verwandter Systeme I. Monatshefte für Math. Phys. 38, 173-198. Tr. in From Frege to Gödel, ed. J. v. Heijenoort, pp. 596-616. Cambridge Mass.: Harvard U.P. 1967. [138]

Grätzer, G. (1979) Universal Algebra (2nd ed.) New York: Springer-Verlag. MR 80g:08001. [6]

Grätzer, G., Jónsson, B. & Lasker, H. (1973) The amalgamation property in equational classes of modular lattices. Pacific J. Math. 45, 507-524. MR 51:3014 Joel Berman. [53]

Gregory, J. (1976) Higher Souslin trees and the generalized continuum hypothesis. J. Symbolic Logic 41, 663-671. MR 58:5208. [150]

Grilliot, T.J. (1972) Omitting types: application to recursion theory. J. Symbolic Logic 37, 81-89. MR 49:8839 J.C. Owings Jr. [130, 168]

Grossberg, R. (198-) Omitting types theorem for L(Q) for successors of singulars. (preprint) [274]

Grossberg, R. & Shelah, S. (1983) On universal locally finite groups. Israel J. Math. 44, 289-302. [280]

Grzegorczyk, A., Mostowski, A. & Ryll-Nardzewski, C. (1961) Definability of sets in models of axiomatic theories. Bull. Acad. Polon. Sci. Sér. Sci. Math. Astronom. Phys. 9, 163-167. MR 29:1138 P. Axt. [168]

Hajnal, A. (1961) Proof of a conjecture of S. Ruziewicz. Fund. Math. 50, 123-8. MR 24:A1833 P. Erdős. [169]

Hall, P. (1959 a) On the finiteness of certain soluble groups. Proc. London Math. Soc. 9, 595-622. MR 22:1618 K. Gruenberg. [82]

Hall, P. (1959 b) Some constructions for locally finite groups. J. London Math. Soc. 34, 305-319. Zbl 88 p. 23 B.H. Neumann. [81]

Halpern, J.D. (see under Pincus, D.)

Hausdorff, F. (1914) Grundzüge der Mengenlehre. Leipzig: Veit. [130]

Hechler, S.H. (1974) On the existence of certain cofinal subsets of ${}^{\omega}\omega$. In Axiomatic Set Theory II, ed. T.J. Jech, pp. 155-173. Providence R.I.: Amer. Math. Soc. MR 50:12716 John K. Truss. [141]

Henkin, L. (1949) The completeness of the first-order functional calculus. J. Symbolic Logic 14, 159-166. MR 11 p. 487 J.C.C. McKinsey. [18, 34]

Henkin, L. (1956) Two concepts from the theory of models. J. Symbolic Logic 21, 28-32. MR 17 p. 816 G. Kreisel. [81]

Henkin, L. (1957) A generalization of the concept of ω-completeness. J. Symbolic Logic 22, 1-14. MR 20:1626 A. Robinson. [167]

Henkin, L. (1963) An extension of the Craig-Lyndon interpolation theorem. J. Symbolic Logic 28, 201-216. MR 30:1927 R.C. Lyndon. [18, 168]

Henrard, P. (1971) Une théorie sans modèle générique. C.R. Acad. Sci. Paris Sér. A-B 272, A293-A294. MR 44:48 K.J. Barwise.

Henrard, P. (1973) Le "forcing-compagnon" sans "forcing". C.R. Acad. Sci. Paris Sér. A-B 276, A821-A822. MR 47:3165 A.B. Slomson. [82, 131]

Hensel, K. (1907) Über die arithmetischen Eigenschaften der Zahlen. Jahresber. der D.M.V. 16, 299-319, 388-393, 474-496. [80]

Henson, C.W. (1972) Countable homogeneous relational structures and $\aleph_0$-categorical theories. J. Symbolic Logic 37, 494-500. MR 48: 94 M. Benda. [130]

Hickin, K.K. (1978) Complete universal locally finite groups. Trans. Amer. Math. Soc. 239, 213-227. MR 58:902 B.A.F. Wehrfritz. [168]

Hickin, K.K. (198-) A.c. groups: extensions, maximal subgroups, and automorphisms. (preprint) [81]

Hickin, K.K. & Macintyre, A. (1980) Algebraically closed groups: embeddings and centralizers. In Word Problems II, The Oxford Book, ed. S.I. Adian et al., pp. 141-155. Amsterdam: North-Holland. MR 82c:20065 U. Felgner. [81]

Higman, G. (1961) Subgroups of finitely presented groups. Proc. Roy. Soc. London Ser. A 262, 455-475. MR 24:A152 O.H. Kegel. [59, 81]

Higman, G., Neumann, B.H. & Neumann, H. (1949) Embedding theorems for groups. J. London Math. Soc. 24, 247-254. MR 11 p. 322 F.W. Levi. [36, 80, 81]

Higman, G. & Stone, A.H. (1954) On inverse systems with trivial limits. J. London Math. Soc. 29, 233-236. MR 15 p. 773 D. Zelinsky. [273]

Hilbert, D. (1893) Über die vollen Invariantensysteme. Math. Ann. 42, 313-373. [80]

Hilbert, D. (1931) Die Grundlegung der elementaren Zahlenlehre. Math. Ann. 104, 485-494. [138]

Hirschfeld, J. (1976) Another approach to infinite forcing. Ann. Sci. Univ. Clermont Math. 13, 81-86. MR 57:16045 James H. Schmerl.

Hirschfeld, J. (1978) Examples in the theory of existential completeness. J. Symbolic Logic 43, 650-658. MR 81i:03043. [130]

Hirschfeld, J. (1980) Finite forcing, existential types and complete types. J. Symbolic Logic 45, 93-102. MR 81b:03037 M.A. Taĭtslin. [131]

Hirschfeld, J. & Wheeler, W.H. (1975) Forcing, Arithmetic, Division Rings. Lecture Notes in Math. 454. Berlin: Springer-Verlag. MR 52:10412 G. Cherlin. [81, 130, 131, 169]

Hodges, W. (1976) On the effectivity of some field constructions. Proc. London Math. Soc. 32, 133-162. MR 55:10252 G. Cherlin. [80]

Hodges, W. (1980) Interpreting number theory in nilpotent groups. Arch. Math. Logik Grundlag. 20, 103-111. MR 82c:03047 U. Felgner. [131, 169]

Hodges, W. (198- a) Finite extensions of finite groups. Proc. Conference of Assoc. for Symbolic Logic, Aachen 1983. [81, 277]

Hodges, W. (198- b) On constructing many non-isomorphic algebras. Proc. Conference in Univ. Algebra, Darmstadt 1983. [249]

Hodges, W. (198- c) Groupes nilpotents existentiellement clos de classe fixée. Mem. Soc. Math. de France (to appear). [131]

Hodges, W. (198- d) Model Theory. Cambridge U.P. (in preparation). [93]

Hutchinson, J.E. (1976) Elementary extensions of countable models of set theory. J. Symbolic Logic 41, 139-145. MR 53:12944 John W. Dawson Jr. [209]

Jacobson, N. (1974) Basic Algebra I. San Francisco: W.H. Freeman & Co. MR 50:9457 M.F. Smiley. [36, 51]

Jacobson, N. (1980) Basic Algebra II. San Francisco: W.H. Freeman & Co. MR 81g:00001 M.F. Smiley. [47, 48]

Jech, T. (1978) Set Theory. New York: Academic Press. MR 80a:03062 E. Mendelson. [146, 147, 148]

Jónsson, B. (1956) Universal relational systems. Math. Scand. 4, 193-208. MR 20:3091 S. Ginsburg. [169]

Jónsson, B. (1960) Homogeneous universal relational systems. Math. Scand. 8, 137-142. MR 23:A2328 S. Ginsburg. [169]

Jónsson, B. (see also under Grätzer, G.)

Kaiser, K. (1969) Über eine Verallgemeinerung der Robinsonschen Modellvervollständigung I. Z. Math. Logik Grundlag. Math. 15, 37-48. MR 39:48 L.W. Szczerba. [81]

Kargapolov, M.I. & Merzljakov, Ju.I. (1979) Fundamentals of the Theory of Groups. New York: Springer-Verlag. MR 80k:20002. [15, 112]
Karp, C.R. (1964) Languages with Expressions of Infinite Length. Amsterdam: North-Holland. MR 31:1178 E. Engeler. [168]
Karp, C.R. (1965) Finite-quantifier equivalence. In The Theory of Models, Proc. 1963 Internat. Symposium at Berkeley, ed. J.W. Addison et al., pp. 407-412. Amsterdam: North-Holland. MR 35:36 J.E. Fenstad. [130]
Kaufmann, M. (1977) A rather classless model. Proc. Amer. Math. Soc. 62, 330-333. MR 57:16058 James H. Schmerl. [169, 210]
Kaufmann, M. (1979) A new omitting types theorem for L(Q). J. Symbolic Logic 44, 507-521. MR 80m:03073 M. Dubiel. [249]
Kaufmann, M. (1981 a) On existence of Σ_n end extensions. In Logic Year 1979-80, University of Connecticut, ed. M. Lerman et al., Lecture Notes in Math. 859, pp. 92-103. Berlin: Springer-Verlag. MR 82h:03031 P. Štěpánek. [209, 210]
Kaufmann, M. (1981 b) Filter logics: filters on ω_1. Ann. Math. Logic 20, 155-200. MR 84d:03044. [248]
Kaufmann, M. (1983 a) Set theory with a filter quantifier. J. Symbolic Logic 48, 263-287.
Kaufmann, M. (1983 b) Blunt and topless end extensions of models of set theory. J. Symbolic Logic 48, 1053-1073. [210, 273]
Kaufmann, M. (1984) Filter logics on ω. J. Symbolic Logic 49, 241-256. [248]
Kaufmann, M. (198-) The quantifier "there exist uncountably many" and some of its relatives. In Barwise & Feferman (198-). [248]
Kaufmann, M. & Kranakis, E. (1984) Definable ultrapowers and ultrafilters over admissible ordinals. Z. Math. Logik Grundlag. Math. 30, 97-118.
Kaufmann, M. (see also under Barwise, K.J.)
Keisler, H.J. (1966) Some model theoretic results for ω-logic. Israel J. Math. 4, 249-261. MR 36:4974 J.C. Owings Jr. [209]
Keisler, H.J. (1970) Logic with the quantifier "There exist uncountably many". Ann. Math. Logic 1, 1-93. MR 41:8217 K.J. Barwise. [16, 209, 229, 248, 249, 274]
Keisler, H.J. (1971) Model Theory for Infinitary Logic. Amsterdam: North-Holland. MR 49:8855 Andreas Blass. [14, 166, 168]
Keisler, H.J. (1973) Forcing and the omitting types theorem. In Studies in Model Theory, ed. M.D. Morley, MAA Studies in Math. Vol. 8, pp. 96-133. Buffalo NY: Math. Assoc. Amer. MR 49:2340 James H. Schmerl. [34, 82]
Keisler, H.J. (1974) Models with tree structures. In Proceedings of the Tarski Symposium, ed. L. Henkin et al., Proc. Symposia in Pure Math. XXV, pp. 331-348. Providence RI: Amer. Math. Soc. MR 50:9576 Keith Devlin. [274]
Keisler, H.J. (1977) Fundamentals of model theory. In Handbook of Mathematical Logic, ed. J. Barwise, pp. 47-103. Amsterdam: North-Holland. [16, 167]
Keisler, H.J. & Morley, M. (1968) Elementary extensions of models of set theory. Israel J. Math. 6, 49-65. MR 38:5611 J.E. Rubin. [209]
Keisler, H.J. (see also under Bruce, K.B. and Chang, C.C.)
Харлампович, О.Г. (Kharlampovich, O.G.) (1981) Конечно определенная разрешимая группа с неразрешимой проблемой равенства (A finitely presented soluble group with unsolvable word problem). Изв. Акад. Начк СССР Сер. Мат. (Izv. Akad. Nauk

SSSR Ser. Mat.) 45 (4), 852-873. [82]
Kirby, L.A.S. (see under Paris, J.B.)
Knight, J.F. (1978) Prime and atomic models. J. Symbolic Logic 43, 385-393. MR 58:21578 Guus Broesterhuizen. [168, 278]
Komjáth, P. (see under Baumgartner, J.E.)
Kotlarski, H. (1978) Some remarks on well-ordered models. Fund. Math. 99, 123-132. MR 57:12201 W. Marek.
Kotlarski, H. (1980) On Skolem ultrapowers and their non standard variants. Z. Math. Logik Grundlag. Math. 26, 227-236. MR 81m:03040 Gabriel Lolli. [210]
Kranakis, E. (1982) Definable ultrafilters and end extensions of constructible sets. Z. Math. Logik Grundlag. Math. 28, 395-412. [210]
Kranakis, E. (see also under Kaufmann, M.)
Kreisel, G. (see under Gandy, R.O.)
Kripke, S.A. (1965) Semantical analysis of intuitionistic logic I. In Formal Systems and Recursive Functions, ed. J.N. Crossley & M.A.E. Dummett, pp. 92-130. Amsterdam: North-Holland. MR 34: 1184 R.E. Vesley. [34]
Krivine, J.L. (1964) Anneaux préordonnés. J. Analyse Math. 12, 307-326. MR 31:213 R.C. Lyndon. [80]
Kueker, D.W. (1975) Back-and-forth arguments and infinitary logics. In Infinitary Logic: In Memoriam Carol Karp, ed. D.W. Kueker, Lecture Notes in Math. 492, pp. 17-71. MR 57:2905 Fred Halpern. [130, 278]
Kueker, D.W. (see also under Baldwin, J.T.)
Kurepa, G. (1938) Ensembles linéaires et une classe de tableaux ramifiées (tableaux de M. Aronszajn). Publ. Math. Univ. Belgrade 6/7, 129-160. [273]
Lachlan, A.H. (198-) On countable stable structures which are homogeneous for a finite relational language. (preprint) [130]
Lachlan, A.H. (see also under Cherlin, G.L.)
Lascar, D. & Poizat, B. (1979) An introduction to forking. J. Symbolic Logic 44, 330-350. MR 80k:03030 Guus Broesterhuizen. [81]
Lasker, H. (see under Grätzer, G.)
Lee, V. & Nadel, M. (1975) On the number of generic models. Fund. Math. 90, 105-114. MR 53:7723 Nigel J. Cutland. [131]
Leinen, F. (1984) Existentiell abgeschlossene L𝔛-gruppen. Dissertation, Albert-Ludwigs-Universität, Freiburg im Breisgau. [277]
Levin, F. (see under Baumslag, B.)
Lindström, P. (1964) On model-completeness. Theoria 30, 183-196. MR 31: 3317 C.-C. Chang. [80, 81]
Lipshitz, L. & Saracino, D. (1973) The model companion of the theory of commutative rings without nilpotent elements. Proc. Amer. Math. Soc. 38, 381-387. MR 55:12510. [210]
Łoś, J. (1955) On the extending of models I. Fund. Math. 42, 38-54. MR 17 p. 224 G. Kreisel. [80]
Loullis, G. (1978) Infinite forcing for Boolean valued models. Bull. Soc. Math. Grèce 19, 155-182. MR 80f:03042 Lawrence Neff Stout.
Lyndon, R.C. & Schupp, P.E. (1977) Combinatorial Group Theory. Berlin: Springer-Verlag. MR 58:28182 Ian M. Chiswell. [60, 61, 76, 185]
MacDowell, R. & Specker, E.P. (1961) Modelle der Arithmetik. In Infinitistic Methods, pp. 257-263. Oxford: Pergamon Press. MR 27:2425 H. Putnam. [209]

MacHenry, T.S. (1960) The tensor product and the 2nd nilpotent product of groups. Math. Z. 73, 134-145. MR 22:11027a F. Haimo. [113]

Macintyre, A. (1972 a) Omitting quantifier-free types in generic structures. J. Symbolic Logic 37, 512-520. MR 49:37 John T. Baldwin. [82]

Macintyre, A. (1972 b) On algebraically closed groups. Ann. of Math. 96, 53-97. MR 47:6477 U. Felgner. [59, 81, 82, 209]

Macintyre, A. (1973 a) Martin's axiom applied to existentially closed groups. Math. Scand. 32, 46-56. MR 49:4770 U. Felgner.

Macintyre, A. (1973 b) On existentially complete Lie algebras. Notices Amer. Math. Soc. 21, A379. [131]

Macintyre, A. (1975) A note on axioms for infinite-generic structures. J. London Math. Soc. 9, 581-584. MR 51:121 James H. Schmerl. [169]

Macintyre, A. (1976) Existentially closed structures and Jensen's principle ◊. Israel J. Math. 25, 202-210. MR 58:211 O.V. Belegradek. [168, 209]

Macintyre, A. (1977) Model completeness. In Handbook of Mathematical Logic, ed. J. Barwise, pp. 139-180. Amsterdam: North-Holland. [80]

Macintyre, A. (1979) Combinatorial problems for skew fields, I, Analogue of Britton's Lemma, and results of Adjan-Rabin type. Proc. London Math. Soc. 39, 211-236. MR 81h:03092 U. Felgner. [68]

Macintyre, A. (1980) Ramsey quantifiers in arithmetic. In Model Theory of Algebra and Arithmetic, Proceedings, Karpacz, Poland 1979, ed. L. Pacholski et al., Lecture Notes in Math. 834, pp. 186-210. MR 83j:03099. [249]

Macintyre, A. & Saracino, D. (1974) On existentially complete nilpotent Lie algebras. Notices Amer. Math. Soc. 21, A379. [131]

Macintyre, A. (see also under Hickin, K.K. and van den Dries, L.P.D.)

McKay, W.B. (1963) Building Construction I. London: Longman. [17]

McKinsey, J.C.C. (1943) The decision problem for some classes of sentences without quantifiers. J. Symbolic Logic 8, 61-76. MR 5 p. 85 O. Frink. [80]

Macpherson, H.D. (1983) Enumeration of orbits of infinite permutation groups. D.Phil. Thesis, Oxford University. [168]

Magidor, M. & Malitz, J. (1977 a) Compact extensions of L(Q), Ia. Ann. Math. Logic 11, 217-261. MR 56:11746 M. Yasuhara. [249]

Magidor, M. & Malitz, J. (1977 b) Compactness and transfer for a fragment of L^2. J. Symbolic Logic 42, 261-268. MR 58:16129 Lee Badger.

Magnus, W. (1974) Noneuclidean Tesselations and their Groups. New York: Academic Press. MR 50:4774 E. Vinberg. [69]

Maier, B.J. (1981) Existentiell abgeschlossene lokal endliche p-Gruppen. Arch. Math. (Basel) 37, 113-128. MR 83e:20005 U. Felgner [277]

Maier, B.J. (1983) On existentially closed and generic nilpotent groups. Israel J. Math. 46, 170-188. [131]

Maier, B.J. (1984) Existentially closed torsion-free nilpotent groups of class three. J. Symbolic Logic 49, 220-230. [131]

Maier, B.J. (198-) Amalgame nilpotenter Gruppen der Klasse zwei. Publ. Math. Debrecen (to appear). [131]

Makkai, M. (1969) On the model theory of denumerably long formulas with finite strings of quantifiers. J. Symbolic Logic 34, 437-459. MR 41:45 E.G.K. Lopez-Escobar. [168]

Makkai, M. (see also under Barwise, K.J.)

Makowsky, J.A. (198-) Abstract embedding relations. In Barwise & Feferman (198-). [169]

Mal'cev, A.I. (Mal'tsev, A.I.) (1936) Untersuchungen aus dem Gebiete der mathematischen Logik. Mat. Sb. 1 (43), 323-336. Tr. in Mal'cev (1971) pp. 1-14. [34]

Мальцев, А.И. (Mal'tsev, A.I.) (1940) Об изоморфном представлении бесконечных групп матрицами. Mat. Sb. 8 (50), 405-422. Tr. as On the faithful representation of infinite groups by matrices. Amer. Math. Soc. Translations (2) 45 (1965) 1-18. MR 2 p. 216 L. Zippin. [81]

Мальцев, А.И. (Mal'tsev, A.I.) (1956) Квасипримитивные классы абстрактных алгебр. Докл. Акад. Наук СССР 108, 187-189. Tr. as Quasi-primitive classes of abstract algebras, in Mal'cev (1971) pp. 27-31. MR 18 p. 107 R.A. Good. [80]

Мальцев, А.И. (Mal'tsev, A.I.) (1958) Структурная характеристика некоторых классов алгебр. Докл. Акад. Наук СССР 120, 29-32. Tr. as The structural characterization of certain classes of algebras, in Mal'cev (1971) pp. 56-60. MR 20:5154 R.A. Good. [80]

Мальцев, А.И. (Mal'tsev, A.I.) (1960) Об одном соответствии между кольцами и группами. Мат. Сб. 50 (92) 257-266. Tr. as A correspondence between rings and groups, in Mal'cev (1971) pp. 124-137. MR 22:9448 E. Mendelson. [131]

Mal'cev, A.I. (Mal'tsev, A.I.) (1971) The Metamathematics of Algebraic Systems, Collected Papers: 1936-1967, trans. B.F. Wells III. Amsterdam: North-Holland. MR 50:1877.

Malitz, J. (1979) Introduction to Mathematical Logic. New York: Springer-Verlag. MR 81h:03002 John W. Dawson Jr. [16, 18]

Malitz, J. (1983) Downward transfer of satisfiability for sentences of $L^{1,1}$. J. Symbolic Logic 48, 1146-1150. [249]

Malitz, J. & Reinhardt, W. (1972) Maximal models in the language with quantifier "there exist uncountably many". Pacific J. Math. 40, 139-155. MR 47:1574 J.A. Makowsky. [248]

Malitz, J. & Rubin, M. (1980) Compact fragments of higher order logic. In Mathematical Logic in Latin America, ed. A.I. Arruda et al., pp. 219-238. Amsterdam: North-Holland. MR 82k:03055 M. Makkai. [249]

Malitz, J. (see also under Magidor, M.)

Manevitz, L.M. (1976) Robinson forcing is not absolute. Israel J. Math. 25, 211-232. MR 56:15407 G. Cherlin.

Mekler, A.H. (see under van den Dries, L.P.D.)

Merzljakov, Ju.I. (see under Kargapolov, M.I.)

Mez, H.-C. (1982) Existentially closed linear groups. J. Algebra 76, 84-98. MR 84a:20050 A.Kh. Kushkuleĭ. [81]

Mez, H.-C. (see also under Eklof, P.C.)

Millar, T. (1981) Vaught's theorem recursively revisited. J. Symbolic Logic 46, 397-411. MR 82d:03053 J.M. Plotkin.

Miller, C.F. III (unpublished) [82]

Mills, G. (1978) A model of Peano arithmetic with no elementary end extension. J. Symbolic Logic 43, 563-567. MR 58:10425 John W. Dawson Jr. [210]

Mills, G. & Paris, J.B. (1984) Regularity in models of arithmetic. J. Symbolic Logic 49, 272-280. [209]

Mitchell, W. (1972) Aronszajn trees and the independence of the transfer property. Ann. Math. Logic 5, 21-46. MR 47:1612 J.M. Plotkin. [273]

Morgenstern, C.F. (1979) On amalgamations of languages with Magidor-Malitz quantifiers. J. Symbolic Logic 44, 549-558. MR 80k:03035 G. Fuhrken.

Morgenstern, C.F. (1982) On generalized quantifiers in arithmetic. J. Symbolic Logic 47, 187-190. [249]

Morley, M. (1965) Omitting classes of elements. In The Theory of Models, Proc. 1963 Internat. Symposium at Berkeley, ed. J.W. Addison et al., pp. 265-273. Amsterdam: North-Holland. MR 34:1189 J.R. Shoenfield. [93]

Morley, M. & Vaught, R. (1962) Homogeneous universal models. Math. Scand. 11, 37-57. MR 27:37 B. Jónsson. [169]

Morley, M. (see also under Keisler, H.J.)

Mostowski, A. (1957) On a generalization of quantifiers. Fund. Math. 44, 12-36. MR 19 p. 724 G. Kreisel. [248]

Nadel, M. (see under Lee, V.)

Neumann, B.H. (1943) Adjunction of elements to groups. J. London Math. Soc. 18, 4-11. MR 5 p. 58 R. Baer. [35, 80]

Neumann, B.H. (1952) A note on algebraically closed groups. J. London Math. Soc. 27, 247-249. MR 13 p. 721 D.G. Higman. [81]

Neumann, B.H. (1973) The isomorphism problem for algebraically closed groups. In Word Problems, ed. W.W. Boone et al., pp. 553-562. Amsterdam: North-Holland. [81]

Neumann, B.H. & Neumann, H. (1959) Embedding theorems for groups. J. London Math. Soc. 34, 465-479. MR 29:1267 W. Kappe. [61]

Neumann, B.H. (see also under Higman, G.)

Neumann, H. (see under Higman, G. and Neumann, B.H.)

Orey, S. (1956) On ω-consistency and related properties. J. Symbolic Logic 21, 246-252, MR 18 p. 632 A. Rose. [167]

Oxtoby, J.C. (1971) Measure and Category. New York: Springer-Verlag. MR 52:14213 I. Chiţescu. [34]

Pabion, J.R. (see under Cusin, R.)

Paris, J.B. & Kirby, L.A.S. (1978) Σ_n-collection schemas in arithmetic. In Logic Colloquium '77, Proc. Colloquium Wrocław 1977, ed. A. Macintyre et al., pp. 199-209. MR 81e:03056 R. Parikh. [209]

Paris, J.B. (see also under Mills, G.)

Parsons, C. (1970) On a number theoretic choice schema and its relation to induction. In Intuitionism and Proof Theory, ed. A. Kino et al., pp. 459-473. Amsterdam: North-Holland. MR 43:6050 G. Kreisel. [209]

Peirce, C.S. (1885) On the algebra of logic. Amer. J. Math. 7, 180-202. [212]

Phillips, R.G. (1974) Omitting types in arithmetic and conservative extensions. In Victoria Symposium on Nonstandard Analysis, Univ. Victoria 1972, ed. A. Hurd & P. Loeb, Lecture Notes in Math. 369, pp. 195-202. Berlin: Springer-Verlag. MR 57:16059 J.M. Plotkin. [209]

Pierce, K.R. (see under Glass, A.M.W.)

Pincus, D. (1972) Zermelo-Fraenkel consistency results by Fraenkel-Mostowski methods. J. Symbolic Logic 37, 721-743. MR 49:2374 Paul E. Howard. [80]

Pincus, D. & Halpern, J.D. (1981) Partitions of products. Trans. Amer. Math. Soc. 267, 549-568. MR 83b:03058 James H. Schmerl. [141]

Podewski, K.-P. & Reineke, J. (1979) Algebraically closed commutative local rings. J. Symbolic Logic 44, 89-94. MR 80c:03039 Paul C. Eklof. [80, 81]

Podewski, K.-P. & Reineke, J. (198-) Algebraically closed commutative indecomposable rings. Algebra Universalis (to appear).

Poincaré, H. (1913) Dernières Pensées. Paris: Flammarion. [35]

Point, F. (198-) Finite generic models of T_{UH}, for certain model companionable theories T. J. Symbolic Logic (to appear). [131]

Poizat, B. (198-) Théorie des Modèles. (in preparation) [16]

Poizat, B. (see also under Lascar, D.)

Poland, J. (see under van den Dries, L.P.D.)

Potthoff, K. (1981) Einführung in die Modelltheorie und ihre Anwendungen. Darmstadt: Wissenschaftliche Buchgesellschaft. [16]

Pouzet, M. (1972 a) Modèle universel d'une théorie n-complète. C.R. Acad. Sci. Paris Sér. A-B 274, A433-A436. MR 45:1745a J.L. Bell. [130]

Pouzet, M. (1972 b) Modèle universel d'une théorie n-complète: Modèle uniformément préhomogène. C.R. Acad. Sci. Paris Sér. A-B 274, A695-A698. MR 45:1745b J.L. Bell. [130, 168]

Pouzet, M. (1972 c) Modèle universel d'une théorie n-complète: Modèle préhomogène. C.R. Acad. Sci. Paris Ser. A-B 274, A813-A816. MR 45:1745c J.L. Bell. [130]

Pouzet, M. (1973) Extensions complètes d'une théorie forcing complète. Israel J. Math. 16, 212-215. MR 48:10797 J.E. Fenstad.

Rabin, M.O. (1959) Arithmetical extensions with prescribed cardinality. Nederl. Akad. Wetensch. Indag. Math. 21, 439-446. MR 21:5564 E. Mendelson. [210]

Rabin, M.O. (1962) Non-standard models and independence of the induction axiom. In Essays on the Foundations of Mathematics, ed. Y. Bar-Hillel et al., pp. 287-299. Amsterdam: North-Holland. MR 28:4999 G. Kreisel. [80, 209]

Rado, R. (see under Erdős, P.)

Rasiowa, H. & Sikorski, R. (1950) A proof of the completeness theorem of Gödel. Fund. Math. 37, 193-200. MR 12 p. 661 I.L. Novak. [168]

Reineke, J. (1982) On algebraically closed models of theories of commutative rings. In Logic, Methodology and Philosophy of Science VI, Proc. Sixth Internat. Congress of Logic, Methodology and Philosophy of Science, Hanover 1979, pp. 223-234. Amsterdam: North-Holland. MR 84c:03064 G. Cherlin.

Reineke, J. (see also under Podewski, K.-P.)

Reinhardt, W. (see under Malitz, J.)

Robinson, A. (1951) On axiomatic systems which possess finite models. In Methodos, pp. 140-149. Milan: Editrice La Fiaccola. Reprinted in Robinson (1979) pp. 322-331. [82]

Robinson, A. (1956) Complete Theories. Amsterdam: North-Holland. MR 17 p. 817 P.R. Halmos. [80, 81, 131]

Robinson, A. (1971 a) Forcing in model theory. In Symposia Math. V, INDAM, Rome 1969/70, pp. 69-82. London: Academic Press. Reprinted in Robinson (1979) pp. 205-218. MR 43:4651 J.E. Rubin. [80. 82, 131]

Robinson, A. (1971 b) Infinite forcing in model theory. In Proc. Second Scand. Logic Symposium, Oslo 1970, pp. 317-340. Amsterdam: North-Holland. Reprinted in Robinson (1979) pp. 243-266. MR 50:9574 Andreas Blass. [80, 169]

Robinson, A. (1971 c) On the notion of algebraic closedness for noncommutative groups and fields. J. Symbolic Logic 36, 441-

444. Reprinted in Robinson (1979) pp. 478-481. MR 45:43 J.M. Plotkin. [80, 157, 169]

Robinson, A. (1973 a) Model theory as a framework for algebra. In Studies in Model Theory, ed. M.D. Morley, MAA Studies in Math. Vol. 8, pp. 134-157. Buffalo NY: Math. Assoc. Amer. Reprinted in Robinson (1979) pp. 60-83. MR 49:2365 W. Taylor.

Robinson, A. (1973 b) Nonstandard arithmetic and generic arithmetic. In Proc. Fourth Internat. Congress for Logic, Methodology and Philos. Sci., Bucharest 1971, pp. 137-154. Amsterdam: North-Holland. Reprinted in Robinson (1979) pp. 280-297. MR 56:5283 G. Cherlin.

Robinson, A. (1979) Selected Papers of Abraham Robinson, Vol. I, Model Theory and Algebra, ed. H.J. Keisler. Amsterdam: North-Holland. MR 80h:01039a E.G. Straus. [133]

Robinson, A. (see also under Barwise, K.J. and Fisher, E.R.)

Rogers, H. Jr. (1967) Theory of Recursive Functions and Effective Computability. New York: McGraw-Hill. MR 37:61 R.L. Goodstein. [60, 66, 79]

Rosser, J.B. (see under Church, A.)

Rotman, J.J. (1973) The Theory of Groups (2nd ed.). Boston: Allyn & Bacon. MR 56:451 R.C. Lyndon. [52, 61, 62]

Rowlands Hughes, D. (see under Bacsich, P.D.)

Rubin, M. (1983) A Boolean algebra with few subalgebras, interval Boolean algebras and retractiveness. Trans. Amer. Math. Soc. 278, 65-89. [210, 279]

Rubin, M. & Shelah, S. (1983) On the expressibility hierarchy of Magidor-Malitz quantifiers. J. Symbolic Logic 48, 542-557. [249, 280]

Rubin, M. (see also under Malitz, J.)

Russell, B. (1919) Introduction to Mathematical Philosophy. London: George Allen and Unwin. [169]

Ryll-Nardzewski, C. (1959) On the categoricity in power $\aleph_0$. Bull. Acad. Polon. Sci. Sér. Sci. Math. 7, 545-548. MR 22:2543 H.B. Curry. [167]

Ryll-Nardzewski, C. (see also under Grzegorczyk, A.)

Sabbagh, G. (1971) Sous-modules purs, existentiellement clos et élémentaires. C.R. Acad. Sci. Paris Sér. A-B 272, A1289-A1292. MR 46:3296 I.L. Gaal. [81, 131, 169]

Sabbagh, G. (1975) Sur les groupes qui ne sont pas réunion d'une suite croissante de sous-groupes propres. C.R. Acad. Sci. Paris Sér. A-B 280, 763-766. MR 51:3270 Paul C. Eklof. [81]

Sabbagh, G. (1976) Caractérisation algébrique des groupes de type finit ayant un problème de mots résoluble (théorème de Boone-Higman, travaux de B.H. Neumann et Macintyre). In Seminaire Bourbaki 1974/5, Exposés 453-470, Exp. 457, Lecture Notes in Math. 514, pp. 61-80. Berlin: Springer-Verlag. MR 56:5258 R.C. Lyndon. [81]

Sabbagh, G. (see also under Eklof, P.C.)

Sacerdote, G.S. (1974) Projective model completeness. J. Symbolic Logic 39, 117-123. MR 49:8854 G. Cherlin.

Sacerdote, G.S. (1975 a) Infinite coforcing in model theory. Adv. in Math. 17, 261-280. MR 57:9528 J.M. Plotkin.

Sacerdote, G.S. (1975 b) Projective model theory and coforcing. In Model Theory and Algebra: A Memorial Tribute to Abraham Robinson, ed. D.H. Saracino & V.B. Weispfenning, Lecture Notes in Math. 498, pp. 276-306. Berlin: Springer-Verlag. MR 53:10580 G. Cherlin.

Sacks, G.E. (1963) Degrees of Unsolvability. Princeton NJ: Princeton U.P. MR 32:4013 K. Appel. [91]

Sacks, G.E. (1972) Saturated Model Theory. Reading Mass.: W.A. Benjamin. MR 53:2668 P. Štěpánek. [16, 81, 186, 261]

Saracino, D. (1973) Model companions for $\aleph_0$-categorical theories. Proc. Amer. Math. Soc. 39, 591-598. MR 47:4786 John T. Baldwin. [168]

Saracino, D. (1974 a) m-existentially complete structures. Colloq. Math. 30, 7-13. MR 49:10542 G. Cherlin.

Saracino, D. (1974 b) Wreath products and existentially complete solvable groups. Trans. Amer. Math. Soc. 197, 327-339. MR 49:7137 Paul C. Eklof. [131]

Saracino, D. (1976) Existentially complete nilpotent groups. Israel J. Math. 25, 241-248. MR 56:11780 Paul C. Eklof. [131]

Saracino, D. (1978) Existentially complete torsion-free nilpotent groups. J. Symbolic Logic 43, 126-134. MR 58:212 Paul C. Eklof. [131]

Saracino, D. (1983) Amalgamation bases for nil-2 groups. Algebra Universalis 16, 47-62. [131]

Saracino, D. & Wood, C. (1979) Periodic existentially closed nilpotent groups. J. Algebra 58, 189-207. MR 80m:03071 Piero Mangani. [131]

Saracino, D. & Wood, C. (1982) QE nil-2 groups of exponent 4. J. Algebra 76, 337-352. MR 83i:03052 U. Felgner. [131]

Saracino, D. & Wood, C. (1983) Finitely generic abelian lattice-ordered groups. Trans. Amer. Math. Soc. 277, 113-123. [131]

Saracino, D. & Wood, C. (1984) Nonexistence of a universal countable commutative ring. Comm. Algebra 12, 1171-1173. [130]

Saracino, D. (see also under Lipschitz, L. and Macintyre, A.)

Saxl, J. (see under Baldwin, J.T.)

Schenkman, E. (1965) Group Theory. Princeton NJ: Van Nostrand. MR 33: 5702 C.W. Curtis. [112]

Schlipf, J. (see under Barwise, K.J.)

Schmerl, J.H. (1972) An elementary sentence which has ordered models. J. Symbolic Logic 37, 521-530. Zbl 278:02041 G. Fuhrken. [273]

Schmerl, J.H. (1973) Peano models with many generic classes. Pacific J. Math. 46, 523-536. MR 50:6831 M. Boffa. Correction, Pacific J. Math. 92 (1981) 195-198. [209]

Schmerl, J.H. (1974) A partition property characterizing cardinals hyperinaccessible of finite type. Proc. Amer. Math. Soc. 188, 281-291. MR 49:2386 N.H. Williams. [169, 273]

Schmerl, J.H. (1975) The number of equivalence classes of existentially complete structures. In Model Theory and Algebra: A Memorial Tribute to Abraham Robinson, ed. D.H. Saracino & V.B. Weispfenning, Lecture Notes in Math. 498, pp. 170-171. Berlin: Springer-Verlag. MR 54:2450 William H. Wheeler. [130]

Schmerl, J.H. (1976) On κ-like structures which embed stationary and closed unbounded subsets. Ann. Math. Logic 11, 289-314. MR 55:7763 G. Fuhrken. [248]

Schmerl, J.H. (1977) Theories with only a finite number of existentially complete models. Israel J. Math. 28, 350-356. MR 57:16052 A.B. Slomson. [130]

Schmerl, J.H. (1980) Decidability and $\aleph_0$-categoricity of theories of partially ordered sets. J. Symbolic Logic 45, 585-611. [131]

Schmerl, J.H. (1981) Recursively saturated, rather classless models of Peano arithmetic. In Logic Year 1979-80, The University of Connecticut, ed. M. Lerman et al., Lecture Notes in Math. 859, pp. 268-282. Berlin: Springer-Verlag. MR 83b:03039 Matt Kaufmann. [209]

Schmerl, J.H. (198-) Peano arithmetic and hyper-Ramsey logic. (preprint) [249, 280]

Schmerl, J.H. & Shelah, S. (1972) On power-like models for hyperinaccessible cardinals. J. Symbolic Logic 37, 531-537. MR 47:6474 P. Štěpánek. [273]

Schmerl, J.H. & Simpson, S.G. (1982) On the role of Ramsey quantifiers in first order arithmetic. J. Symbolic Logic 47, 423-435. MR 83j:03062 Klaus Potthoff. [249, 280]

Schreier, O. (see under Artin, E.)

Schupp, P.E. (see under Lyndon, R.C.)

Scott, D.S. (1965) Logic with denumerably long formulas and finite strings of quantifiers. In The Theory of Models, Proc. 1963 Internat. Symposium at Berkeley, ed. J.W. Addison et al., pp. 329-341. Amsterdam: North-Holland. MR 34:32 E. Engeler. [130]

Scott, W.R. (1951) Algebraically closed groups. Proc. Amer. Math. Soc. 2, 118-121. MR 12 p. 671 G. Higman. [80]

Seese, D.G. (see under Baudisch, A.)

Shelah, S. (1972) A note on model complete models and generic models. Proc. Amer. Math. Soc. 34, 509-514. MR 45:3188 K.J. Barwise.

Shelah, S. (1975 a) Categoricity in $\aleph_1$ of sentences in $L_{\omega_1,\omega}(Q)$. Israel J. Math. 20, 127-148. MR 52:83 F.R. Drake. [168, 249]

Shelah, S. (1975 b) A compactness theorem for singular cardinals, free algebras, Whitehead problem and transversals. Israel J. Math. 21, 319-349. MR 52:10410 Paul C. Eklof. [16, 168]

Shelah, S. (1975 c) The lazy model-theoretician's guide to stability. Logique et Analyse 1975 volume, 241-308. Reprinted in Six Days of Model Theory, ed. P. Henrard, pp. 9-76. Albeuve, Switzerland: Castella. MR 58:27447 John T. Baldwin. [279]

Shelah, S. (1977) Existentially-closed groups in $\aleph_1$ with special properties. Bull. Soc. Math. Grèce 18, 17-27. MR 80j:03047 Piero Mangani. [168]

Shelah, S. (1978 a) Classification Theory and the Number of Non-isomorphic Models. Amsterdam: North-Holland. MR 81a:03030 Daniel Lascar. [93, 130, 168, 248]

Shelah, S. (1978 b) End extensions and numbers of countable models. J. Symbolic Logic 43, 550-562. MR 80b:03037 V. Harnik. [210, 279]

Shelah, S. (1978 c) Models with second order properties I, Boolean algebras with no definable [*should be* undefinable] automorphisms. Ann. Math. Logic 14, 57-72. MR 80b:03047a Martin Weese. [262]

Shelah, S. (1978 d) Models with second order properties II, Trees with no undefined branches. Ann. Math. Logic 14, 73-87. MR 80b:03047b Martin Weese. [169, 210]

Shelah, S. (1979) Boolean algebras with few endomorphisms. Proc. Amer. Math. Soc. 74, 135-142. MR 82i:06017 D. Monk. [274]

Shelah, S. (1980 a) Models with second order properties III, Omitting types for L(Q). Arch. Math. Logik Grundlag. 21, 1-11. MR 83a:03031 Martin Weese. [168, 249, 274]

Shelah, S. (1980 b) On a problem of Kurosh, Jonsson groups, and applications. In Word Problems II, The Oxford Book, ed. S.I. Adian et al., pp. 373-394. Amsterdam: North-Holland. MR 81j:20047 G. Cherlin. [280]

Shelah, S. (1981) On uncountable Boolean algebras with no uncountable pairwise comparable or incomparable sets of elements. Notre Dame J. Formal Logic 22, 301-308. MR 83d:03060 J.L. Bell. [210, 274]

Shelah, S. (1982) Proper Forcing. Lecture Notes in Math. 940. Berlin: Springer-Verlag. [273]

Shelah, S. (1983 a) Constructions of many complicated uncountable structures and Boolean algebras. Israel J. Math. 45, 100-146. [210]

Shelah, S. (1983 b,c) Classification theory for non-elementary classes I, The number of uncountable models of $\psi \in L_{\omega_1,\omega}$, parts A, B. Israel J. Math. 46, 212-240, 241-273. [16, 168, 249, 250, 279]

Shelah, S. (1983 d) Models with second order properties IV, A general method and eliminating diamonds. Ann. Pure Appl. Logic 25, 183-212. [169, 209, 274, 281]

Shelah, S. (198- a) Classification theory for non-elementary classes II, abstract elementary classes. [169]

Shelah, S. (198- b) Uncountable constructions. Israel J. Math. (to appear). [168, 210]

Shelah, S. (198- c) Models with second order properties V. (in preparation) [169, 274]

Shelah, S. (198- d) Classification theory for non-elementary classes I, part C. Israel J. Math. (to appear) [250, 279]

Shelah, S. & Steinhorn, C. (198-) The non-axiomatizability of $L(Q^2_{\aleph_1})$ by finitely many schemas. (preprint) [246]

Shelah, S. & Ziegler, M. (1979) Algebraically closed groups of large cardinality. J. Symbolic Logic 44, 522-532. MR 80j:03048 Paul C. Eklof. [94]

Shelah, S. (see also under Bonnet, R., Galvin, F., Giorgetta, D., Grossberg, R., Rubin, M. and Schmerl, J.H.)

Shoenfield, J.R. (1967) Mathematical Logic. Reading Mass.: Addison-Wesley. MR 37:1224 G. Kreisel. [9, 11, 13, 18]

Shoenfield, J.R. (1971) A theorem on quantifier elimination. In Symposia Math. V, INDAM, Rome 1969-70, pp. 173-176. London: Academic Press. MR 42:7497 K.J. Barwise. [81]

Sikorski, R. (see under Rasiowa, H.)

Simmons, H. (1972) Existentially closed structures. J. Symbolic Logic 37, 293-310. MR 51:12518 James H. Schmerl. [80]

Simmons, H. (1973) An omitting types theorem with an application to the construction of generic structures. Math. Scand. 33, 46-54. MR 48:10798 James H. Schmerl. [82]

Simmons, H. (1975 a) The complexity of T^f and omitting types in F_T. In Model Theory and Algebra, A Memorial Tribute to Abraham Robinson, ed. D.H. Saracino & V.B. Weispfenning, Lecture Notes in Math. 498, pp. 403-407. Berlin: Springer-Verlag. MR 53:7766 M. Boffa.

Simmons, H. (1975 b) Counting countable e.c. structures. Logique et Analyse 1975 volume, 309-357. Reprinted in Six Days of Model Theory, ed. P. Henrard, pp. 77-125. Albeuve, Switzerland: Castella. MR 57:12203 John Cowles. [130, 131, 169]

Simmons, H. (1976) Large and small existentially closed structures. J. Symbolic Logic 41, 379-390. MR 54:87 John T. Baldwin. [130]

Simmons, H. (see also under Fisher, E.R.)

Simpson, S.G. (see under Schmerl, J.H.)

Skolem, T. (1934) Über die Nichtcharakterisierbarkeit der Zahlenreihe mittels endlich oder abzählbar unendlich vieler Aussagen mit ausschliesslich Zahlenvariablen. Fund. Math. 23, 150-161. [209]

Skolem, T. (1955) Peano's axioms and models of arithmetic. In Mathematical Interpretation of Formal Systems, pp. 1-14. Amsterdam: North-Holland. MR 17 p. 699 E. Mendelson. [209]

Слободской, А.М. (Slobodskoĭ, A.M.) (1981) Неразрешимость универсальной теории конечных групп. Алгебра и Логика 20, 207-230. Tr. as Undecidability of the universal theory of finite groups. Algebra and Logic 20, 139-156. MR 83h:03062 O.V. Belegradek. [277]

Slomson, A.B. (see under Bell, J.L.)

Smullyan, R.M. (1963) A unifying principal in quantification theory. Proc. Nat. Acad. Sci. USA 49, 828-832. MR 27:2410 G. Kreisel. [34, 168]

Specker, E.P. (1949) Sur un problème de Sikorski. Colloq. Math. 2, 9-12. MR 12 p. 597 F. Bagemihl. [274]

Specker, E.P. (see also under Gaifman, H. and MacDowell, R.)

Steinhorn, C. (see under Shelah, S.)

Steinitz, E. (1910) Algebraische Theorie der Körper. J. Reine Angew. Math. 137, 167-309. [80]

Stone, A.H. (see under Higman, G.)

Svenonius, L. (1959) $\aleph_0$-categoricity in first-order predicate calculus. Theoria 25, 82-94. MR 25:1986a R.C. Lyndon. [168]

Tait, W.W. (see under Gandy, R.O.)

Тайцлин, М.А. (Taĭtslin, M.A.) (1973) Экзистенциально замкнутые регулярные коммутативные полугруппы. Алгебра и Логика 12, 689-703. Tr. as Existentially closed regular commutative semigroups. Algebra and Logic 12, 394-401. MR 52:10413 A.I. Omarov.

Тайцлин, М.А. (Taĭtslin, M.A.) (1977) Экзистенциально замкнутые коммутативные полугруппы (Existentially closed commutative semigroups). Fund. Math. 94, 231-243. MR 55:5436 A.I. Omarov.

Taĭtslin, M.A. (see also under Belyaev, V.Ya.)

Tarski, A. (1954) Contributions to the theory of models I, II. Nederl. Akad. Wetensch. Indag. Math. 16, 572-581, 582-588. MR 16 p. 554 A. Robinson. [80]

Taylor, W. (see under Fuhrken, G.)

Трофимов, М.Ю. (Trofimov, M.Yu.) (1975) Об определимости в алгебраически замкнутых системах. Алгебра и Логика 14, 320-327. Tr. as On definability in algebraically closed systems. Algebra and Logic 14, 198-202. MR 55:2559 O.V. Belegradek. [81]

Tulipani, S. (198-) On the universal theory of classes of finite models. (preprint) [82]

Tuschik, P. (see under Baudisch, A.)

van den Dries, L.P.D., Glass, A.M.W., Macintyre, A., Mekler, A.H. & Poland, J. (1982) Elementary equivalence and the commutator subgroup. Glasgow Math. J. 23, 115-117. [34]

van der Waerden, B.L. (1950) Modern Algebra II. New York: Frederick Ungar Publ. Co. [43]

van Praag, P. (see under Boffa, M.)

Vaught, R.L. (1961) Denumerable models of complete theories. In Infinitistic Methods, Proc. Symp. Foundations of Math., Warsaw 1959, pp. 303-321. Oxford: Pergamon Press. MR 32:4011 G. Fuhrken. [130, 168, 209]

Vaught, R.L. (1964) The completeness of logic with the added quantifier "there are uncountably many". Fund. Math. 54, 303-304. MR 29:3364 A. Mostowski. [248]
Vaught, R.L. (see also under Morley, M.)
Weese, M. (see under Baudisch, A.)
Weispfenning, V. (1976) Negative-existentially complete structures and definability in free extensions. J. Symbolic Logic 41, 95-108. MR 54:2451 William H. Wheeler.
Weispfenning, V. (1977) Nullstellensätze - a model theoretic framework. Z. Math. Logik Grundlag. Math. 23, 539-545. MR 57:16056 Sauro Tulipani.
Wheeler, W.H. (1976) Model-companions and definability in existentially complete structures. Israel J. Math. 25, 305-330. MR 56:15413 W. Hodges.
Wheeler, W.H. (1978) A characterization of companionable, universal theories. J. Symbolic Logic 43, 402-429. MR 58:10400 B.I. Zil'ber. [81]
Wheeler, W.H. (1979) Amalgamation and elimination of quantifiers for theories of fields. Proc. Amer. Math. Soc. 77, 243-250. MR 80j:03049 Paul C. Eklof. [80]
Wheeler, W.H. (see also under Fisher, E.R. and Hirschfeld, J.)
Wiegold, J. (1959) Nilpotent products of groups with amalgamations. Publ. Math. Debrecen 6, 131-168. MR 21:3478 R.R. Struik. [113, 131]
Wood, C. (1972) Forcing for infinitary languages. Z. Math. Logik Grundlag. Math. 18, 385-402. MR 48:100 W.H. Wheeler.
Wood, C. (see also under Saracino, D.)
Yasuhara, M. (1974) The amalgamation property, the universal-homogeneous models, and the generic models. Math. Scand. 34, 5-36. MR 51:7860 M. Makkai.
Ziegler, M. (1980) Algebraisch abgeschlossene Gruppen. In Word Problems II, The Oxford Book, ed. S.I. Adian et al., pp. 449-576. Amsterdam: North-Holland. MR 82b:20004 M. Yasuhara. [34, 81, 82, 130, 131, 169, 249]
Ziegler, M. (see also under Shelah, S.)

INDEX